Environmental Hydraulics of Open Channel Flows

Environmental Hydraulics of Open Channel Flows

Hubert Chanson
ME, ENSHM Grenoble, INSTN, PhD (Cant), DEng (Qld)
Eur Ing, MIEAust, MIAHR
13th Arthur Ippen awardee (IAHR)
Reader in Environmental Fluid Mechanics and
Water Engineering
The University of Queensland, Australia
E-mail: h.chanson@uq.edu.au
Web site: http://www.uq.edu.au/~e2hchans/

ELSEVIER
BUTTERWORTH
HEINEMANN

AMSTERDAM BOSTON HEIDELBERG LONDON NEW YORK OXFORD
PARIS SAN DIEGO SAN FRANCISCO SINGAPORE SYDNEY TOKYO

Elsevier Butterworth-Heinemann
Linacre House, Jordan Hill, Oxford OX2 8DP
200 Wheeler Road, Burlington, MA 01803

First published 2004

British Library Cataloguing in Publication Data
A catalogue record for this book is available from the British Library

Library of Congress Cataloguing in Publication Data
A catalogue record for this book is available from the Library of Congress

ISBN: 0 7506 6165 8

For information on all Elsevier Butterworth-Heinemann publications
visit our web site at http://books.elsevier.com

Typeset by Charon Tec Pvt. Ltd, Chennai, India
Printed and bound in Great Britain

Working together to grow
libraries in developing countries

www.elsevier.com | www.bookaid.org | www.sabre.org

ELSEVIER   BOOK AID International   Sabre Foundation

Contents

Preface

Rivers play a major role in shaping the landscapes of our planet (Table P.1, Fig. P.1). Extreme flow rates may vary from zero in drought periods to huge amount of waters in flood periods. For example, the maximum observed flood discharge of the Amazon River at Obidos was about 370 000 m³/s (Herschy 2002). This figure may be compared with the average annual discharges of the Congo River (41 000 m³/s at the mouth) and of the Murray-Darling River (0.89 m³/s at the mouth) (Table P.1). Even arid, desertic regions are influenced by fluvial action when periodic flood waters surge down dry watercourses (Fig. P.1(a)).

Hydraulic engineers have had an important role to contribute although the technical challenges are gigantic, often involving multiphase flows and interactions between fluids and biological life. These engineers were at the forefront of science for centuries. For example, the arts of tapping groundwater developed early in the Antiquity in Armenia and Persia, the Roman aqueducts, and the Grand canal navigation system in China. In the author's opinion, the extreme complexity of hydraulic engineering is closely linked with:

1. *The geometric scale of water systems*: e.g. from $<10\,\mathrm{m}^2$ for a soil erosion pattern (e.g. rill) to over $1000\,\mathrm{km}^2$ for a river catchment area typically, and ocean surface area over $1 \times 10^6\,\mathrm{km}^2$.
2. *The broad range of relevant time scales*: e.g. $<1\,\mathrm{s}$ for a breaking wave, about $1 \times 10^4\,\mathrm{s}$ for tidal processes, about $1 \times 10^8\,\mathrm{s}$ for reservoir siltation, and about $1 \times 10^9\,\mathrm{s}$ for deep sea currents.

Table P.1 Characteristics of the world's longest rivers

River system (1)	Length (km) (2)	Catchment area (km²) (3)	Average annual discharge (m³/s) (4)	Average sediment transport rate (tons/day) (5)
Amazon-Ucayali-Apurimac (South America)	6400	6 000 000	180 000	1 300 000
Congo (Africa)	4700	3 700 000	41 000	–
Yangtze (Asia)	6300	1 808 500	31 000	–
Yenisey-Baikal-Selenga (Asia)	5540	2 580 000	19 800	–
Parana (South America)	4880	2 800 000	17 293	–
Mississippi-Missouri-Red Rock (North America)	5971	3 100 00	17 000	–
Ob-Irtysh (Asia)	5410	2 975 000	12 700	–
Amur-Argun (Asia)	4444	1 855 000	10 900	–
Volga (Europe)	3530	1 380 000	8050	–
Nile (Africa)	6650	3 349 000	3100	–
Huang Ho (Yellow River) (Asia)	5464	752 000	1840	4 400 000
Murray-Darling (Australia)	3370	1 072 905	0.89	–

Average annual discharge: at the river mouth.

(a)

(b)

Fig. P.1 Photographs of natural rivers. (a) Small flood in the Gascoyne River, Carnarvon, WA (Australia) (courtesy of Gascoyne Development Commission and Robert Panasiewicz). The Gascoyne River has catchment area of about 67 770 km^2 and it extends 630 km inland. Average annual rainfall is <250 mm throughout the basin and this is an ephemeral river. There are typically one to two flow periods per year following seasonal rainfall or cyclone activity, but it may fails to flow at all once every 5 or 6 years. (b) Tingalpa Creek, Redlands Qld (Australia) on 21 January 2003 at high tide at about 9 km from the river mouth, looking upstream.

3. *The variability of river flows* from zero (dry river bed during droughts) to gigantic floods.
4. *The complexity of basic fluid mechanics*, with governing equations characterized by non-linearity, natural fluid instabilities, interactions between water, solid, air and biological life and;
5. *Man's (and Life's) total dependence* on water.

DISCUSSION

Armed conflicts around water systems have been plenty. In the Bible, a wind-setup effect allowed Moses and the Hebrews to cross shallow-water lakes and marshes during their exodus. Droughts were artificially introduced: e.g. during the siege of the ancient city of Khara Khoto (Black City) in AD 1372, the Chinese army diverted the Ezen River[1] supplying water to the city.[2] Man-made flooding[3] of an army or a city was carried out by the Assyrians (Babylon, Iraq BC 689), the Spartans (Mantinea, Greece BC 385–384), the Chinese (Huai River, AD 514–515).[4] A related case was the air raids of the dam buster campaign conducted by the British in 1943. Artifical flooding created by dyke destruction played a role in several wars: e.g. the war between the cities of Lagash and Umma (Assyria) around BC 2500 was fought for the control of irrigation systems and dykes.

The 21st Century is facing political instabilities centred around water systems, and freshwater system issues might be the focal point of future armed conflicts. For example, the Tigris and Euphrates River catchments and the Mekong River. The scope of the relevant problems is broad and complex: e.g. water quality, pollution, flooding and drought. An example is the disaster of the Aral Sea with the formation of the permanently-dry isthmus between the northern small Aral Sea and the southern big Aral Sea since 1987 (Waltham and Sholji 2001).

This book was developed to introduce students, professionals and managers to the challenges of open channel flows and environmental hydraulics. After a concise introduction (Part 1), the second section (Part 2) deals with mixing and dispersion of matter in natural river systems. Part 3 presents an introduction to unsteady open channel flows, and the interactions between flowing water and its surroundings are discussed in Part 4.

Mixing and dispersion of contaminants in natural systems are developed in Part 2. Applications include release of organic and nutrient-rich waste water into the ecosystem (e.g. from treated sewage effluent), smothering of seagrass and coral, storm water runoff during flood events, and injection of heated water from an industrial discharge (e.g. at a cooling power plant). For example, during an accidental release of waste occurs in a stream, the water resource scientist needs to predict the arrival time of the contaminant cloud, the peak concentration of solute and the duration of the pollution. Basic theory of molecular diffusion and advection is extended to turbulent advective diffusion in channels.

Gradually varied flow calculations are developed in Part 3. First the basic equations of one-dimensional unsteady open channel flows are presented. That is, the Saint-Venant

[1] Also called Hei He River (Black River) by the Chinese.
[2] Located in the Gobi Desert, Khara Khoto was ruled by the Mongol King Khara Bator (Webster 2002).
[3] By building an upstream dam and destroying it.
[4] It may be added the aborted attempt to blow up Ordunte Dam, during the Spanish Civil War, by the troops of General Franco, and the anticipation of German Dam destruction at the German–Swiss border to stop the crossing of the Rhine River by the Allied Forces in 1945.

equations and the method of characteristics in Chapter 11*. Later simple applications are developed. The propagation of waves, and positive and negative surges is presented in Chapter 12, while the dam break wave problem is discussed in Chapter 13. Simple numerical models are presented and explained in Chapter 14.

There are strong interactions between turbulent water flows and the surrounding environment. Part 4 introduces the basic concepts of the transport of solids (Chapter 16), and of the mixing of air and water at free surfaces (Chapter 17).

At the beginning of the book, the reader will find the table of contents, a list of symbols and a glossary of technical terms and names. After the conclusion, a detailed list of references is presented. The last section presents a correction form. Readers who find an error or mistake are welcome to record the error on the page and to send a copy to the author. Corrections and updates will be posted on the Internet at: http://www.uq.edu.au/~e2hchans/reprints/book7.htm

Discussion

The lecture material is based upon the author's experience at the University of Queensland, and at other universities. It is designed primarily for undergraduate students in civil, environmental and hydraulic engineering. The author has taught Part 1 in Years 2 and 3, and Parts 2 and 3 as parts of advanced undergraduate electives in Year 4. Some material of Part 4 is usually introduced in the advanced hydraulics elective subject, and the course is further developed at postgraduate levels.

The author wants to stress, however, that field studies are a necessary complement to traditional lectures in environmental hydraulics. In the context of undergraduate subjects, design applications in classroom are restricted to simple flow situations and boundary conditions for which the basic equations can be solved analytically or with simple models. Fieldwork activities (Fig. P.2) are essential to illustrate real professional situations, and the complex interactions between all engineering and non-engineering constraints.

The author has organized undergraduate fieldworks in hydraulic engineering for more than 10 years involving more than 1000 undergraduate students. Figure P.2 illustrates recent examples. Figure P.2(a) shows mixing and dispersion class students conducting an ecological assessment of the estuarine zone of a small subtropical creek. For 12 h, students surveyed hydrodynamics, water quality parameters, fish populations, bird behaviours and wildlife sightings at four sites (Chanson *et al.* 2003). They concluded their works with a group report and an oral presentation in front of student peers, lecturers, professionals and local community groups. Figure P.2(b) shows hydraulic design students in front of the fully silted Korrumbyn Creek Dam disused since 1926. The dam and reservoir were accessed after a 45-min bushwalk guided by National Park and Wildlife rangers in the dense sub-tropical rainforest of Mount Warning National Park (NSW). The fieldworks was focused on sediment processes in the catchment. Students surveyed both upper and lower catchments, the fully silted reservoir and discussed its possible use as touristic attraction and potential source of aggregate for the local construction industry. Figure P.2(c) presents the civil design students surveying a flood plain in the heart of Brisbane. Students working in groups surveyed eight sections of the creek including culverts and wide flood plains. Each group conducted hydraulic computations for design and less-than-design flow rates, and prepared newer designs for a larger flood.

Anonymous student feedback demonstrated the very significant role of fieldworks in the teaching of hydraulic engineering (Chanson 2004c). Seventy-eight per cent of students

---

* It is acknowledged that, in Chapter 11, the basic derivation of Saint-Venant equations and method of characteristics presents some similarities with sections of another textbook (Chanson, 2004b).

(a)

(b)

Fig. P.2 Photographs of undergraduate student field trips. (a) Mixing in estuary fieldwork (39 students) at Eprapah Creek on 4 April 2003, students conducting sampling tests in the mangrove (courtesy of Ms H. Joyce). (b) Field study on 4 September 2002 with hydraulic design class (24 students), students in front of the fully silted Korrumbyn Creek Dam in a dense sub-tropical rainforest.

(c)

(d)

Fig. P.2 (*Contd*) (c) Civil design students (73 students) surveying a flood plain in 2002 (courtesy of L. Cheung). (d) Group bonding at the end of 12 h of estuarine field study (4 April 2003) (courtesy of Ms H. Joyce).

believed strongly and very strongly that 'fieldwork is an important component of the subject'. Eighty-four per cent of students agreed strongly and very strongly that 'all things considered, fieldworks and site visits are the vital components of civil and environmental engineering curricula'. Ninety-six per cent of students believed that 'fieldworks play a vital role to comprehend real-word engineering' and 100% of interviewed employers stressed that fieldworks under academic supervision was a basic requirement for civil and environmental engineering graduates. Lecturers and professionals should not be complaisant with university hierarchy and administration clerks to cut costs by eliminating field studies. Although the preparation and organization of fieldworks with large class sizes are a major effort, the outcome is very rewarding for the students and the lecturer. From his own experience, the author has had great pleasure in bringing his students to hydraulics fieldwork for more than a decade and to experience first hand their personal development (Fig. P.2(d)).

Internet resources
General resources

Gallery of photographs	http://www.uq.edu.au/~e2hchans/photo.html
Reprints of research papers	http://www.uq.edu.au/~e2hchans/reprints.html
Internet technical resources	http://www.uq.edu.au/~e2hchans/url_menu.html
NASA Earth observatory	http://earthobservatory.nasa.gov/
NASA rain, wind and air-sea gas exchange research	http://bliven2.wff.nasa.gov/index.htm
USACE inlets online	http://www.oceanscience.net/inletsonline/
Estuaries in South Africa	http://www.upe.ac.za/cerm/
Whirlpools	http://www.uq.edu.au/~e2hchans/whirlpl.html

Mixing and dispersion in rivers

Rivers seen from space	http://www.athenapub.com/rivers1.htm
Aerial photographs of rivers	ftp://geology.wisc.edu/pub/air

Acknowledgments

The author wants to thank Prof. Colin J. Apelt, University of Queensland, for his help, support and assistance all along the academic career of the author, and Dr Jean Cunge who presented some superb lectures. The author thanks particularly his friend Prof. Shin-ichi Aoki, Toyohashi University of Technology (Japan) for his valuable advice and comments.

The author thanks his research students who conducted relevant experimental work: Ms Chantal Donnelly, Dr Carlos Gonzalez, Ms Karen Hickox, Mr Chung Hwee Jerry Lim, Mr Mamuro Maruyama, Ms Claire Quinlan, Mr Chye-Guan Sim, Mr Frankie Tan, Mr York-Wee Tan and Dr Luke Toombes.

The author wishes to express his gratitude to the followings who made available some photographs of interest:

Acres International, Canada;
Mr Amir Aghakoochak, Iran;
Michael Armitage, University of Sheffield, UK;
Dr Marie Augendre, Université de Lyon 2, France;
Dr Antje Bornschein, University of Dresden, Germany;
Mr and Mrs Chanson, France;
Consortium for Estuarine Research and Management (CERM), South Africa;
Coastal and Hydraulics Laboratory, US Army Corps of Engineers;
Prof. Andre Fourie, University of Witwatersrand, South Africa;
Gascoyne Development Commission, WA, Australia;
Dr Michael R. Gourlay, Brisbane, Australia;
Gary & Rhonda Higgins, Northern Territory, Australia;
Lim Hiok Hwa, Department of Irrigation and Drainage, Sarawak, Indonesia;
Dr Eric Jones, Proudman Oceanographic Laboratory, UK;
Pr J. Knauss, Münich University of Technology, Germany;
Ms Sasha Kurz, Brisbane, Australia;
Ms Nathalie Lemiere, Sequana-Normandie, France;
Mr Jerry Lim, Singapore;
Dr Pedro Lomonaco, University of Cantabria, Spain;
Dr Lou Maher, University of Wisconsin, USA;
Dr John Macintosh, Water Solutions, Australia;
Dr Richard Manasseh, CSIRO, Australia;
Mr Dennis Murphy, USA;
Prof. Okada, Mt Usu Vulcano Observatory. Hokudai Faculty of Science, Japan;
Mr Robert Panasiewicz, Gascoyne Development Commission, Australia;
Prof. D. Howell Peregrine, University of Bristol, UK;
Mr Bruno de Quinsonas, Le Touvet, France;
Mr Marq Redeker, Ruhrverband, Germany;

The Santa Clarita Valley Historical Society, California, USA;
Mr Chye-Guan Sim, Singapore;
Daniel Stephens, USA;
Mr Frankie Tan, Singapore;
Mr York-Wee Tan, Singapore;
Tonkin and Taylor, New Zealand;
Mr Didier Toulouze, Fréjus, France;
US Army Corps of Engineers, Portland district;
US Naval Historical Center, USA;
Waterways Scientific Services, Queensland Environmental Protection Agency, Australia;
Prof. Steven J. Wright, University of Michigan, USA.

The author thanks the following people in providing relevant experimental data:

Prof. S. Aoki, Toyohashi University of Technology, Japan;
Dr I. Ramsay, Queensland Environmental Protection Agency, Australia;
Dr Y. Yasuda, Nihon University, Japan.

The author thanks also the following people in providing additional information: Prof. Shin-ichi Aoki (Japan); Dr Richard Brown, QUT (Australia); Dr Antje Bornschein, University of Dresden (Germany); Mr and Mrs Chanson (France); Dr Stephen Coleman, University of Auckland (New Zealand); Dr Peter Cummings (Australia); Mr John Ferris, Qld EPA (Australia); John Grimston (New Zealand); Dr Eric Jones, Proudman Oceanographic Laboratory (UK); Prof. Iwao Ohtsu, Nihon University (Japan); Robert Panasiewicz, Gascoyne Development Commission (Australia); Dr Ian Ramsay, Qld EPA (Australia); John Remi (Canada); Mr M. Tomkins (Australia); Dr Youichi Yasuda, Nihon University (Japan).

The help and assistance of the following colleagues must be acknowledged: Prof. C.J. Apelt and Dr P. Nielsen.

At last but not the least, the author thanks all the people (including colleagues, former students, students and professionals) who gave him information, feedback and comments on his lecture material. In particular, some material on the Saint-Venant equations and the method of characteristics derived from Dr Jean Cunge's lecture notes.

About the author

Hubert Chanson is a Reader in Environmental Fluid Mechanics and Water Engineering at the University of Queensland since 1990. He was born in 1961 in Paris, France. He lives in Brisbane, Australia, with his wife Ya-Hui (Karen) Chou and their children Bernard and Nicole. He received a degree of 'Ingénieur Hydraulicien' from the Hydraulic Engineering School of Grenoble, France (ENSHMG) in 1983 and a postgraduate degree of 'Ingénieur Génie Atomique' from the Nuclear Engineering Institute of Saclay (INSTN) in 1984. He worked for the industry in France as an R&D Engineer at the Atomic Energy Commission from 1984 to 1986, and as a computer professional in fluid mechanics for Thomson-CSF between 1989 and 1990. From 1986 to 1988, he studied at the University of Canterbury (New Zealand) as part of a PhD project. He was awarded a Doctor of Engineering from the University of Queensland in 1999 for outstanding research achievements in gas–liquid bubbly flows. In 2003, the International Association for Hydraulic engineering and Research (IAHR) presented him the 13th Arthur Ippen Award for outstanding achievements in hydraulic engineering. This award is regarded as the highest achievement in hydraulic research.

His research interests cover design of hydraulic structures, experimental investigations of two-phase flows, coastal hydrodynamics, water quality modelling, environmental management and natural resources. He authored several books: *Hydraulic Design of Stepped Cascades, Channels, Weirs and Spillways* (Pergamon, 1995), *Air Bubble Entrainment in Free-Surface Turbulent Shear Flows* (Academic Press, 1997), *The Hydraulics of Open Channel Flows: An Introduction* (Butterworth-Heinemann, 1999) and *The Hydraulics of Stepped Chutes and Spillways* (Balkema, 2001). He co-authored another book *Fluid Mechanics for Ecologists* (IPC Press, 2002). His textbook *The Hydraulics of Open Channel Flows: An Introduction* has already been translated into Chinese (Hydrology Bureau of Yellow River Conservancy Committee) and Spanish (McGraw Hill Interamericana), and the second edition was recently released (Elsevier, 2004). His publication record includes over 200 international refereed papers and his work was cited over 1000 times since 1990. Hubert Chanson has been active also as consultant for both governmental agencies and private organizations.

Hubert Chanson has been awarded six fellowships from the Australian Academy of Science. In 1995 he was a Visiting Associate Professor at National Cheng Kung University (Taiwan, ROC) and he was a Visiting Research Fellow at Toyohashi University of Technology (Japan) in 1999 and 2001. In 2004, he was a Visiting Research Fellow at the Laboratoire Central des Ponts et Chaussées (France), at Université de Bretagne Occidentale (France) and at McGill University (Canada).

Hubert Chanson was the Invited Keynote Lecturer at the *1998 ASME Fluids Engineering Symposium on Flow Aeration* (Washington DC), first *International Conference of the International Federation for Environmental Management System IFEMS '01* (Tsurugi, Japan 2001), *6th International Conference on Civil Engineering ICCE '03* (Isfahan, Iran 2003), *30th IAHR Biennial Congress* (Thessaloniki, Greece 2003) and *International Conference on*

Hydraulics of Dams and River Structures HDRS '04 (Tehran 2004). He gave invited lectures at the *Workshop on Flow Characteristics around Hydraulic Structures* (Nihon University, Japan 1998), International Workshop on Hydraulics of Stepped Spillways (ETH-Zürich, Switzerland 2000) and *29th IAHR Biennial Congress* (Beijing, China 2001). He lectured several short courses in Australia and overseas (e.g. France, Japan, Taiwan).

His Internet home page is http://www.uq.edu.au/~e2hchans. He developed a gallery of photographs web site http://www.uq.edu.au/~e2hchans/photo.html, that received more than 100 000 hits since inception, and a series of world-known technical Internet resources.[1] Reprints of his research papers may be downloaded from: http://www.uq.edu.au/~e2hchans/reprints.html.

[1]http://www.uq.edu.au/~e2hchans/url_menu.html

To Nicole, Ya-Hui

and Bernard

Glossary

Abutment   Part of the valley side against which the dam is constructed. Artificial abutments are sometimes constructed to take the thrust of an arch where there is no suitable natural abutment.

Académie des Sciences de Paris   The Académie des Sciences, Paris, is a scientific society, part of the Institut de France formed in 1795 during the French Revolution. The academy of sciences succeeded the Académie Royale des Sciences, founded in 1666 by Jean-Baptiste Colbert.

Acid   A sour compound that is capable, in solution, of reacting with a base to form a salt and has a pH <7.

Acidity   Having marked acid properties, more broadly having a pH < 7.

Accretion   Increase of channel bed elevation resulting from the accumulation of sediment deposits.

Adiabatic   Thermodynamic transformation occurring without loss nor gain of heat.

Advection   Movement of a mass of fluid which causes change in temperature or in other physical or chemical properties of fluid.

Aeration device (or *aerator*)   Device used to introduce artificially air within a liquid. Spillway aeration devices are designed to introduce air into high-velocity flows. Such aerators include basically a deflector and air is supplied beneath the deflected waters. Downstream of the aerator, the entrained air can reduce or prevent cavitation erosion.

Afflux   Rise of water level above normal level (i.e. natural flood level) on the upstream side of a culvert or of an obstruction in a channel.

Aggradation   Raise in channel bed elevation caused by deposition of sediment material. Another term is accretion.

Air   Mixture of gases comprising the atmosphere of the Earth. The principal constituents are nitrogen (78.08%) and oxygen (20.95%). The remaining gases in the atmosphere include argon, carbon dioxide, water vapour, hydrogen, ozone, methane, carbon monoxide, helium, krypton, ...

Air concentration   Concentration of un-dissolved air defined as the volume of air per unit volume of air and water. It is also called the void fraction.

Air entrainment   Entrapment and entrainment of un-dissolved air into a water flow. It is also called the air bubble entrainment and self-aeration.

Alembert (d')   Jean le Rond d'Alembert (1717–1783) was a French mathematician and philosopher. He was a friend of Leonhard Euler and Daniel Bernoulli. In 1752 he published his famous d'Alembert's paradox for an ideal-fluid flow past a cylinder (Alembert 1752).

Algual bloom   Dense aquatic population of microscopic organisms and alguae produced by an abundance of nutrient salts in surface water, coupled with adequate sunlight for photosynthesis. The bloom depletes that water oxygen content, poison aquatic animals and waterfowl irritate the skin and respiratory tract of humans.

Alkalinity   Having marked basic properties (as a hydroxide or carbonate of an alkali metal); more broadly having a pH >7.

Alternate depth   In open channel flow, for a given flow rate and channel geometry, the relationship between the specific energy and flow depth indicates that, for a given specific energy, there is no real solution (i.e. no possible flow), one solution (i.e. critical flow) or two solutions for the flow depth. In the latter case, the two flow depths are called alternate depths. One corresponds to a subcritical flow and the second to a supercritical flow.

Analytical model   System of mathematical equations which are the algebraic solutions of the fundamental equations.

Apelt   C.J. Apelt is an Emeritus Professor in Civil Engineering at the University of Queensland (Australia).

Apron   The area at the downstream end of a weir to protect against erosion and scouring by water.

Aqueduct   A conduit for conveying a large quantity of flowing waters. The conduit may include canals, siphons, pipelines.

Arch dam   Dam in plan dependent on arch action for its strength.

Arched dam   Gravity dam which is curved in plan. Alternatives include 'curved-gravity dam' and 'arch-gravity dam'.

Archimedes   Greek mathematician and physicist. He lived between BC 290–280 and BC 212 (or 211). He spent most of his life in Syracuse (Sicily, Italy) where he played a major role in the defence of the city against the Romans. His treaty 'On Floating Bodies' is the first-known work on hydrostatics, in which he outlined the concept of buoyancy.

Aristotle   Greek philosopher and scientist (BC 384–322), student of Plato. His work 'Meteorologica' is considered as the first comprehensive treatise on atmospheric and hydrological processes.

Armouring   Progressive coarsening of the bed material resulting from the erosion of fine particles. The remaining coarse material layer forms an armour, protecting further bed erosion.

Assyria   Land to the North of Babylon comprising, in its greatest extent, a territory between the Euphrates and the mountain slopes East of the Tigris. The Assyrian Kingdom lasted from about BC 2300 to BC 606.

Atomic number   The atomic number (of an atom) is defined as the number of units of positive charge in the nucleus. It determines the chemical properties of an atom.

Atomic weight   Ratio of the average mass of a chemical element's atoms to some standard. Since 1961 the standard unit of atomic mass has been 1/12 the mass of an atom of the isotope carbon-12.

Avogadro number   Number of elementary entities (i.e. molecules) in 1 mol of a substance: $6.0221367 \times 10^{23}\,\text{mol}^{-1}$. Named after the Italian physicist Amedeo Avogadro.

Backwater   In a tranquil flow motion (i.e. subcritical flow) the longitudinal flow profile is controlled by the downstream flow conditions: e.g. an obstacle, a structure, a change of cross-section. Any downstream control structure (e.g. bridge piers, weir) induces a backwater effect. More generally the term backwater calculations or backwater profile refers to the calculation of the longitudinal flow profile. The term is commonly used for both supercritical and subcritical flow motion.

Backwater calculation   Calculation of the free-surface profile in open channels. The first successful calculations were developed by the Frenchman J.B. Bélanger who used a finite difference step method for integrating the equations (Bélanger 1828).

Bagnold   Ralph Alger Bagnold (1896–1990) was a British geologist and a leading expert on the physics of sediment transport by wind and water. During World War II, he founded the Long Range Desert Group and organized long-distance raids behind enemy lines across the Libyan Desert.

Bakhmeteff   Boris Alexandrovitch Bakhmeteff (1880–1951) was a Russian hydraulician. In 1912, he developed the concept of specific energy and energy diagram for open channel flows.

Barrage   French word for dam or weir, commonly used to described large dam structure in English.

Barré de Saint-Venant   Adhémar Jean Claude Barré de Saint-Venant (1797–1886), French engineer of the 'Corps des Ponts-et-Chaussées', developed the equation of motion of a fluid particle in terms of the shear and normal forces exerted on it (Barré de Saint-Venant 1871a, b).

Barrel   For a culvert, central section where the cross-section is minimum. Another term is the throat.

Bathymetry   Measurement of water depth at various places in water (e.g. river, ocean).

Bazin   Henri Emile Bazin was a French hydraulician (1829–1917) and engineer, member of the French 'Corps des Ponts-et-Chaussées' and later of the Académie des Sciences de Paris. He worked as an Assistant of Henri P.G. Darcy at the beginning of his career.

Bed form   Channel bed irregularity that is related to the flow conditions. Characteristic bed forms include ripples, dunes and antidunes.

Bed load   Sediment material transported by rolling, sliding and saltation motion along the bed.

Bélanger   Jean-Baptiste Ch. Bélanger (1789–1874) was a French hydraulician and professor at the Ecole Nationale Supérieure des Ponts et Chaussées (Paris). He suggested first the application of the momentum principle to hydraulic jump flow (Bélanger 1828). In the same book, he presented the first 'backwater' calculation for open channel flow.

Bélanger equation   Momentum equation applied across a hydraulic jump in a horizontal channel (named after J.B.C. Bélanger).

Bélidor   Bertrand Forêt de Bélidor (1693–1761) was a Teacher at the Ecole Nationale des Ponts et Chaussées. His treatise *Architecture Hydraulique* (Bélidor 1737–1753) was a well-known hydraulic textbook in Europe during the 18th and 19th Centuries.

Benthic   Related to processes occurring at the bottom of the waters.

Bernoulli   Daniel Bernoulli (1700–1782) was a Swiss mathematician, physicist and botanist who developed the Bernoulli equation in his *Hydrodynamica, de Viribus et Motibus Fluidorum* textbook (first draft in 1733, first publication in 1738, Strasbourg).

Bessel   Friedrich Wilhelm Bessel (1784–1846) was a German astronomer and mathematician. In 1810 he computed the orbit of Halley's comet. As a mathematician he introduced the Bessel functions (or circular functions) which have found wide use in physics, engineering and mathematical astronomy.

Bidone   Giorgio Bidone (1781–1839) was an Italian hydraulician. His experimental investigations on the hydraulic jump were published between 1820 and 1826.

Biesel   Francis Biesel (1920–1993) was a French hydraulic engineer and a pioneer of computational hydraulics.

Biochemical oxygen demand   The biochemical oxygen demand (BOD) is the amount of oxygen used by micro-organisms in the process of breaking down organic matter in water.

Blasius   H. Blasius (1883–1970) was German scientist, student and collaborator of L. Prandtl.

BOD   See *Biochemical oxygen demand*.

Boltzmann   Ludwig Eduard Boltzmann (1844–1906) was an Austrian physicist.

Boltzmann constant   Ratio of the universal gas constant ($8.3143 \, \text{K} \, \text{J}^{-1} \text{mol}^{-1}$) to the Avogadro number ($6.0221367 \times 10^{23} \, \text{mol}^{-1}$). It equals: $1.380662 \times 10^{-23} \, \text{J} \, \text{K}$.

Borda   Jean-Charles de Borda (1733–1799) was a French mathematician and military engineer. He achieved the rank of Capitaine de Vaisseau and participated to the US War of Independence with the French Navy. He investigated the flow through orifices and developed the Borda mouthpiece.

Borda mouthpiece   A horizontal re-entrant tube in the side of a tank with a length such that the issuing jet is not affected by the presence of the walls.

Bore   A surge of tidal origin is usually termed a bore (e.g. the Mascaret in the Seine River, France).

Bossut   Abbé Charles Bossut (1730–1804) was a French ecclesiastic and experimental hydraulician, author of a hydrodynamic treaty (Bossut 1772).

Bottom outlet   Opening near the bottom of a dam for draining the reservoir and eventually flushing out reservoir sediments.

Boundary layer   Flow region next to a solid boundary where the flow field is affected by the presence of the boundary and where friction plays an essential part. A boundary layer flow is characterized by a range of velocities across the boundary layer region from zero at the boundary to the free-stream velocity at the outer edge of the boundary layer.

Boussinesq   Joseph Valentin Boussinesq (1842–1929) was a French hydrodynamicist and Professor at the Sorbonne University (Paris). His treatise *Essai Sur la Théorie des Eaux Courantes* (1877) remains an outstanding contribution in hydraulics literature.

Boussinesq coefficient   Momentum correction coefficient named after J.V. Boussinesq who first proposed it (Boussinesq 1877).

Boussinesq–Favre wave   An undular surge (see *Undular surge*).

Bowden   Prof. Kenneth F. Bowden contributed to the present understanding of dispersion in estuaries and coastal zones.

Boys   P.F.D. du Boys (1847–1924) was a French hydraulic engineer. He made a major contribution to the understanding of sediment transport and bed-load transport (Boys 1879).

Braccio   Ancient measure of length (from the Italian 'braccia'). One braccio equals 0.6096 m (or 2 ft).

Brackish water Water with a salinity less than about 25 ppt. In open oceans, the salinity of surface waters is about 35 ppt.

Braised channel Stream characterized by random interconnected channels separated by islands or bars. By comparison with islands, bars are often submerged at large flows.

Bresse Jacques Antoine Charles Bresse (1822–1883) was a French applied mathematician and hydraulician. He was Professor at the Ecole Nationale Supérieure des Ponts et Chaussées, Paris as successor of J.B.C. Bélanger. His contribution to gradually varied flows in open channel hydraulics is considerable (Bresse 1860).

Broad-crested weir A weir with a flat long crest is called a broad-crested weir when the crest length over the upstream head is >1.5–3. If the crest is long enough, the pressure distribution along the crest is hydrostatic, the flow depth equals the critical flow depth and the weir can be used as a critical depth meter.

Buat Comte Pierre Louis George du Buat (1734–1809) was a French military engineer and hydraulician. He was a friend of Abbé C. Bossut. Du Buat is considered as the pioneer of experimental hydraulics. His textbook (Buat 1779) was a major contribution to flow resistance in pipes, open channel hydraulics and sediment transport.

Bubble Small volume of gas within a liquid (e.g. air bubble in water). The term bubble is used also for a thin film of liquid inflated with gas (e.g. soap bubble) or a small air globule in a solid (e.g. gas inclusion during casting). More generally the term air bubble describes a volume of air surrounded by liquid interface(s).

Buoyancy Tendency of a body to float, to rise or to drop when submerged in a fluid at rest. The physical law of buoyancy (or Archimedes' principle) was discovered by the Greek mathematician Archimedes. It states that any body submerged in a fluid at rest is subjected to a vertical (or buoyant) force. The magnitude of the buoyant force is equal to the weight of the fluid displaced by the body.

Buoyant jet Submerged jet discharging a fluid lighter or heavier than the mainstream flow. If the jet's initial momentum is negligible, it is called a *buoyant plume*.

Buttress dam A special type of dam in which the water face consists of a series of slabs or arches supported on their air faces by a series of buttresses.

Byewash Ancient name for a spillway: i.e. channel to carry waste waters.

Candela SI unit for luminous intensity, defined as the intensity in a given direction of a source emitting a monochromatic radiation of frequency 540×10^{12} Hz and which has a radiant intensity in that direction of 1/683 W per unit solid angle.

Carnot Lazare N.M. Carnot (1753–1823) was a French military engineer, mathematician, general and statesman who played a key role during the French Revolution.

Carnot Sadi Carnot (1796–1832), eldest son of Lazare Carnot, was a French scientist who worked on steam engines and described the Carnot cycle relating to the theory of heat engines.

Cartesian coordinate One of three coordinates that locate a point in space and measure its distance from one of three intersecting coordinate planes measured parallel to that one of three straight-line axes that is the intersection of the other two planes. It is named after the French mathematician René Descartes.

Cascade (1) A steep stream intermediate between a rapid and a water fall. The slope is steep enough to allow a succession of small drops but not sufficient to cause the water to drop vertically (i.e. waterfall). (2) A man-made channel consisting of a series of steps: e.g. a stepped fountain, a staircase chute, a stepped sewer.

Cataract A series of rapids or waterfalls. It is usually termed for large rivers: e.g. the six cataracts of the Nile River between Karthum and Aswan.

Catena d'Acqua (Italian term for 'chain of water') variation of the cascade developed during the Italian Renaissance. Water is channelled down the centre of an architectural ramp contained on both sides by stone carved into a scroll pattern to give a chain-like appearance. Waters flow as a supercritical regime with regularly spaced increase and decrease of channel width, giving a sense of continuous motion highlighted by shock wave patterns at the free surface. One of the best examples is at Villa Lante, Italy. The stonework was carved into crayfish, the emblem of the owner, Cardinal Gambara.

Cauchy Augustin Louis de Cauchy (1789–1857) was a French engineer from the 'Corps des Ponts-et-Chaussées'. He devoted himself later to mathematics and he taught at Ecole Polytechnique, Paris, and at the Collège de France. He worked with Pierre-Simon Laplace and J. Louis Lagrange. In fluid mechanics, he contributed greatly to the analysis of wave motion.

Cavitation Formation of vapour bubbles and vapour pockets within a homogeneous liquid caused by excessive stress (Franc *et al.* 1995). Cavitation may occur in low-pressure regions where the liquid has been accelerated (e.g. turbines, marine propellers, baffle blocks of dissipation basin). Cavitation modifies the hydraulic characteristics of a system, and it is characterized by damaging erosion, additional noise, vibrations and energy dissipation.

Celsius Anders Celsius (1701–1744) was a Swedish astronomer who invented the Celsius thermometer scale (or centigrade scale) in which the interval between the freezing and boiling points of water is divided into 100°.

Celsius degree (or *degree centigrade*) Temperature scale based on the freezing and boiling points of water, 0 and 100°C respectively.

Chadar Type of narrow sloping chute peculiar to Islamic gardens and perfected by the Mughal gardens in Northern India (e.g. at Nishat Bagh). These stone channels were used to carry water from one terrace garden down to another. A steep slope ($\alpha \sim 20$–$35°$) enables sunlight to be reflected to the maximum degree. The chute bottom is very rough to enhance turbulence and free-surface aeration. The discharge per unit width is usually small, resulting in thin sheets of aerated waters.

Chézy Antoine Chézy (1717–1798) (or Antoine de Chézy) was a French engineer and member of the French 'Corps des Ponts-et-Chaussées'. He designed canals for the water supply of the city of Paris. In 1768 he proposed a resistance formula for open channel flows called the Chézy equation. In 1798, he became the Director of the Ecole Nationale Supérieure des Ponts et Chaussées after teaching there for many years.

Chézy coefficient Resistance coefficient for open channel flows first introduced by the Frenchman A. Chézy. Although it was thought to be a constant, the coefficient is a function of the relative roughness and Reynolds number.

Chimu Indian of a Yuncan tribe dwelling near Trujillo on the North-West coast of Peru. The Chimu empire lasted from AD 1250–1466. It was overrun by the Incas in 1466.

Chlorophyll One of the most important classes of pigments involved in photosynthesis. Chlorophyll is found in virtually all photosynthetic organisms. It absorbs energy from light that is then used to convert carbon dioxide to carbohydrates. Chlorophyll occurs in several distinct forms: chlorophyll a and chlorophyll b are the major types found in higher plants and green algae. High concentrations of cholophyll a occur in algal bloom.

Choke In open channel flow, a channel contraction might obstruct the flow and induce the appearance of critical flow conditions (i.e. control section). Such a constriction is sometimes called a 'choke'.

Choking flow Critical flow in a channel contraction. The term is used for both open channel flow and compressible flow.

Chord length (1) The chord or chord length of an airfoil is the straight-line distance joining the leading and trailing edges of the foil. (2) The chord length of a bubble (or bubble chord length) is the length of the straight line connecting the two intersections of the air bubble free surface with the leading tip of the measurement probe (e.g. conductivity probe, conical hot-film probe) as the bubble is transfixed by the probe tip.

Clausius Rudolf Julius Emanuel Clausius (1822–1888) was a German physicist and thermodynamicist. In 1850 he formulated the Second Law of Thermodynamics.

Clay Earthy material that is plastic when moist and that becomes hard when baked or fired. It is composed mainly of fine particles of a group of hydrous alumino-silicate minerals (particle sizes <0.05 mm usually).

Clean-air turbulence Turbulence experienced by aircraft at high altitude above the atmospheric boundary layer. It is a form of Kelvin–Helmholtz instability occurring when a destabilizing pressure gradient of the fluid become large relative to the stabilizing pressure gradient.

Clepsydra Greek name for water clock.

Cofferdam Temporary structure enclosing all or part of the construction area so that construction can proceed in dry conditions. A diversion cofferdam diverts a stream into a pipe or channel.

Cohesive sediment Sediment material of very small sizes (i.e. $<50\,\mu m$) for which cohesive bonds between particles (e.g. intermolecular forces) are significant and affect the material properties.

Colbert Jean-Baptiste Colbert (1619–1683) was a French statesman. Under King Louis XIV, he was the Minister of Finances, the Minister of 'Bâtiments et Manufactures' (buildings and industries) and the Minister of the Marine.

Conjugate depth In open channel flow, another name for sequent depth.

Control Considering an open channel, subcritical flows are controlled by the downstream conditions. This is called a 'downstream flow control'. Conversely supercritical flows are controlled only by the upstream flow conditions (i.e. 'upstream flow control').

Control section In an open channel, cross-section where critical flow conditions take place. The concept of 'control' and 'control section' are used with the same meaning.

Control surface This is the boundary of a control volume.

Control volume This refers to a region in space and is used in the analysis of situations where flow occurs into and out of the space.

Convection Transport (usually) in the direction normal to the flow direction induced by hydrostatic instability: e.g. flow past a heated plate.

Coriolis Gustave Gaspard Coriolis (1792–1843) was a French mathematician and engineer of the 'Corps des Ponts-et-Chaussées' who first described the Coriolis force (i.e. effect of motion on a rotating body).

Coriolis coefficient Kinetic energy correction coefficient named after G.G. Coriolis who introduced first the correction coefficient (Coriolis 1836).

Couette M. Couette was a French scientist who measured experimentally the viscosity of fluids with a rotating viscosimeter (Couette 1890).

Couette flow Flow between parallel boundaries moving at different velocities, named after the Frenchman M. Couette. The most common Couette flows are the cylindrical Couette flow used to measure dynamic viscosity and the two-dimensional Couette flow between parallel plates.

Couette viscosimeter This system consisting of two co-axial cylinders of radii, r_1 and r_2 rotating in opposite direction, used to measure the viscosity of the fluid placed in the space between the cylinders. In a steady state, the torque transmitted from one cylinder to another per unit length equals:

$$\frac{4\pi\mu\omega_o r_1^2 r_2^2}{r_2^2 - r_1^2}$$

where ω_o is the relative angular velocity and μ is the dynamic viscosity of the fluid.

Courant Richard Courant (1888–1972) was an American mathematician born in Germany who made significant advances in the calculus of variations.

Courant number Dimensionless number characterizing the stability of explicit finite difference schemes.

Craya Antoine Craya was a French hydraulician and Professor at the University of Grenoble.

Creager profile Spillway shape developed from a mathematical extension of the original data of Bazin in 1886–1888 (Creager 1917).

Crest of spillway Upper part of a spillway. The term 'crest of dam' refers to the upper part of an uncontrolled overflow.

Crib (1) Framework of bars or spars for strengthening. (2) Frame of logs or beams to be filled with stones, rubble or filling material and sunk as a foundation or retaining wall.

Crib dam Gravity dam built-up of boxes, cribs, crossed timbers or gabions, and filled with earth or rock.

Critical depth This is the flow depth for which the mean specific energy is minimum.

Critical flow conditions In open channel flows, the flow conditions such as the specific energy (of the mean flow) is minimum are called the critical flow conditions. With commonly used Froude number definitions, the critical flow conditions occur for $Fr = 1$. If the flow is critical, small changes in specific energy cause large changes in flow depth. In practice, critical flow over a long reach of channel is unstable.

Culvert Covered channel of relatively short length installed to drain water through an embankment (e.g. highway, railroad, dam).

Cyclopean dam Gravity masonry dam made of very large stones embedded in concrete.

Danel Pierre Danel (1902–1966) was a French hydraulician and engineer. One of the pioneers of modern hydrodynamics, he worked from 1928 to his death for Neyrpic known prior to 1948 as 'Ateliers Neyret-Beylier-Piccard et Pictet'.

Darcy Henri Philibert Gaspard Darcy (1805–1858) was a French civil engineer. He studied at Ecole Polytechnique between 1821 and 1823, and later at the Ecole Nationale Supérieure des Ponts et Chaussées (Brown 2002). He performed numerous experiments of flow resistance in pipes (Darcy 1858) and in open channels (Darcy and Bazin 1865), and of seepage flow in porous media (Darcy 1856). He gave his name to the Darcy–Weisbach friction factor and to the Darcy law in porous media.

Darcy law Law of groundwater flow motion which states that the seepage flow rate is proportional to the ratio of the head loss over the length of the flow path. It was discovered by H.P.G. Darcy (1856) who showed that, for a flow of liquid through a porous medium, the flow rate is directly proportional to the pressure difference.

Darcy–Weisbach friction factor Dimensionless parameter characterizing the friction loss in a flow. It is named after the Frenchman H.P.G. Darcy and the German J. Weisbach.

Debris Debris comprise mainly large boulders, rock fragments, gravel-sized to clay-sized material, tree and wood material that accumulate in creeks.

Degradation Lowering in channel bed elevation resulting from the erosion of sediments.

Density-stratified flows Flow field affected by density stratification caused by temperature variations in lakes, estuaries and oceans. There is a strong feedback process: i.e. mixing is affected by density stratification, which depends in turn upon mixing.

Descartes René Descartes (1596–1650) was a French mathematician, scientist, and philosopher. He is recognized as the father of modern philosophy. He stated: 'cogito ergo sum' ('I think therefore I am').

Diffusion The process whereby particles of liquids, gases or solids intermingle as the result of their spontaneous movement caused by thermal agitation and in dissolved substances move from a region of higher concentration to one of lower concentration. The term turbulent diffusion is used to describe the spreading of particles caused by turbulent agitation.

Diffusion coefficient Quantity of a substance that in diffusing from one region to another passes through each unit of cross-section per unit of time when the volume concentration is unity. The unit of the diffusion coefficient is m^2/s.

Diffusivity Another name for the diffusion coefficient.

Dimensional analysis Organization technique used to reduce the complexity of a study, by expressing the relevant parameters in terms of numerical magnitude and associated units, and grouping them into dimensionless numbers. The use of dimensionless numbers increases the generality of the results.

Dispersion Longitudinal scattering of particles by the combined effects of shear and diffusion.

Dissolved oxygen content Mass concentration of dissolved oxygen in water. It is a primary indicator of water quality: e.g. oxygenated water is considered to be of good quality.

Diversion channel Waterway used to divert water from its natural course.

Diversion dam Dam or weir built across a river to divert water into a canal. It raises the upstream water level of the river but does not provide any significant storage volume.

DOC See *Dissolved oxygen content*.

Drag reduction Reduction of the skin friction resistance in fluids in motion. In a broader sense, reduction in flow resistance (skin friction and form drag) in fluids in motion.

Drainage layer Layer of pervious material to relieve pore pressures and/or facilitate drainage: e.g. drainage layer in an earthfill dam.

Drogue (1) Sea anchor. (2) Cylindrical device towed for water sampling by a boat.

Drop (1) Volume of liquid surrounded by gas in a free-fall motion (i.e. dropping). (2) By extension, small volume of liquid in motion in a gas. (3) A rapid change of bed elevation also called step.

Droplet Small drop of liquid.

Drop structure   Single-step structure characterized by a sudden decrease in bed elevation.

Du Boys (or *Duboys*)   See P.F.D. du Boys.

Du Buat (or *Dubuat*)   See P.L.G. du Buat.

Dupuit   Arsène Jules Etienne Juvénal Dupuit (1804–1866) was a French engineer and economist. His expertise included road construction, economics, statics and hydraulics.

Earth dam   Massive earthen embankment with sloping faces and made watertight.

Ebb   Reflux of the tide towards the sea. That is, the flow motion between a high tide and a low tide. The ebb flux is maximum at mid-tide. (The opposite is the *flood*.)

Ecole Nationale Supérieure des Ponts et Chaussées, Paris   French civil engineering school founded in 1747. The direct translation is: 'National School of Bridge and Road Engineering'. Among the directors, there were the famous hydraulicians A. Chézy and G. de Prony. Other famous professors included B.F. de Bélidor, J.B.C. Bélanger, J.A.C. Bresse, G.G. Coriolis and L.M.H. Navier.

Ecole Polytechnique, Paris   Leading French engineering school founded in 1794 during the French Révolution under the leadership of Lazare Carnot and Gaspard Monge. It absorbed the state artillery school in 1802 and was transformed into a military school by Napoléon Bonaparte in 1804. Famous professors included Augustin Louis Cauchy, Jean Baptiste Joseph Fourier, Siméon-Denis Poisson, Jacques Charles François Sturm, among others.

Eddy viscosity   Another name for the momentum exchange coefficient. It is also called 'eddy coefficient' by Schlichting (1979). (See *Momentum exchange coefficient*.)

Effluent   Waste water (e.g. industrial refuse, sewage) discharged into the environment, often serving as a pollutant.

Ekman   V. Walfrid Ekman (1874–1954) was a Swedish oceanographer best known for his studies of the dynamics of ocean currents.

Embankment   Fill material (e.g. earth, rock) placed with sloping sides and with a length greater than its height.

Ephemeral channel   A river that is usually not flowing above ground except during the rainy season. Ephemeral channels are also called arroyo, wadi, wash, dry wash, oued or coulee (coulée).

Escalier d'Eau   See *Water staircase*.

Estuary   Water passage where the tide meets a river flow. An estuary may be defined as a region where salt water is diluted with fresh water.

Euler   Leonhard Euler (1707–1783) was a Swiss mathematician and physicist, and a close friend of Daniel Bernoulli.

Eulerian method   Study of a process (e.g. dispersion) from a fixed reference in space. For example, velocity measurements at a fixed point. (A different method is the *Lagrangian method*.)

Eutrophication   Process by which a body of water becomes enriched in dissolved nutrients (e.g. phosphorus, nitrogen) that stimulate the growth of aquatic plant life, often resulting in the depletion of dissolved oxygen.

Explicit method   Calculation containing only independent variables; numerical method in which the flow properties at one point are computed as functions of known flow conditions only.

Extrados   Upper side of a wing or exterior curve of a foil. The pressure distribution on the extrados must be smaller than that on the intrados to provide a positive lift force.

Face   External surface which limits a structure: e.g. air face of a dam (i.e. downstream face), water face (i.e. upstream face) of a weir.

Favre   H. Favre (1901–1966) was a Swiss professor at ETH-Zürich. He investigated both experimentally and analytically positive and negative surges. Undular surges are sometimes called Boussinesq Favre waves. Later he worked on the theory of elasticity.

Fawer jump   Undular hydraulic jump.

Fetch   The fetch, or fetch length, is the unobstructed distance over which the wind acts on the water body. Fetch is an important factor in wind wave development, with increasing wave height with increasing fetch up to a maximum of 1600 km. The wave height does not increase with increasing fetch beyond that distance.

Fick   Adolf Eugen Fick was a 19th Century German physiologist who developed the diffusion equation for neutral particle (Fick 1855).

Finite differences   Approximate solutions of partial differential equations, which consists essentially of replacing each partial derivative by a ratio of differences between two immediate values e.g., $\partial v/\partial t \approx \delta v/\delta t$. The method was first introduced by Runge (1908).

Fischer   Hugo B. Fischer (1937–1983) was a Professor at the University of California, Berkeley. He earned his BSc, MS and PhD at the California Institute of Technology. He was a Professor of Civil Engineering at the University of California, Berkeley from 1966 until 1983. Fischer was a recognized authority in the salt-water intrusion, water pollution, heat dispersion in waterways, and the mixing in rivers and oceans (e.g. Fischer *et al.* 1979). He died in a glider accident in May 1983.

Fixed-bed channel   The bed and sidewalls are non-erodible. Neither erosion nor accretion occurs.

Flashboard   A board or a series of boards placed on or at the side of a dam to increase the depth of water. Flashboards are usually lengths of timber, concrete or steel placed on the crest of a spillway to raise the upstream water level.

Flash flood   Flood of short duration with a relatively high peak flow rate.

Flashy   Term applied to rivers and streams whose discharge can rise and fall suddenly, and is often unpredictable.

Flettner   Anton Flettner (1885–1961) was a German engineer and inventor. In 1924 he designed a rotor ship based on the Magnus effect. Large vertical circular cylinders were mounted on the ship. They were mechanically rotated to provide circulation and to propel the ship. More recently a similar system was developed for the ship 'Alcyone' of Jacques-Yves Cousteau.

Flip bucket   A flip bucket or ski-jump is a concave curve at the downstream end of a spillway, to deflect the flow into an upward direction. Its purpose is to throw the water clear of the hydraulic structure and to induce the disintegration of the jet in air.

Flood   (1) High-water stage in which the river overflows its banks. (2) The flux of the rising tide. In coastal zones, the flood tide is the rising tide. (The opposite of the flood is the *ebb*.)

Fog   Small water droplets near ground level forming a cloud sufficiently dense to reduce drastically visibility. The term fog refers also to clouds of smoke particles or ice particles.

Forchheimer   Philipp Forchheimer (1852–1933) was an Austrian hydraulician who contributed significantly to the study of groundwater hydrology.

Fortier   André Fortier was a French scientist and engineer. He became later Professor at the Sorbonne, Paris.

Fourier   Jean Baptiste Joseph Fourier (1768–1830) was a French mathematician and physicist known for his development of the Fourier series. In 1794 he was offered a professorship of mathematics at the Ecole Normale in Paris and was later appointed at the Ecole Polytechnique. In 1798 he joined the expedition to Egypt lead by (then) General Napoléon Bonaparte. His research in mathematical physics culminated with the classical study 'Théorie Analytique de la Chaleur' (Fourier 1822) in which he enunciated his theory of heat conduction.

Free surface   Interface between a liquid and a gas. More generally a free surface is the interface between the fluid (at rest or in motion) and the atmosphere. In two-phase gas–liquid flow, the term 'free surface' includes also the air–water interface of gas bubbles and liquid drops.

Free-surface aeration   Natural aeration occurring at the free surface of high velocity flows is referred to as free-surface aeration or self-aeration.

French Revolution (Révolution Française)   Revolutionary period that shook France between 1787 and 1799. It reached a turning point in 1789 and led to the destitution of the monarchy in 1791. The constitution of the First Republic was drafted in 1790 and adopted in 1791.

Frontinus   Sextus Julius Frontinus (AD 35–103 or 104) was a Roman engineer and soldier. After AD 97, he was 'curator aquarum' in charge of the water supply system of Rome. He dealt with discharge measurements in pipes and canals. In his analysis he correctly related the proportionality between discharge and cross-sectional area. His book '*De Aquaeductu Urbis Romae*' (*Concerning the Aqueducts of the City of Rome*) described the operation and maintenance of Rome water supply system.

Froude   William Froude (1810–1879) was a English naval architect and hydrodynamicist who invented the dynamometer and used it for the testing of model ships in towing tanks. He was assisted by his son

Robert Edmund Froude who, after the death of his father, continued some of his work. In 1868, he used Reech's law of similarity to study the resistance of model ships.

Froude number The Froude number is proportional to the square root of the ratio of the inertial forces over the weight of fluid. The Froude number is used generally for scaling free-surface flows, open channels and hydraulic structures. Although the dimensionless number was named after William Froude, several French researchers used it before. Dupuit (1848) and Bresse (1860) highlighted the significance of the number to differentiate the open channel flow regimes. Bazin (1865a) confirmed experimentally the findings. Ferdinand Reech introduced the dimensionless number for testing ships and propellers in 1852. The number is called the Reech–Froude number in France.

G.K. formula Empirical resistance formula developed by the Swiss engineers E. Ganguillet and W.R. Kutter in 1869.

Gabion A gabion consists of rockfill material enlaced by a basket or a mesh. The word 'gabion' originates from the Italian 'gabbia' cage.

Gabion dam Crib dam built-up of gabions.

Gas transfer Process by which gas is transferred into or out of solution: i.e. dissolution or desorption respectively.

Gate Valve or system for controlling the passage of a fluid. In open channels the two most common types of gates are the underflow gate and the overflow gate.

Gauckler Philippe Gaspard Gauckler (1826–1905) was a French engineer and member of the French 'Corps des Ponts-et-Chaussées'. He re-analysed the experimental data of Darcy and Bazin (1865), and in 1867 he presented a flow resistance formula for open channel flows (Gauckler–Manning formula) sometimes called improperly the Manning equation (Gauckler 1867). Later he became Directeur des Antiquités et des Beaux-Arts (Director of Anquities and Fine Arts) for the French Republic in Tunisia and he directed an extensive survey of Roman hydraulic works in Tunisia.

Gay-Lussac Joseph-Louis Gay-Lussac (1778–1850) was a French chemist and physicist.

Ghaznavid (or Ghaznevid) one of the Moslem dynasties (10–12 Centuries) ruling South-Western Asia. Its capital city was at Ghazni (Afghanistan).

Gradually varied flow It is characterized by relatively small changes in velocity and pressure distributions over a short distance (e.g. long waterway).

Gravity dam Dam which relies on its weight for stability. Normally the term 'gravity dam' refers to masonry or concrete dam.

Grille d'eau (French for 'water screen') a series of water jets or fountains aligned to form a screen. An impressive example is 'les Grilles d'Eau' designed by A. Le Nôtre at Vaux-le Vicomte, France.

Gulf Stream Warm ocean current flowing in the North Atlantic north-eastward. The Gulf Stream is part of a general clockwise-rotating system of currents in the North Atlantic.

Hartree Douglas R. Hartree (1897–1958) was an English physicist. He was a Professor of Mathematical Physics at Cambridge. His approximation to the Schrödinger equation is the basis for the modern physical understanding of the wave mechanics of atoms. The scheme is sometimes called the Hartree–Fock method after the Russian physicist V. Fock who generalized Hartree's scheme.

Hasmonean Designing the family or dynasty of the Maccabees, in Israel. The Hasmonean Kingdom was created following the uprising of the Jews in BC 166.

Helmholtz Hermann Ludwig Ferdinand von Helmholtz (1821–1894) was a German scientist who made basic contributions to physiology, optics, electrodynamics and meteorology.

Hennin Georg Wilhelm Hennin (1680–1750) was a young Dutchman hired by the tsar Peter the Great to design and build several dams in Russia (Danilveskii 1940, Schnitter 1994). He went to Russia in 1698 and stayed until his death in April 1750.

Hero of Alexandria Greek mathematician (1st Century AD) working in Alexandria, Egypt. He wrote at least 13 books on mathematics, mechanics and physics. He designed and experimented the first steam engine. His treatise *Pneumatica* described Hero's fountain, siphons, steam-powered engines, a water organ, and hydraulic and mechanical water devices. It influenced directly the waterworks design during the Italian Renaissance. In his book *Dioptra*, Hero stated rightly the concept of continuity for incompressible flow: the discharge being equal to the area of the cross-section of the flow times the speed of the flow.

Himyarite Important Arab tribe of antiquity dwelling in Southern Arabia (BC 700 to AD 550).

Hohokams Native Americans in South-West America (Arizona), they build several canal systems in the Salt River Valley during the period BC 350 to AD 650. They migrated to Northern Mexico around AD 900 where they build other irrigation systems.

Hokusai Katsushita Japanese painter and wood engraver (1760–1849). His *Thirty-Six Views of Mount Fuji* (1826–1833) are world known.

Huang Chun-Pi One of the greatest masters of Chinese painting in modern China (1898–1991). Several of his paintings included mountain rivers and waterfalls: e.g. 'Red trees and waterfalls', 'The house by the water-falls', 'Listening to the sound of flowing waters', 'Water-falls'.

Humboldt Alexander von Humboldt (1769–1859) was a German explorer who was a major figure in the classical period of physical geography and biogeography.

Humboldt current Flows off the west coast of South America. It was named after Alexander von Humboldt who took measurements in 1802 that showed the coldness of the flow. It is also called Peru current.

Hydraulic diameter This is defined as the equivalent pipe diameter: i.e. four times the cross-sectional area divided by the wetted perimeter. The concept was first expressed by the Frenchman P.L.G. du Buat (1779).

Hydraulic fill dam Embankment dam constructed of materials which are conveyed and placed by suspension in flowing water.

Hydraulic jump Transition from a rapid (supercritical flow) to a slow flow motion (subcritical flow). Although the hydraulic jump was described by Leonardo da Vinci, the first experimental investigations were published by Giorgio Bidone in 1820. The present theory of the jump was developed by Bélanger (1828) and it has been verified experimentally numerous researchers (e.g. Bakhmeteff and Matzke 1936).

Hyperconcentrated flow Sediment-laden flow with large suspended sediment concentrations (i.e. typically >1% in volume). Spectacular hyperconcentrated flows are observed in the Yellow River basin (China) with volumetric concentrations >8%.

Ideal fluid Frictionless and incompressible fluid. An ideal fluid has zero viscosity: i.e. it cannot sustain shear stress at any point.

Idle discharge Old expression for spill or waste water flow.

Implicit method Calculation in which the dependent variable and the one or more independent variables are not separated on opposite sides of the equation; numerical method in which the flow properties at one point are computed as functions of both independent and dependent flow conditions.

Inca South-American Indian of the Quechuan tribes of the highlands of Peru. The Inca civilization dominated Peru between AD 1200 and 1532. The domination of the Incas was terminated by the Spanish conquest.

Inflow (1) Upstream flow. (2) Incoming flow.

Inlet (1) Upstream opening of a culvert, pipe or channel. (2) A tidal inlet is a narrow water passage between peninsulas or islands.

Intake Any structure in a reservoir through which water can be drawn into a waterway or pipe. By extension, upstream end of a channel.

Interface Surface forming a common boundary of two phases (e.g. gas–liquid interface) or two fluids.

International system of units See Système international d'unités.

Intrados Lower side of a wing or interior curve of a foil.

Invert (1) Lowest portion of the internal cross-section of a conduit. (2) Channel bed of a spillway. (3) Bottom of a culvert barrel.

Inviscid flow This is a non-viscous flow.

Ippen Arthur Thomas Ippen (1907–1974) was Professor in Hydrodynamics and Hydraulic Engineering at MIT (USA). Born in London of German parents, educated in Germany (Technische Hochschule in Aachen), he moved to USA in 1932, where he obtained the MS and PhD degrees at the California Institute of Technology. There he worked on high-speed free-surface flows with Theodore von Karman. In 1945 he was appointed at MIT until his retirement in 1973.

Irrotational flow This is defined as a zero vorticity flow. Fluid particles within a region have no rotation. If a frictionless fluid has no rotation at rest, any later motion of the fluid will be irrotational. In irrotational flow each element of the moving fluid undergoes no net rotation, with respect to chosen coordinate axes, from one instant to another.

JHRC Jump height rating curve.

JHRL Jump height rating level.

Jet d'eau French expression for water jet. The term is commonly used in architecture and landscape.

Jevons W.S. Jevons (1835–1882) was an English chemist and economist. His work on salt finger intrusions (Jevons 1858) was a significant contribution to the understanding of double-diffusive convection. He performed his experiments in Sydney, Australia, 23 years prior to Rayleigh's experiments (Rayleigh 1883)

Karman Theodore von Karman (or von Kármán) (1881–1963) was a Hungarian fluid dynamicist and aerodynamicist who worked in Germany (1906–1929) and later in USA. He was a student of Ludwig Prandtl in Germany. He gave his name to the vortex shedding behind a cylinder (Karman vortex street).

Karman constant (or von Karman constant) This is the 'universal' constant of proportionality between the Prandtl mixing length and the distance from the boundary. Experimental results indicate that K = 0.40.

Kelvin (Lord) William Thomson (1824–1907), Baron Kelvin of Largs, was a British physicist. He contributed to the development of the Second Law of Thermodynamics, the absolute temperature scale (measured in Kelvin), the dynamical theory of heat, fundamental work in hydrodynamics, …

Kelvin–Helmholtz instability Instability at the interface of two ideal fluids in relative motion. The instability can be caused by a destabilizing pressure gradient of the fluid (e.g. clean-air turbulence) or free-surface shear (e.g. fluttering fountain). It is named after H.L.F. Helmoltz who solved first the problem (Helmholtz 1868) and Lord Kelvin (1871).

Kennedy Prof. John Fisher Kennedy (1933–1991) was a hydraulic professor at the University of Iowa. He succeeded Hunter Rouse as Head of the Iowa Institute of Hydraulic Research.

Keulegan Garbis Hovannes Keulegan (1890–1989) was an Armenian mathematician who worked as hydraulician for the US Bureau of Standards since its creation in 1932.

Kuroshio It is a strong surface oceanic current of flowing north-easterly in North Pacific, between the Philippines and the east coast of Japan. It travels at rates ranging between 0.05 and 0.3 m/s and it is also called Japan current. Known to European geographers as early as 1650, it is called Kuroshio (Black Current) because it appears a deeper blue than surrounding seas by Captain James Cook.

Lagrange Joseph-Louis Lagrange (1736–1813) was a French mathematician and astronomer. During the 1789 Revolution, he worked on the Committee to Reform the Metric System. He was a Professor of Mathematics at the École Polytechnique from the start in 1795.

Lagrangian method Study of a process in a system of coordinates moving with an individual particle. For example, the study of ocean currents with buoys. (A different method is the *Eulerian method*.)

Laminar flow This is characterized by fluid particles moving along smooth paths in laminas or layers, with one layer gliding smoothly over an adjacent layer. Laminar flows are governed by Newton's law of viscosity which relates the shear stress to the rate of angular deformation:

$$\tau = \mu \, \frac{\partial V}{\partial y}$$

Langevin Paul Langevin (1879–1946) was a French physicist, specialist in magnetism, ultrasonics, and relativity. In 1905 Einstein identified Brownian motion as due to imbalances in the forces on a particle resulting from molecular impacts from the liquid. Shortly thereafter, Paul Langevin formulated a theory in which the minute fluctuations in the position of the particle were due explicitly to a random force. His approach had great utility in describing molecular fluctuations in other systems, including non-equilibrium thermodynamics.

Laplace Pierre-Simon Laplace (1749–1827) was a French mathematician, astronomer and physicist. He is best known for his investigations into the stability of the solar system.

LDA velocimeter   Laser Doppler anemometer system.

Left abutment   Abutment on the left-hand side of an observer when looking downstream.

Left bank (left wall)   Looking downstream, the left bank or the left channel wall is on the left.

Leonardo da Vinci   Italian artist (painter and sculptor) who extended his interest to medicine, science, engineering and architecture (AD 1452–1519).

Lining   Coating on a channel bed to provide water tightness, to prevent erosion or to reduce friction.

Lumber   Timber sawed or split into boards, planks or staves.

McKay   Prof. Gordon M. McKay (1913–1989) was Professor in Civil Engineering at the University of Queensland.

Mach   Ernst Mach (1838–1916) was an Austrian physicist and philosopher. He established important principles of optics, mechanics and wave dynamics.

Mach number   See *Sarrau–Mach number*.

Magnus   H.G. Magnus (1802–1870) was a German physicist who investigated the so-called Magnus effect in 1852.

Magnus effect   A rotating cylinder, placed in a flow, is subjected to a force acting in the direction normal to the flow direction: i.e. a lift force which is proportional to the flow velocity times the rotation speed of the cylinder. This effect, called the Magnus effect, has a wide range of applications (Swanson 1961).

Manning   Robert Manning (1816–1897) was the Chief Engineer of the Office of Public Works, Ireland. In 1889, he presented two formulae (Manning 1890). One was to become the so-called 'Gauckler–Manning formula' but Robert Manning did prefer to use the second formula that he gave in his paper. It must be noted that the Gauckler–Manning formula was proposed first by the Frenchman P.G. Gauckler (1867).

Mariotte   Abbé Edme Mariotte (1620–1684) was a French physicist and plant physiologist. He was the Member of the Académie des Sciences de Paris and wrote a fluid mechanics treaty published after his death (Mariotte 1686).

Mascaret   Tidal bore in French.

Masonry dam   Dam constructed mainly of stone, brick or concrete blocks jointed with mortar.

Meandering channel   Alluvial stream characterized by a series of alternating bends (i.e. meanders) as a result of alluvial processes.

MEL culvert   See *Minimum energy loss culvert*.

Metric system   See Système métrique.

Minimum energy loss culvert   Culvert designed with very smooth shapes to minimize energy losses. The design of an MEL culvert is associated with the concept of constant total head. The inlet and outlet must be streamlined in such a way that significant form losses are avoided (Apelt 1983).

Mixing   Process by which contaminants combine into a more or less uniform whole by diffusion or dispersion.

Mixing length   The mixing length theory is a turbulence theory developed by L. Prandtl, first formulated in 1925 (Prandtl 1925). Prandtl assumed that the mixing length is the characteristic distance travelled by a particle of fluid before its momentum is changed by the new environment.

Mochica   (1) South American civilization (AD 200–1000) living in the Moche River Valley, Peru along the Pacific coastline. (2) Language of the Yuncas.

Mole   Mass numerically equal in grams to the relative mass of a substance (i.e. 12 g for carbon-12). The number of molecules in 1 mol of gas is 6.0221367×10^{23} (i.e. Avogadro number).

Momentum exchange coefficient   In turbulent flows the apparent kinematic viscosity (or kinematic eddy viscosity) is analogous to the kinematic viscosity in laminar flows. It is called the momentum exchange coefficient, the eddy viscosity or the eddy coefficient. The momentum exchange coefficient is proportional to the shear stress divided by the strain rate. It was first introduced by the Frenchman J.V. Boussinesq (1877, 1896).

Monge   Gaspard Monge (1746–1818), Comte de Péluse, was a French mathematician who invented descriptive geometry and pioneered the development of analytical geometry. He was a prominent figure during the French Revolution, helping to establish the Système Métrique and the École Polytechnique, and being Minister for the Navy and colonies between 1792 and 1793.

Moor   (1) Native of Mauritania, a region corresponding to parts of Morocco and Algeria. (2) Moslem of native North African races.

Morning-Glory spillway   Vertical discharge shaft, more particularly the circular hole form of a drop inlet spillway. The shape of the intake is similar to a Morning-Glory flower (American native plant (*Ipomocea*)). It is sometimes called a tulip intake.

Mud   Slimy and sticky mixture of solid material and water.

Munk   Walter H. Munk was an American geophysicist and oceanographer who expanded Sverdrup's work on ocean circulation.

Mughal (or Mughul or Mogul or Moghul)   Name or adjective referring to the Mongol conquerors of India and to their descendants. The Mongols occupied India from 1526 up to the 18th Century although the authority of the Mughal Emperor became purely nominal after 1707. The fourth emperor, Jahangir (1569–1627), married a Persian Princess Mehr-on Nesa who became known as Nur Jahan. His son Shah Jahan (1592–1666) built the famous Taj Mahal between 1631 and 1654 in memory of his favourite wife Arjumand Banu better known by her title: Mumtaz Mahal or Taj Mahal.

Nabataean   Habitant from an ancient kingdom to the East and South-East of Palestine that included the Neguev Desert. The Nabataean Kingdom lasted from around BC 312 to AD 106. The Nabataeans built a large number of soil-and-retention dams. Some are still in use today.

Nappe flow   Flow regime on a stepped chute where the water bounces from one step to the next one as a succession of free-fall jets.

Navier   Louis Marie Henri Navier (1785–1835) was a French engineer who primarily designed bridges but also extended Euler's equations of motion (Navier 1823).

Navier–Stokes equation   Momentum equation applied to a small control volume of incompressible fluid. It is usually written in vector notation. The equation was first derived by L. Navier in 1822 and S.D. Poisson in 1829 by a different method. It was derived later in a more modern manner by A.J.C. Barré de Saint-Venant in 1843 and G.G. Stokes in 1845.

Neap tide   Tide of minimum range occurring at the first and the third quarters of the moon. (The opposite is the *spring tide*.)

Negative surge   A negative surge results from a sudden change in flow that decreases the flow depth. It is a retreating wave front moving upstream or downstream.

Nephelometric turbidity units   Units of water turbidity. It is a measure of how light is scattered by suspended particulate material in the water. Waters with a turbidity level of >5 NTU are not safe for recreational use or human consumption. Waters with levels >25 NTU cannot sustain aquatic life.

Newton   Sir Isaac Newton (1642–1727) was an English mathematician and physicist. His contributions in optics, mechanics and mathematics were fundamental.

Nikuradse   J. Nikuradse was a German engineer who investigated experimentally the flow field in smooth and rough pipes (Nikuradse 1932, 1933).

Non-uniform equilibrium flow   The velocity vector varies from place to place at any instant: steady non-uniform flow (e.g. flow through an expanding tube at a constant rate) and unsteady non-uniform flow (e.g. flow through an expanding tube at an increasing flow rate).

Normal depth   Uniform equilibrium open channel flow depth.

Normal flow conditions   At uniform equilibrium in an open channel, the momentum principle states that the boundary shear force (i.e. flow resistance) equals exactly the gravity force component in the flow directions. These conditions are called normal flow conditions or uniform equilibrium flow conditions.

NTU   Units of turbidity measurement. See *Nephelometric turbidity units*.

Nutrient   Substance that an organism must obtain from its surroundings for growth and the sustainment of life. Nitrogen and phosphorus are important nutrients for plant growth. High levels of nitrogen and phosphorus may cause excessive growth, weed proliferation and algal bloom, leading to eutrophication.

Obvert   Roof of the barrel of a culvert. Another name is soffit.

One-dimensional flow   Neglects the variations and changes in velocity and pressure transverse to the main flow direction. An example of one-dimensional flow can be the flow through a pipe.

One-dimensional model   Model defined with one spatial coordinate, the variables being averaged in the other two directions.

Organic compound   Class of chemical compounds in which one or more atoms of carbon are linked to atoms of other elements (e.g. hydrogen, oxygen, nitrogen).

Organic matter   Substance derived from living organisms.

Outflow   Downstream flow.

Outlet   (1) Downstream opening of a pipe, culvert or canal. (2) Artificial or natural escape channel.

pH   Measure of acidity and alkalinity of a solution. It is a number on a scale on which a value of 7 represents neutrality, lower numbers indicate increasing acidity and higher numbers increasing alkalinity. On the pH scale, each unit represents a 10-fold change in acidity or alkalinity.

Pascal   Blaise Pascal (1623–1662) was a French mathematician, physicist and philosopher. He developed the modern theory of probability. Between 1646 and 1648, he formulated the concept of pressure and showed that the pressure in a fluid is transmitted through the fluid in all directions. He measured also the air pressure both in Paris and on the top of a mountain overlooking Clermont-Ferrand (France).

Pascal   Unit of pressure named after the Frenchman B. Pascal: 1 pascal equals a newton per square metre.

Pelton turbine (or wheel)   Impulse turbine with one to six circular nozzles that deliver high-speed water jets into air which then strike the rotor blades shaped like scoop and known as bucket. A simple bucket wheel was designed by Sturm in the 17th Century. The American Lester Allen Pelton patented the actual double-scoop (or double-bucket) design in 1880.

Pervious zone   Part of the cross-section of an embankment comprising material of high permeability.

Photosynthesis   This is the process by which green plants and certain other organisms transform light energy into chemical energy. Photosynthesis in green plants harnesses sunlight energy to convert carbon dioxide, water and minerals, into organic compounds and gaseous oxygen.

Pitot   Henri Pitot (1695–1771) was a French mathematician, astronomer and hydraulician. He was a member of the French Académie des Sciences from 1724. He invented the Pitot tube to measure flow velocity in the Seine River (first presentation in 1732 at the Académie des Sciences de Paris).

Pitot tube   Device to measure flow velocity. The original Pitot tube consisted of two tubes, one with an opening facing the flow. L. Prandtl developed an improved design (e.g. Howe 1949) which provides the total head, piezometric head and velocity measurements. It is called a Prandtl–Pitot tube and more commonly a Pitot tube.

Pitting   Formation of small pits and holes on surfaces due to erosive or corrosive action (e.g. cavitation pitting).

Plato   Greek philosopher (about BC 428–347) who influenced greatly Western philosophy.

Plunging jet   Liquid jet impacting (or impinging) into a receiving pool of liquid.

Poiseuille   Jean-Louis Marie Poiseuille (1799–1869) was a French physician and physiologist who investigated the characteristics of blood flow. He carried out experiments and formulated first the expression of flow rates and friction losses in laminar fluid flow in circular pipes (Poiseuille 1839).

Poiseuille flow   Steady laminar flow in a circular tube of constant diameter.

Poisson   Siméon Denis Poisson (1781–1840) was a French mathematician and scientist. He developed the theory of elasticity, a theory of electricity and a theory of magnetism.

Pororoca   Tidal bore of the Amazon River.

Positive surge   A positive surge results from a sudden change in flow that increases the depth. It is an abrupt wave front. The unsteady flow conditions may be solved as a quasi-steady flow situation.

Potential flow   Ideal-fluid flow with irrotational motion.

Prandtl   Ludwig Prandtl (1875–1953) was a German physicist and aerodynamicist who introduced the concept of boundary layer (Prandtl 1904) and developed the turbulent 'mixing length' theory. He was Professor at the University of Göttingen.

Preissmann   Alexandre Preissmann (1916–1990) was born and educated in Switzerland. From 1958, he worked on the development of hydraulic mathematical models at Sogreah in Grenoble.

Prismatic   A prismatic channel has an unique cross-sectional shape independent of the longitudinal distance along the flow direction. For example, a rectangular channel of constant width is prismatic.

Prony   Gaspard Clair François Marie Riche de Prony (1755–1839) was a French mathematician and engineer. He succeeded A. Chézy as Director General of the Ecole Nationale Supérieure des Ponts et Chaussées, Paris during the French Revolution.

Radial gate   Underflow gate for which the wetted surface has a cylindrical shape.

Rankine   William J.M. Rankine (1820–1872) was a Scottish engineer and physicist. His contribution to thermodynamics and steam engine was important. In fluid mechanics, he developed the theory of sources and sinks, and used it to improve ship hull contours. One ideal-fluid flow pattern, the combination of uniform flow, source and sink, is named after him: i.e. flow past a Rankine body.

Rapidly varied flow   This flow is characterized by large changes over a short distance (e.g. sharp-crested weir, sluice gate, hydraulic jump).

Rayleigh   John William Strutt, Baron Rayleigh (1842–1919) was an English scientist who made fundamental findings in acoustics and optics. His works are the basics of wave propagation theory in fluids. He received the Nobel Prize for Physics in 1904 for his work on the inert gas argon.

Reech   Ferdinand Reech (1805–1880) was a French naval instructor who proposed first the Reech–Froude number in 1852 for the testing of model ships and propellers.

Rehbock   Theodor Rehbock (1864–1950) was a German hydraulician and Professor at the Technical University of Karlsruhe. His contribution to the design of hydraulic structures and physical modelling is important.

Renaissance   Period of great revival of art, literature and learning in Europe in the 14–16 Centuries.

Reynolds   Osborne Reynolds (1842–1912) was a British physicist and mathematician who expressed first the Reynolds number (Reynolds 1883) and later the Reynolds stress (i.e. turbulent shear stress).

Reynolds number   Dimensionless number proportional to the ratio of the inertial force over the viscous force. In pipe flows, the Reynolds number is commonly defined as:

$$Re = \rho \, \frac{VD}{\mu}$$

Rheology   Science describing the deformation of fluid and matter.

Riblet   Series of longitudinal grooves. Riblets are used to reduce skin drag (e.g. on aircraft, ship hull). The presence of longitudinal grooves along a solid boundary modifies the bottom shear stress and the turbulent bursting process. Optimum groove width and depth are about 20–40 times the laminar sublayer thickness (i.e. about 10–20 μm in air, 1–2 mm in water).

Richardson   Lewis Fry Richardson (1881–1953) was a British meteorologist who pioneered mathematical weather forecasting. It is believed that he took interest in the dispersion of smoke from shell explosion while he was an ambulance driver on the World War I battle front, leading to his classical publications (Richardson 1922, 1926).

Richardson number   Dimensionless number characterizing density stratification, commonly used to predict the occurrence of fluid turbulence and the destruction of density currents in water or air. A common definition is:

$$Ri = \frac{g}{\rho} \, \frac{\partial \rho / \partial y^2}{(\partial V / \partial y)^2}$$

Richelieu   Armand Jean du Plessis (1585–1642), Duc de Richelieu and French Cardinal, was the Prime Minister of King Louis XIII of France from 1624 to his death.

Riemann   Bernhard Georg Friedrich Riemann (1826–1866) was a German mathematician.

Right abutment   Abutment on the right-hand side of an observer when looking downstream.

Right bank (right wall)   Looking downstream, the right bank or the right channel wall is on the right.

Riquet   Pierre Paul Riquet (1604–1680) was the Designer and Chief Engineer of the Canal du Midi built between 1666 and 1681. The canal provided an inland route between the Atlantic Ocean and the Mediterranean Sea across Southern France.

Rockfill   Material composed of large rocks or stones loosely placed.

Rockfill dam   Embankment dam in which >50% of the total volume comprises compacted or dumped pervious natural stones.

Roller   In hydraulics, large-scale turbulent eddy: e.g. the roller of a hydraulic jump.

Roller compacted concrete (RCC)   Roller compacted concrete is defined as a no-slump consistency concrete that is placed in horizontal lifts and compacted by vibratory rollers. RCC has been commonly used as construction material of gravity dams since the 1970s.

Roll wave   On steep slopes free-surface flows become unstable. The phenomenon is usually clearly visible at low flow rates. The waters flow down the chute in a series of wave fronts called roll waves.

Rouse   Hunter Rouse (1906–1996) was an eminent hydraulician who was Professor and Director of the Iowa Institute of Hydraulic Research at the University of Iowa (USA).

SAF   St Anthony's Falls hydraulic laboratory at the University of Minnesota (USA).

Sabaen   Ancient name of the people of Yemen in Southern Arabia. Renowned for the visit of the Queen of Sabah (or Sheba) to the King of Israel around BC 950 and for the construction of the Marib dam (BC 115 to AD 575). The fame of the Marib Dam was such that its final destruction in AD 575 was recorded in the Koran.

Saltation   (1) Action of leaping or jumping. (2) In sediment transport, particle motion by jumping and bouncing along the bed.

Saint-Venant   See *Barré de Saint-Venant*.

Salinity   Amount of dissolved salts in water. The definition of the salinity is based on the electrical conductivity of water relative to a specified solution of KCl and H_2O (Bowie *et al.* 1985). In surface waters of open oceans, the salinity ranges from 33 to 37 ppt typically.

Salt   (1) Common salt, or sodium chloride (NaCl), is a crystalline compound that is abundant in Nature. (2) When mixed, acids and bases neutralize one another to produce salts, that is substances with a salty taste and none of the characteristic properties of either acids or bases.

Sarrau   French Professor at Ecole Polytechnique, Paris, who first introduced the Sarrau–Mach number (Sarrau 1884).

Sarrau–Mach number   Dimensionless number proportional to the ratio of inertial forces over elastic forces. Although the number is commonly named after E. Mach who introduced it in 1887, it is often called the Sarrau number after Prof. Sarrau who first highlighted the significance of the number (Sarrau 1884). The Sarrau–Mach number was once called the Cauchy number as a tribute to Cauchy's contribution to wave motion analysis.

Scalar   A quantity that has a magnitude described by a real number and no direction. A scalar means a real number rather than a vector.

Scale effect   Discrepancy between model and prototype resulting when one or more dimensionless parameters have different values in the model and prototype.

Scour   Bed material removal caused by the eroding power of the flow.

Sea water   Sea water is a complex mixture of 96.5% water, 2.5% salts and smaller amounts of other substances. The most abundant salts are sodium chloride (NaCl, 29.536 ppt), sulphate (SO_4, 2.649 ppt), magnesium (Mg, 1.272 ppt), calcium (Ca, 0.400 ppt) and potassium (K, 0.380 ppt) (Riley and Skirrow 1965, Open University Course Team 1995).

Secchi   Peitro Angelo Secchi was an Italian astronomer. In 1865, he initiated experiments using disks of various sizes and colors to determine water clarity.

Secchi disk   A simple weighted disk with a 20–40 cm diameter that is divided into four quadrates alternating black and white colours. The disk is lowered into turbid waters until it can no longer be seen and lifted back up until it is seen again. The average of the two depths gives the clarity of the water.

Secondary current   It is a flow generated at right angles to the primary current. It is a direct result of the Reynolds stresses and exists in any non-circular conduits (Liggett 1994, pp. 256–259). In natural rivers, they are significant at bends, and between a flood plain and the main channel.

Sediment   Any material carried in suspension by the flow or as bed load which would settle to the bottom in absence of fluid motion.

Sediment load   Material transported by a fluid in motion.

Sediment transport   Transport of material by a fluid in motion.

Sediment transport capacity   Ability of a stream to carry a given volume of sediment material per unit time for given flow conditions. It is the sediment transport potential of the river.

Sediment yield   Total sediment outflow rate from a catchment, including bed load and suspension.

Seepage   Interstitial movement of water that may take place through a dam, its foundation or abutments.

Seiche   Rhythmic water oscillations caused by resonnance in a lake, harbour or estuary. This results in the formation standing waves. In first approximation, the resonnance period[1] of a water body with characteristic surface length L and depth d is about: $2L/\sqrt{gd}$.

Sennacherib (or *Akkadian Sin-Akhkheeriba*)   King of Assyria (BC 705–681), son of Sargon II (who ruled during BC 722–705). He build a huge water supply for his capital city Nineveh (near the actual Mossul, Iraq) in several stages. The latest stage comprised several dams and over 75 km of canals and paved channels.

Separation   In a boundary layer, a deceleration of fluid particles leading to a reversed flow within the boundary layer is called a separation. The decelerated fluid particles are forced outwards and the boundary layer is separated from the wall. At the point of separation, the velocity gradient normal to the wall is zero:

$$\left(\frac{\partial V_x}{\partial y}\right)_{y=0} = 0$$

Separation point   In a boundary layer, intersection of the solid boundary with the streamline dividing the separation zone and the deflected outer flow. The separation point is a stagnation point.

Sequent depth   In open channel flow, the solution of the momentum equation at a transition between supercritical and subcritical flow gives two flow depths (upstream and downstream flow depths). They are called sequent depths.

Sewage   Refused liquid or waste matter carried off by sewers. It may be a combination of water-carried wastes from residences and industries together with groundwater, surface water and storm water.

Sewer   An artificial subterranean conduit to carry off water and waste matter.

Shear flow   The term shear flow characterizes a flow with a velocity gradient in a direction normal to the mean flow direction: e.g. in a boundary layer flow along a flat plate, the velocity is zero at the boundary and equals the free-stream velocity away from the plate. In a shear flow, momentum (i.e. per unit volume: ρV) is transferred from the region of high velocity to that of low velocity. The fluid tends to resist the shear associated with the transfer of momentum.

Shear stress   In a shear flow, the shear stress is proportional to the rate of transfer of momentum. In laminar flows, Newton's law of viscosity states:

$$\tau = \mu \frac{\partial V}{\partial y}$$

where τ is the shear stress, μ is the dynamic viscosity of the flowing fluid, V is the velocity and y is the direction normal to the flow direction. For large shear stresses, the fluid can no longer sustain the viscous shear stress and turbulence spots develop. After apparition of turbulence spots, the turbulence expands rapidly to the entire shear flow. The apparent shear stress in turbulent flow is expressed as:

$$\tau = \rho(\nu + \nu_T) \frac{\partial V}{\partial y}$$

where ρ is the fluid density, ν is the kinematic viscosity (i.e. $\nu = \mu\rho$), and ν_T is a factor depending upon the fluid motion and called the eddy viscosity or momentum exchange coefficient in turbulent flow.

Shock waves   With supercritical flows, a flow disturbance (e.g. change of direction, contraction) induces the development of shock waves propagating at the free surface across the channel (e.g. Ippen and Harleman 1956). Shock waves are called also lateral shock waves, oblique hydraulic jumps, Mach waves, crosswaves, diagonal jumps.

---

[1] Sometimes called sloshing motion period.

Side-channel spillway   A side-channel spillway consists of an open spillway (along the side of a channel) discharging into a channel running along the foot of the spillway and carrying the flow away in a direction parallel to the spillway crest (e.g. Arizona-side spillway of the Hoover Dam, USA).

Similitude   Correspondence between the behaviour of a model and that of its prototype, with or without geometric similarity. The correspondence is usually limited by scale effects.

Siphon   Pipe system discharging waters between two reservoirs or above a dam in which the water pressure becomes sub-atmospheric. The shape of a simple siphon is close to an omega (i.e. Ω-shape). Inverted siphons carry waters between two reservoirs with pressure larger than atmospheric. Their design follows approximately a U-shape. Inverted siphons were commonly used by the Romans along their aqueducts to cross valleys.

Siphon-spillway   Device for discharging excess water in a pipe over the dam crest.

Skimming flow   Flow regime above a stepped chute for which the water flows as a coherent stream in a direction parallel to the pseudo-bottom formed by the edges of the steps. The same term is used to characterize the flow regime of large discharges above rockfill and closely spaced large roughness elements.

Slope   (1) Side of a hill. (2) Inclined face of a canal (e.g. trapezoidal channel). (3) Inclination of the channel bottom from the horizontal.

Sluice gate   Underflow gate with a vertical sharp edge for stopping or regulating flow.

Soffit   Roof of the barrel of a culvert. Another name is obvert.

Specific energy   Quantity proportional to the energy per unit mass, measured with the channel bottom as the elevation datum, and expressed in metres of water. The concept of specific energy, first developed by B.A. Bakhmeteff in 1912, is commonly used in open channel flows.

Spillway   Opening built into a dam or the side of a reservoir to release (to spill) excess flood waters.

Splitter   Obstacle (e.g. concrete block, fin) installed on a chute to split the flow and to increase the energy dissipation.

Spray   Water droplets flying or falling through air: e.g. spray thrown up by a waterfall.

Spring tide   Tide of greater-than-average range around the times of new and full moon. (The opposite of a spring tide is the *neap tide*.)

Stage–discharge curve   Relationship between discharge and free-surface elevation at a given location along a stream.

Stagnation point   This is defined as the point where the velocity is zero. When a streamline intersects itself, the intersection is a stagnation point. For irrotational flow a streamline intersects itself at right angle at a stagnation point.

Staircase   Another adjective for 'stepped': e.g. a staircase cascade is a stepped cascade.

Stall   Aerodynamic phenomenon causing a disruption (i.e. separation) of the flow past a wing associated with a loss of lift.

Steady flow   Occurs when conditions at any point of the fluid do not change with the time:

$$\frac{\partial V}{\partial t} = 0 \qquad \frac{\partial \rho}{\partial t} = 0 \qquad \frac{\partial P}{\partial t} = 0 \qquad \frac{\partial T}{\partial t} = 0$$

Stilling basin   Structure for dissipating the energy of the flow downstream of a spillway, outlet work, chute or canal structure. In many cases, a hydraulic jump is used as the energy dissipator within the stilling basin.

Stokes   George Gabriel Stokes (1819–1903), British mathematician and physicist, is known for his research in hydrodynamics and a study of elasticity.

Stommel   Henry Melson Stommel (1920–1992) was an American oceanographer and meteorologist, internationally known during the 1950s for his theories on circulation patterns in the Atlantic Ocean.

Stop-logs   Form of sluice gate comprising a series of wooden planks, one above the other, and held at each end.

Storm water   Excess water running off the surface of a drainage area during and immediately following a period of rain. In urban areas, waters drained off a catchment area during or after a heavy rainfall are usually conveyed in man-made storm waterways.

Storm waterway Channel built for carrying storm waters.

Straub L.G. Straub (1901–1963) was Professor and Director of the St Anthony Falls Hydraulics Laboratory at the University of Minnesota (USA).

Stream function Vector function of space and time which is related to the velocity field as: $\vec{V} = -\overrightarrow{\text{curl}}\,\vec{\phi}$. The stream function exists for steady and unsteady flow of incompressible fluid as it does satisfy the continuity equation. The stream function was introduced by the French mathematician Lagrange.

Streamline It is the line drawn so that the velocity vector is always tangential to it (i.e. no flow across a streamline). When the streamlines converge the velocity increases. The concept of streamline was first introduced by the Frenchman J.C. de Borda.

Streamline maps It should be drawn so that the flow between any two adjacent streamlines is the same.

Stream tube This is a filament of fluid bounded by streamlines.

Subcritical flow In open channel the flow is defined as subcritical if the flow depth is larger than the critical flow depth. In practice, subcritical flows are controlled by the downstream flow conditions.

Subsonic flow A compressible flow with a Sarrau–Mach number less than unity: i.e. the flow velocity is less than the sound celerity.

Supercritical flow In open channel, when the flow depth is less than the critical flow depth, the flow is supercritical and the Froude number is >1. Supercritical flows are controlled from upstream.

Supersonic flow A compressible flow with a Sarrau–Mach number larger than unity: i.e. the flow velocity is larger than the sound celerity.

Surface tension Property of a liquid surface displayed by its acting as if it were a stretched elastic membrane. Surface tension depends primarily upon the attraction forces between the particles within the given liquid and also upon the gas, solid or liquid in contact with it. The action of surface tension is to increase the pressure within a water droplet or within an air bubble. For a spherical bubble of diameter d_{ab}, the increase of internal pressure necessary to balance the tensile force caused by surface tension equals: $\Delta P = 4\sigma/d_{ab}$ where σ is the surface tension.

Surfactant (or *surface active agent*) Substance that, when added to a liquid, reduces its surface tension thereby increasing its wetting property (e.g. detergent).

Surge A surge in an open channel is a sudden change of flow depth (i.e. abrupt increase or decrease in depth). An abrupt increase in flow depth is called a positive surge while a sudden decrease in depth is termed a negative surge. A positive surge is also called (improperly) a 'moving hydraulic jump' or a 'hydraulic bore'.

Surge wave Results from a sudden change in flow that increases (or decreases) the depth.

Suspended load Transported sediment material maintained into suspension.

Sverdrup Harald Ulrik Sverdrup (1888–1957) was a Norwegian meteorologist and oceanographer known for his studies of the physics, chemistry, and biology of the oceans. He explained the equatorial countercurrents and helped develop a method of predicting surf and breakers.

Sverdrup Volume discharge units in oceanic circulation: 1 Sverdrup = $1 \times 10^6\,\text{m}^3/\text{s}$.

Swash In coastal engineering, the swash is the rush of water up a beach from the breaking waves.

Swash line The upper limit of the active beach reached by highest sea level during big storms.

Système International d'Unités International System of Units adopted in 1960 based on the metre–kilogram–second (MKS) system. It is commonly called SI unit system. The basic seven units are: for length, the metre; for mass, the kilogram; for time, the second; for electric current, the ampere; for luminous intensity, the candela; for amount of substance, the mole; for thermodynamic temperature, the kelvin.

Système Métrique International decimal system of weights and measures which was adopted in 1795 during the French Révolution. Between 1791 and 1795, the Académie des Sciences de Paris prepared a logical system of units based on the metre for length and the kilogram for mass. The standard metre was defined as 1×10^7 times a meridional quadrant of earth. The gram was equal to the mass of $1\,\text{cm}^3$ of pure water at the temperature of its maximum density (i.e. 4°C) and 1 kilogram equals 1000 grams. The litre was defined as the volume occupied by a cube of $1 \times 10^3\,\text{cm}^3$.

TWRC Tail water rating curve.

TWRL   Tail water rating Level.

Tainter gate   This is a radial gate.

Tailwater depth   Downstream flow depth.

Tailwater level   Downstream free-surface elevation.

Taylor   Sir Geoffrey Ingram Taylor (1886–1975) was a British fluid dynamicist based in Cambridge. He established the basic developments of shear dispersion (Taylor 1953, 1954). His great-father was the British mathematician George Boole (1815–1864) who established modern symbolic logic and Boolean algebra.

Thompson   Sir Benjamin Thompson (1753–1814), also known as Count Rumford, proposed in 1797 that evaporation in the Northern Hemisphere would cause heavier, saltier water to sink and flow southward, and that a warmer north-bound current would be needed to balance the southern one.

Total head   The total head is proportional to the total energy per unit mass and per gravity unit. It is expressed in metres of water.

Training wall   Sidewall of chute spillway.

Trashrack   Screen comprising metal or reinforced concrete bars located at the intake of a waterway to prevent the progress of floating or submerged debris.

Turbidity   Opacity of water. Turbidity is a measure of the absence of clarity of the water.

Turbulence   Flow motion characterized by its unpredictable behaviour, strong mixing properties and a broad spectrum of length scales.

Turbulent flow   In turbulent flows the fluid particles move in very irregular paths, causing an exchange of momentum from one portion of the fluid to another. Turbulent flows have great mixing potential and involve a wide range of eddy length scales.

Turriano   Juanelo Turriano (1511–1585) was an Italian clockmaker, mathematician and engineer who worked for the Spanish Kings Charles V and later Philip II. It is reported that he checked the design of the Alicante dam for King Philip II.

Two-dimensional flow   All particles are assumed to flow in parallel planes along identical paths in each of these planes. There are no changes in flow normal to these planes. An example of two-dimensional flow can be an open channel flow in a wide rectangular channel.

USACE   United States Army Corps of Engineers.

USBR   United States Bureau of Reclamation.

Ukiyo-e (or *Ukiyoe*)   This is a type of Japanese painting and colour woodblock prints during the period 1803–1867.

Undular hydraulic jump   Hydraulic jump characterized by steady stationary free-surface undulations downstream of the jump and by the absence of a formed roller. The undulations can extend far downstream of the jump with decaying wave lengths, and the undular jump occupies a significant length of the channel. It is usually observed for $1 < Fr_1 < 1.5$–3 (Chanson 1995b). The first significant study of undular jump flow can be attributed to Fawer (1937) and undular jump flows should be called Fawer jump in homage to Fawer's work.

Undular surge   Positive surge characterized by a train of secondary waves (or undulations) following the surge front. Undular surges are sometimes called Boussinesq–Favre waves in homage to the contributions of J.B. Boussinesq and H. Favre.

Uniform equilibrium flow   Occurs when the velocity is identically the same at every point, in magnitude and direction, for a given instant:

$$\frac{\partial V}{\partial s} = 0$$

in which time is held constant and s is a displacement in any direction. That is, steady uniform flow (e.g. liquid flow through a long pipe at a constant rate) and unsteady uniform flow (e.g. liquid flow through a long pipe at a decreasing rate).

Universal gas constant (also called *molar gas constant* or *perfect gas constant*)   Fundamental constant equal to the pressure times the volume of gas divided by the absolute temperature for 1 mol of perfect gas. The value of the universal gas constant is $8.31441 \, \text{J K}^{-1} \text{mol}^{-1}$.

Unsteady flow   The flow properties change with the time.

Uplift   Upward pressure in the pores of a material (interstitial pressure) or on the base of a structure. Uplift pressures led to the destruction of stilling basins and even to the failures of concrete dams (e.g. Malpasset dam break in 1959).

Upstream flow conditions   Flow conditions measured immediately upstream of the investigated control volume.

VNIIG   Institute of Hydrotechnics Vedeneev in St Petersburg (Russia).

VOC   Volatile organic compound.

Valence   Property of an element that determines the number of other atoms with which an atom of the element can combine.

Validation   Comparison between model results and prototype data, to validate the model. The validation process must be conducted with prototype data that are different from that used to calibrate and to verify the model.

Vauban   Sébastien Vauban (1633–1707) was Maréchal de France. He participated to the construction of several water supply systems in France, including the extension of the feeder system of the Canal du Midi between 1686 and 1687, and parts of the water supply system of the gardens of Versailles.

Velocity potential   It is defined as a scalar function of space and time such that its negative derivative with respect to any direction is the fluid velocity in that direction: $\vec{V} = -\overrightarrow{\text{grad}} \, \phi$. The existence of a velocity potential implies irrotational flow of ideal fluid. The velocity potential was introduced by the French mathematician J. Louis Lagrange (1781).

Vena contracta   Minimum cross-sectional area of the flow (e.g. jet or nappe) discharging through an orifice, sluice gate or weir.

Venturi meter   In closed pipes, smooth constriction followed by a smooth expansion. The pressure difference between the upstream location and the throat is proportional to the velocity square. It is named after the Italian physicist Giovanni Battista Venturi (1746–1822).

Villareal de Berriz   Don Pedro Bernardo Villareal de Berriz (1670–1740) was a Basque nobleman. He designed several buttress dams, some of these being still in use (Smith 1971).

Viscosity   Fluid property which characterizes the fluid resistance to shear: i.e. resistance to a change in shape or movement of the surroundings.

Vitruvius   Roman architect and engineer (BC 94–??). He built several aqueducts to supply the Roman capital with water. (*Note*: there are some incertitude on his full name: 'Marcus Vitruvius Pollio' or 'Lucius Vitruvius Mamurra', Garbrecht 1987a.)

Von Karman constant   See *Karman constant*.

WES   Waterways Experiment Station of the US Army Corps of Engineers.

Wadi   Arabic word for a valley which becomes a watercourse in rainy seasons.

Wake region   The separation region downstream of the streamline that separates from a boundary is called a wake or wake region.

Warrie   Australian aboriginal name for 'rushing water'.

Waste waterway   Old name for a spillway, particularly used in irrigation with reference to the waste of waters resulting from a spill.

Wasteweir   A spillway. The name refers to the waste of hydroelectric power or irrigation water resulting from the spill. A 'staircase' wasteweir is a stepped spillway.

Water   Common name applied to the liquid state of the hydrogen–oxygen combination H_2O. Although the molecular structure of water is simple, the physical and chemical properties of H_2O are unusually complicated. Water is a colourless, tasteless and odourless liquid at room temperature. One most important property of water is its ability to dissolve many other substances: H_2O is frequently called the universal solvent. Under standard atmospheric pressure, the freezing point of water is 0°C (273.16 K) and its boiling point is 100°C (373.16 K).

Water clock   Ancient device for measuring time by the gradual flow of water through a small orifice into a floating vessel. The Greek name is Clepsydra.

Waterfall   Abrupt drop of water over a precipice characterized by a free-falling nappe of water. The highest waterfalls are the Angel fall (979 m) in Venezuela (Churún Merú), Tugel fall (948 m) in South Africa, Mtarazi (762 m) in Zimbabwe.

Water mill   Mill (or wheel) powered by water.

Water staircase (or '*Escalier d'Eau*')   This is the common architectural name given to a stepped cascade with flat steps.

Weak jump   A weak hydraulic jump is characterized by a marked roller, no free-surface undulation and low energy loss. It is usually observed after the disappearance of undular hydraulic jump with increasing upstream Froude numbers.

Weber   Moritz Weber (1871–1951) was a German Professor at the Polytechnic Institute of Berlin. The Weber number characterizing the ratio of inertial force over surface tension force was named after him.

Weber number   Dimensionless number characterizing the ratio of inertial forces over surface tension forces. It is relevant in problems with gas–liquid or liquid–liquid interfaces.

Weir   Low river dam used to raise the upstream water level. Measuring weirs are built across a stream for the purpose of measuring the flow.

Weisbach   Julius Weisbach (1806–1871) was a German applied mathematician and hydraulician.

Wen Cheng-Ming   Chinese landscape painter (1470–1559). One of his famous works is the painting of 'Old trees by a cold waterfall'.

WES standard spillway shape   Spillway shape developed by the US Army Corps of Engineers at the Waterways Experiment Station.

Wetted perimeter   Considering a cross-section (usually selected normal to the flow direction), the wetted perimeter is the length of wetted contact between the flowing stream and the solid boundaries. For example, in a circular pipe flowing full, the wetted perimeter equals the circle perimeter.

Wetted surface   In open channel, the term 'wetted surface' refers to the surface area in contact with the flowing liquid.

White waters   Non-technical term used to design free-surface aerated flows. The refraction of light by the entrained air bubbles gives the 'whitish' appearance to the free surface of the flow.

White water sports   Include canoe, kayak and rafting racing down swift-flowing turbulent waters.

Wind setup   Water level rise in the downwind direction caused by wind shear stress. The opposite is a wind setdown.

Wing wall   Sidewall of an inlet or outlet.

Wood   I.R. Wood is an Emeritus Professor in Civil Engineering at the University of Canterbury (New Zealand).

Yen   Professor Ben Chie Yen (1935–2001) was a hydraulic professor at the University of Illinois at Urbana-Champaign, although born and educated in Taiwan.

Yunca   Indian of a group of South American tribes of which the Chimus and the Chinchas are the most important. The Yunca civilization developed a pre-Inca culture on the coast of Peru.

List of symbols

A	flow cross-sectional area (m^2)
A_s	particle cross-sectional area (m^2)
B	open channel free-surface width (m)
B_{max}	inlet lip width (m) of MEL culvert
B_{min}	(1) minimum channel width (m) for onset of choking flow
	(2) barrel width (m) of a culvert
C	(1) celerity (m/s): e.g. celerity of sound in a medium, celerity of a small disturbance at a free surface
	(2) dimensional discharge coefficient
Ca	Cauchy number
$C_{Chézy}$	Chézy coefficient (m$^{1/2}$/s)
C_D	dimensionless discharge coefficient (SI units)
C_L	lift coefficient
C_d	(1) skin friction coefficient (also called drag coefficient)
	(2) drag coefficient
C_{des}	design discharge coefficient (SI units)
C_o	initial celerity (m/s) of a small disturbance
C_p	specific heat at constant pressure (J/kg/K): $C_p = \left(\dfrac{\partial h}{\partial T} \right)_P$
C_s	mean volumetric sediment concentration
$(C_s)_{mean}$	mean sediment suspension concentration
C_{sound}	sound celerity (m/s)
C_v	specific heat at constant volume (J/kg/K)
c_s	sediment concentration
D	circular pipe diameter (m)
D_H	hydraulic diameter (m), or equivalent pipe diameter, defined as:

$$D_H = 4 \frac{\text{cross-sectional area}}{\text{wetted perimeter}} = \frac{4A}{P_w}$$

D_s	sediment diffusivity (m^2/s)
D_t	diffusion coefficient (m^2/s)
D_1, D_2	characteristics of velocity distribution in turbulent boundary layer
d	flow depth (m) measured perpendicular to the channel bed
d_{ab}	air bubble diameter (m)
d_b	brink depth (m)
d_c	critical flow depth (m)
d_{charac}	characteristic geometric length (m)
d_{conj}	conjugate flow depth (m)
d_o	(1) uniform equilibrium flow depth (m): i.e. normal depth
	(2) initial flow depth (m)

d_p pool depth (m)

d_s (1) sediment size (m)

        (2) dam break wave front thickness (m)

d_{tw} tailwater flow depth (m)

d_{50} median grain size (m) defined as the size for which 50% by weight of the material is finer

d_{84} sediment grain size (m) defined as the size for which 84% by weight of the material is finer

d_i characteristic grain size (m), where i = 10, 16, 50, 75, 84, 90

d_* dimensionless particle parameter: $d_* = d_s \sqrt[3]{(\rho_s/\rho - 1)(g/v^2)}$

E mean specific energy (m) defined as: $E = H - z_0$

E local specific energy (m) defined as: $E = \dfrac{P}{\rho g} + (z - z_0) + \dfrac{v^2}{2g}$

Eu Euler number

E_b bulk modulus of elasticity (Pa): $E_b = \rho \dfrac{\partial P}{\partial \rho}$

E_{co} compressibility (1/Pa): $E_{co} = \dfrac{1}{\rho}\dfrac{\partial \rho}{\partial P}$

E_{min} minimum specific energy (m)

e internal energy per unit mass (J/kg)

E total energy (J) of system

F force (N)

$\vec{F}$ force vector

F_b buoyant force (N)

F_d drag force (N)

F_{fric} friction force (N)

F_p pressure force (N)

F_{visc} viscous force (N/m^3)

F_{vol} volume force per unit volume (N/m^3)

F_p' pressure force (N) acting on the flow cross-sectional area

F_p'' pressure force (N) acting on the channel side boundaries

f Darcy friction factor (also called head loss coefficient)

Fr Froude number

F_r ratio of prototype to model forces: $F_r = F_p/F_m$

g gravity constant (m/s^2) in Brisbane, Australia: $g = 9.80$ m/s^2

$g_{centrif}$ centrifugal acceleration (m/s^2)

H (1) mean total head (m): $H = d \cos\theta + z_0 + \alpha \dfrac{V_{mean}^2}{2gd}$ assuming a hydrostatic pressure distribution

        (2) depth-averaged total head (m) defined as: $H = \dfrac{I}{d}\displaystyle\int^d H\, dy$

H_{dam} reservoir height (m) at dam site

H_{des} design upstream head (m)

H_{res} residual head (m)

H_1 upstream total head (m)

H_2 downstream total head (m)

H local total head (m) defined as: $H = \dfrac{P}{\rho g} + z + \dfrac{v^2}{2g}$

h specific enthalpy (i.e. enthalpy per unit mass) (J/kg): $h = e + \dfrac{P}{\rho}$

h	(1) dune bed form height (m)
	(2) step height (m)
i	integer subscript
i	imaginary number: $i = \sqrt{-1}$
JHRL	jump height rating level (m RL)
K	hydraulic conductivity (m/s) of a soil
K_M	empirical coefficient (s) in the Muskingum method
K	von Karman constant (i.e. K = 0.4)
K'	head loss coefficient: $K' = \dfrac{\Delta H}{0.5 V^2/g}$
k	permeability (m^2) of a soil
k_{Bazin}	Bazin resistance coefficient
$k_{Strickler}$	Strickler resistance coefficient (m$^{1/3}$/s)
k_s	equivalent sand roughness height (m)
L	length (m)
L_{crest}	crest length (m)
L_{culv}	culvert length (m) measured in the flow direction
L_d	drop length (m)
L_{inlet}	inlet length (m) measured in the flow direction
L_r	(1) length of roller of hydraulic jump (m)
	(2) ratio of prototype to model lengths: $L_r = L_p/L_m$
l	(1) dune bed form length (m)
	(2) step length (m)
M	momentum function (m^2)
Ma	Sarrau–Mach number
Mo	Morton number
M_r	ratio of prototype to model masses
M	total mass (kg) of system
$\dot{m}$	mass flow rate per unit width (kg/s/m)
$\dot{m}_s$	sediment mass flow rate per unit width (kg/s/m)
N	inverse of velocity distribution exponent
No	Avogadro constant: No = 6.0221367×10^{23} mol^{-1}
Nu	Nusselt number
N_{bl}	inverse of velocity distribution exponent in turbulent boundary layer
$n_{Manning}$	Gauckler–Manning coefficient (s/m$^{1/3}$)
P	absolute pressure (Pa)
P_{atm}	atmospheric pressure (Pa)
$P_{centrif}$	centrifugal pressure (Pa)
P_r	ratio of prototype to model pressures
P_{std}	standard atmosphere (Pa) or normal pressure at sea level
P_v	vapour pressure (Pa)
P_w	wetted perimeter (m)
Po	porosity factor
Q	total volume discharge (m^3/s)
Q_{des}	design discharge (m^3/s)
Q_{max}	maximum flow rate (m^3/s) in open channel for a constant specific energy
Q_r	ratio of prototype to model discharges

Q_h	heat added to a system (J)
q	discharge per meter width (m^2/s)
q_{des}	design discharge (m^2/s) per unit width
q_{max}	maximum flow rate per unit width (m^2/s) in open channel for a constant specific energy
q_s	sediment flow rate per unit width (m^2/s)
q_h	heat added to a system per unit mass (J/kg)
R	invert curvature radius (m)
R	fluid thermodynamic constant (J/kg/K) also called gas constant: $P = \rho R T$ perfect gas law (i.e. Mariotte law)
Re	Reynolds number
Re_*	shear Reynolds number
R_H	hydraulic radius (m) defined as: $R_H = \dfrac{\text{cross-sectional area}}{\text{wetted perimeter}} = \dfrac{A}{P_w}$
Ro	universal gas constant: Ro = 8.3143 J/K/mol
r	radius of curvature (m)
S	sorting coefficient of a sediment mixture: $S = \sqrt{d_{90}/d_{10}}$
S	specific entropy (i.e. entropy per unit mass) (J/K/kg): $dS = \left(\dfrac{dq_h}{T}\right)_{rev}$
S_c	critical slope
S_f	friction slope defined as: $S_f = \dfrac{-\partial H}{\partial s}$
S_o	bed slope defined as: $S_o = -\dfrac{\partial z_o}{\partial s} = \sin\theta$
S_t	transition slope for a multi-cell MEL culvert
s	curvilinear co-ordinate (m) (i.e. distance measured along a streamline and positive in the flow direction)
s	relative density of sediment: $\mathbf{s} = \rho_s/\rho$
T	thermodynamic (or absolute) temperature (K)
T_o	reference temperature (K)
TWRL	tailwater rating level (m RL)
t	time (s)
t_r	ratio of prototype to model times: $t_r = t_p/t_m$
t_s	sedimentation time scale (s)
U	(1) volume force potential (m^2/s^2)
	(2) wave celerity (m/s) for an observed standing on the bank
V	flow velocity (m/s)
v	(local) velocity (m/s)
$\vec{V}$	velocity vector; in Cartesian co-ordinates the velocity vector equals: $\vec{V} = (V_x, V_y, V_z)$
V_c	critical flow velocity (m/s)
V_H	characteristic velocity (m/s) defined in terms of the dam height
V_{mean}	mean flow velocity (m/s): $V_{mean} = Q/A$
V_{max}	maximum velocity (m/s) in a cross-section; in fully developed open channel flow, the velocity is maximum near the free surface
V_r	ratio of prototype to model velocities: $V_r = V_p/V_m$
V_s	average speed (m/s) of sediment motion
V_{srg}	surge velocity (m/s) as seen by an observer immobile on the channel bank

V_o (1) uniform equilibrium flow velocity (m/s)
     (2) initial flow velocity (m/s)

V' depth-averaged velocity (m/s): $V' = \dfrac{1}{d}\displaystyle\int_0^d V\,dy$

V_* shear velocity (m/s) defined as: $V_* = \sqrt{\dfrac{\tau_o}{\rho}}$

Vol   volume (m³)

v_s particle volume (m/³)

W (1) channel bottom width (m)
     (2) channel width (m) at a distance y from the invert

We Weber number

W_p work (J) done by the pressure force

W_s work (J) done by the system by shear stress (i.e. torque exerted on a rotating shaft)

W_t total work (J) done by the system

w_o (1) particle settling velocity (m/s)
     (2) fall velocity (m/s) of a single particle in a fluid at rest

w_s settling velocity (m/s) of a suspension

w_s work done by shear stress per unit mass (J/kg)

Wa Coles wake function

X horizontal co-ordinate (m) measured from spillway crest

X_M empirical coefficient in the Muskingum method

X_r ratio of prototype to model horizontal distances

x Cartesian co-ordinate (m)

$\vec{x}$ Cartesian co-ordinate vector: $\vec{x} = (x, y, z)$

x_s dam break wave front location (m)

Y (1) vertical co-ordinate (m) measured from spillway crest
     (2) free-surface elevation (m): $Y = z_o + d$

y (1) distance (m) measured normal to the flow direction
     (2) distance (m) measured normal to the channel bottom
     (3) Cartesian co-ordinate (m)

$y_{channel}$ channel height (m)

y_s characteristic distance (m) from channel bed

Z_r ratio of prototype to model vertical distances

z (1) altitude or elevation (m) measured positive upwards
     (2) Cartesian co-ordinate (m)

z_{apron} apron invert elevation (m)

z_{crest} spillway crest elevation (m)

z_o (1) reference elevation (m)
     (2) bed elevation (m)

Greek symbols

α Coriolis coefficient or kinetic energy correction coefficient

β momentum correction coefficient (i.e. Boussinesq coefficient)

Δ angle between the characteristics and the x-axis

Δd change in flow depth (m)

ΔE change in specific energy (m)

ΔH	head loss (m): i.e. change in total head
ΔP	pressure difference (Pa)
Δq_s	change in sediment transport rate (m²/s)
Δs	small distance (m) along the flow direction
Δt	small time change (s)
ΔV	change in flow velocity (m/s)
Δx	small distance (m) along the x-direction
Δz_o	change in bed elevation (m)
Δz_o	weir height (m) above natural bed level
δ	(1) sidewall slope
	(2) boundary layer thickness (m)
δ_{ij}	identity matrix element
δ_s	bed-load layer thickness (m)
δt	small time increment (s)
δx	small distance increment (m) along the x-direction
ε_{ij}	velocity gradient element (m/s²)
ϕ	sedimentological size parameter
ϕ_s	angle of repose
Ψ	stream function (m²/s)
γ	specific heat ratio: $\gamma = C_p/C_v$
μ	dynamic viscosity (Pa s)
ν	kinematic viscosity (m²/s): $\nu = \mu/\rho$
ϕ	velocity potential (m²/s)
Π	wake parameter
π	= 3.14159265358979323846264 3
θ	channel slope
ρ	density (kg/m³)
ρ_s	sediment density (kg/m³)
ρ_r	ratio of prototype to model densities
ρ_{sed}	sediment mixture density (kg/m³)
σ	surface tension (N/m)
σ_e	effective stress (Pa)
σ_{ij}	stress tensor element (Pa)
σ_g	geometric standard deviation of sediment size distribution: $\sigma_g = \sqrt{d_{84}/d_{16}}$
τ	shear stress (Pa)
τ_{ij}	shear stress component (Pa) of the i-momentum transport in the j-direction
τ_o	average boundary shear stress (Pa)
$(\tau_o)_c$	critical shear stress (Pa) for onset of sediment motion
τ_o'	skin friction shear stress (Pa)
τ_o''	bed form shear stress (Pa)
τ_1	yield stress (Pa)
τ_*	Shields parameter: $\tau_* = \dfrac{\tau_o}{\rho g(\rho_s/\rho - 1)d_s}$
$(\tau_*)_c$	critical Shields parameter for onset of sediment motion

Subscript

air	air
bl	bed-load

c	critical flow conditions
conj	conjugate flow property
des	design flow conditions
dry	dry conditions
exit	exit flow condition
i	characteristics of section $\{i\}$ (in the numerical integration process)
inlet	inlet flow condition
m	model
mean	mean flow property over the cross-sectional area
mixt	sediment-water mixture
model	model conditions
o	(1) uniform equilibrium flow conditions
	(2) initial flow conditions
outlet	outlet flow condition
p	prototype
prototype	prototype conditions
r	ratio of prototype to model characteristics
s	(1) component in the s-direction
	(2) sediment motion
	(3) sediment particle property
sl	suspended load
t	flow condition at time t
tw	tailwater flow condition
w	water
wet	wet conditions
x	x-component
y	y-component
z	z-component
1	upstream flow conditions
2	downstream flow conditions

Abbreviations

CS	control surface
CV	control volume
D/S	downstream
GVF	gradually varied flow
Hg	mercury
RVF	rapidly varied flow
SI	Système international d'unités (International System of Units)
THL	total head line
U/S	upstream

Reminder

1. At 20°C, the density and dynamic viscosity of water (at atmospheric pressure) are: $\rho_w = 998.2 \, \text{kg/m}^3$ and $\mu_w = 1.005 \times 10^{-3} \, \text{Pa s}$.
2. Water at atmospheric pressure and 20.2°C has a kinematic viscosity of exactly $10^{-6} \, \text{m}^2/\text{s}$.
3. Water in contact with air has a surface tension of about 0.0733 N/m at 20°C.
4. At 20°C and atmospheric pressure, the density and dynamic viscosity of air are about $1.2 \, \text{kg/m}^3$ and $1.8 \times 10^{-5} \, \text{Pa s}$ respectively.

Dimensionless numbers

Ca Cauchy number: $Ca = \dfrac{\rho V^2}{E_b}$

Note: the Sarrau–Mach number equals: $Ma \sim \sqrt{Ca}$

C_d (1) drag coefficient for bottom friction (i.e. friction drag):

$$C_d = \frac{\tau_o}{(1/2)\rho V^2} = \frac{\text{shear stress}}{\text{dynamic pressure}}$$

Note: another notation is C_f (e.g. Comolet 1976).

(2) drag coefficient for a structural shape (i.e. form drag):

$$C_d = \frac{F_d}{(1/2)\rho V^2 A} = \frac{\text{drag force per unit cross-sectional area}}{\text{dynamic pressure}}$$

where A is the projection of the structural shape (i.e. body) in the plane normal to the flow direction

Eu Euler number defined as: $Eu = \dfrac{V}{\sqrt{\Delta P/\rho}}$

Fr Froude number defined as: $Fr = \dfrac{V}{\sqrt{gd_{\text{charac}}}} \propto \sqrt{\dfrac{\text{inertial force}}{\text{weight}}}$

Note: some authors use the notation:

$$Fr = \frac{V^2}{gd_{\text{charac}}} = \frac{\rho V^2 A}{\rho \, gAd_{\text{charac}}} \propto \frac{\text{inertial force}}{\text{weight}}$$

Ma Sarrau–Mach number: $Ma = \dfrac{V}{C}$

Mo Morton number defined as: $Mo = \dfrac{g\mu_w^4}{\rho_w \sigma^3}$

The Morton number is a function only of fluid properties and gravity constant. If the same fluids (air and water) are used in both model and prototype, Mo may replace the Weber, Reynolds or Froude number as: $Mo = \dfrac{We^3}{Fr^2 Re^4}$

Nu Nusselt number: $Nu = \dfrac{Hd_{\text{charac}}}{\lambda} \propto \dfrac{\text{heat transfer by convection}}{\text{heat transfer by conduction}}$

where H is the heat transfer coefficient ($\text{W/m}^2/\text{K}$) and λ is the thermal conductivity ($\text{W/m}^2/\text{K}$)

Re Reynolds number: $Re = \dfrac{V d_{charac}}{\nu} \quad \propto \quad \dfrac{\text{inertial forces}}{\text{viscous forces}}$

Re_* shear Reynolds number: $Re_* = \dfrac{V_* k_s}{\nu}$

We Weber number: $We = \dfrac{V}{\sqrt{\sigma/(\rho d_{charac})}} \quad \propto \quad \sqrt{\dfrac{\text{inertial force}}{\text{surface tension force}}}$

   Note: some authors use the notation: $We = \dfrac{V}{\sqrt{\sigma/(\rho d_{charac})}} \quad \propto \quad \dfrac{\text{inertial force}}{\text{surface tension force}}$

τ_* Shields parameter characterizing the onset of sediment motion:

   $\tau_* = \dfrac{\tau_o}{\rho g (\rho_s/\rho - 1) d_s} \quad \propto \quad \dfrac{\text{destabilizing force moment}}{\text{stabilizing moment of weight force}}$

Notes

The variable d_{charac} is the characteristic geometric length of the flow field: e.g. pipe diameter, flow depth, sphere diameter. Some examples are listed below:

Flow	d_{charac}	Comments
Circular pipe flow	D	Pipe diameter
Flow in pipe of irregular cross-section	D_H	Hydraulic diameter
Flow resistance in open channel flow	D_H	Hydraulic diameter
Wave celerity in open channel flow	D	Flow depth
Flow past a cylinder	D	Cylinder diameter

PART 1
Introduction to Open Channel Flows

River bank erosion at Chenchung, PingTung county, Taiwan about 5 km upstream of the river mouth in December 1999. View from the right bank looking upstream.

1

Introduction

Summary

This introduction chapter briefly reviews the fluid properties and some result for static fluids. Then open channel flows are defined.

1.1 Presentation

The term 'Hydraulics' is related to the application of Fluid Mechanics' principles to water engineering structures, and civil and environmental engineering facilities. We consider open channels in which liquid (i.e. water) flows with a free surface. Examples of open channels are natural streams and rivers. Man-made channels include irrigation and navigation canals, drainage ditches, sewer and culvert pipes running partially full, and spillways.

In open channel flows, the free surface rises and falls in response to perturbations to the flow (e.g. changes in channel slope or width). The location of the free surface is unknown beforehand. The main parameters of a hydraulic study are the geometry of the channel, the properties of the flowing fluid and the flow parameters.

1.1.1 Discussion: hydraulic engineering through history

Hydraulic engineers were at the forefront of science for centuries (Fig. 1.1). For example, although the origins of seepage water was long the subject of speculation, the arts of tapping groundwater developed early in the antiquity. The construction of qanats, which were hand-dug underground water collection tunnels, in Armenia and Persia is considered as one great hydrologic achievement of the ancient world. Roman aqueducts were magnificient water-works and demonstrated the 'savoir-faire' of Roman engineers. The 132-km long Carthage aqueduct was considered one of the marvels of the world by the Muslim poet El Kairouani. Many aqueducts were used, repaired and maintained for centuries and some are still used in parts (e.g. Carthage). A major navigation canal system was the *Grand canal* fed by the Tianping diversion weir in China. Completed in BC 219, the 3.9 m high and 470 m long weir diverted the Xiang River into the South and North canals, allowing navigation between Guangzhou (formerly Canton), Shanghai and Beijing.

(a)

(b)

Fig. 1.1 Ancient hydraulic engineering. (a) Nabataean Dam on the Mamshit stream (also called Mampsis or Kunub) on 10 May 2001 (courtesy of Dennis Murphy). Dam wall built around the end of 1st Century BC, downstream slope of the dam wall. (b) Roman aqueduct in Fréjus, Arches de Sainte Croix, downstream of Chateau Aurélien on 14 September 2000. Looking upstream, note the slight bend in the aqueduct in the background.

The development of hydraulic engineering is closely linked to the beginnings of civil engineering as a separate discipline, and the foundation of the 'Corps des Ponts et Chaussées' (Bridge and Highway Corps) in France in 1716 with the establishment of the 'École Nationale des Ponts et Chaussées' (National School of Bridges and Highways) in 1747. Among the directors of the school were the famous hydraulicians A. Chézy (1717–1798) and G. de Prony (1755–1839). Other famous professors included B.F. de Bélidor (1693–1761), J.B.C. Bélanger (1789–1874), J.A.C. Bresse (1822–1883), G.G. Coriolis (1792–1843) and L.M.H. Navier (1785–1835).

(c)

Fig. 1.1 (c) Storm waterway at Miya-jima (Japan) below Senjò-kaku wooden hall on 19 November 2001. The steep stepped chute ($\theta > 45°$, $h \sim 0.4$ m) was built during the 12th Century AD. The Senjò-kaku wooden hall was built by Kyomori (AD 1168) and left unfinished after his death.

1.2 Fluid properties

The density ρ of a fluid is defined as its mass per unit volume.

All real fluids resist any force tending to cause one layer to move over another, but this resistance is offered only while the movement is taking place. The resistance to the movement of one layer of fluid over an adjoining one is referred to as the viscosity of the fluid. Newton's law of viscosity postulates that, for the straight parallel motion of a given fluid, the tangential stress between two adjacent layers is proportional to the velocity gradient in a direction perpendicular to the layers:

$$\tau = \mu \frac{\mathrm{d}v}{\mathrm{d}y} \tag{1.1}$$

where τ is the shear stress between adjacent fluid layers, μ is the dynamic viscosity of the fluid, v is the velocity and y is the direction perpendicular to the fluid motion.

At the interface between a liquid and a gas, a liquid and a solid or two immiscible liquids, a tensile force is exerted at the surface of the liquid and tends to reduce the area of this surface to the greatest possible extent. The surface tension is the stretching force required to form the film.

Notes

1. Isaac Newton (1642–1727) was an English mathematician.
2. The kinematic viscosity is the ratio of viscosity to mass density:

$$\nu = \frac{\mu}{\rho}$$

3. Basic fluid properties are summarized in Table 1.1. The standard atmosphere or normal pressure at sea level equals 360 mm of mercury (Hg) or 101 325 Pa.

Table 1.1 Fluid properties of air, freshwater and sea water at 20°C and standard atmosphere

Fluid properties (1)	Air (2)	Fresh water (3)	Sea water (4)	Remarks (at °C) (5)
Composition	Nitrogen (78%), oxygen (21%) and other gases (1%)	H_2O	H_2O, dissolved sodium and chloride ions (30 g/kg) and dissolved salts	
Density ρ (kg/m^3)	1.197	998.2	1024	20
Dynamic viscosity μ (Pa s)	18.1×10^{-6}	1.005×10^{-3}	1.22×10^{-3}	20
Surface tension between air and water σ (N/m)	N/A	0.0736	0.076	20
Conductivity (μS/cm)	–	87.7	48 800	25

References: Riley and Skirrow (1965), Open University Course Team (1995) and Chanson *et al.* (2002a).

1.3 Fluid statics

Considering a fluid at rest (Fig. 1.2), the pressure at any point within the fluid follows Pascal's law. For any small control volume, there is no shear stress acting on the control surface. The only forces acting on the control volume of fluid are the gravity and the pressure forces.

In a static fluid, the pressure at one point in the fluid has an unique value, independent of the direction. This is called Pascal's law. The pressure variation in a static fluid follows:

$$\frac{dP}{dz} = -\rho g \tag{1.2}$$

where P is the pressure, z is the vertical elevation positive upwards, ρ is the fluid density and g is the gravity constant.

For a body of fluid at rest with a free surface (e.g. a lake) and with a constant density, the pressure variation equals:

$$P(x, y, z) = P_{atm} - \rho g(z - d) \tag{1.3}$$

where P_{atm} is the atmospheric pressure (i.e. air pressure above the free surface) and d is the reservoir depth (Fig. 1.2).

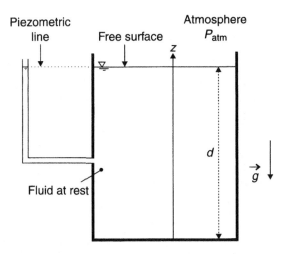

Fig. 1.2 Pressure variation in a static fluid.

Notes

1. Blaise Pascal (1623–1662) was a French mathematician, physicist and philosopher. He developed the modern theory of probability. He also formulated the concept of pressure (between 1646 and 1648) and showed that the pressure in a fluid is transmitted through the fluid in all directions (i.e. Pascal's law).
2. By definition, the pressure always acts normal to a surface. The pressure force has no component tangential to the surface.
3. The pressure force acting on a surface of finite area which is in contact with the fluid is distributed over the surface. The resultant force is obtained by integration:

$$F_p = \int P \, dA$$

where A is the surface area.

   In Fig. 1.2, the pressure force (per unit width) applied on the sidewalls of the tank is:

$$F_p = \frac{1}{2}\rho g d^2 \qquad \text{Pressure force acting on the right wall per unit width}$$

1.4 Open channel flows

An open channel is a waterway, canal or conduit in which a liquid flows with a free surface. An open channel flow describes the fluid motion in open channel (Fig. 1.3). In most applications, the liquid is water and the air above the flow is usually at rest and at standard atmospheric pressure.

Open channel flows are found in Nature as well as in man-made structures. In Nature, rushing waters are encountered in mountain rivers, river rapids and torrents (Fig. 1.3(a)). Tranquil flows are observed in large rivers near their estuaries (Fig. 1.3(b)). Natural rivers

(a)

(b)

Fig. 1.3 Examples of open channel flow. (a) Russel Falls in Tasmania (Australia) (courtesy of Dr R. Manasseh). (b) Eprapah Creek, Queensland (Australia) on 24 November 2003 at sunrise.

(c)

(d)

Fig. 1.3 (c) Hsinwulu River, Taiwan East coast in December 1998. Looking upstream and confluence with tributary on left-foreground. (d) Pont d'Arc, Vallée de l'Ardèche (France) in July 1977. Looking upstream near Vallon-Pont-d'Arc.

have the ability to scour channel beds, to carry sediment materials, and to deposit sediment loads (Fig. 1.3(c) and (d)).

River flow rates may range from extreme values: the hydraulics of droughts and floods are both important.

1.5 Exercises

Give the values (*and* units) of the specified fluid and physical properties:

(a) Density of water at atmospheric pressure and 20°C.
(b) Density of air at atmospheric pressure and 20°C.
(c) Dynamic viscosity of water at atmospheric pressure and 20°C.
(d) Kinematic viscosity of water at atmospheric pressure and 20°C.
(e) Kinematic viscosity of air at atmospheric pressure and 20°C.
(f) Surface tension of air and water at atmospheric pressure and 20°C.
(g) Acceleration of gravity in Brisbane.

In a static fluid, express the pressure variation with depth.

2

Fundamentals of open channel flows

Summary

The general principles of fluid mechanics are applicable to open channel flows. After a brief summary, the hydraulics of short transitions is developed, including the concept of specific energy, critical flow conditions and the hydraulic jump. Then flow resistance calculations at uniform equilibrium are detailed and gradually varied flow (GVF) calculations are developed (i.e. backwater calculations). The application of basic principles to open channel flow situations is summarized at the end.

2.1 Presentation

An open channel is a waterway, canal or conduit in which a liquid flows with a free surface (Fig. 2.1). Classical examples include natural streams (Fig. 2.1(a)–(c)) and man-made waterways (Fig. 2.1(d)–(f)). Figure 2.1(a)–(c) shows small and large rivers with quiet and cascading regimes. Figure 2.1(e) shows culvert during a dry period with some students surveying the structure. Figure 2.1(f) presents the same structure during a medium rainstorm.

Open channel flow describes the fluid motion in open channel. The free surface is at constant atmospheric pressure. The driving force of the fluid motion is a combination of pressure (e.g. beneath a sluice gate) and gravity (e.g. sloping channel). For a given flow rate, the primary unknown is the location of the free surface which is not known beforehand. The free surface rises and falls in response to perturbations to the flow: e.g. changes in channel elevation, reduction in channel width. The main parameters of any hydraulic study are the properties of the fluids (e.g. density, viscosity), the geometry of the channel (e.g. cross-sectional shape, bed slope) and the flow properties (e.g. depth, velocity).

Basic definitions

In an open channel, the pressure distribution is nearly always hydrostatic, unless the curvature of the streamlines is important (e.g. under waves). The mean total head is defined as:

$$H = d \cos \theta + z_o + \frac{V^2}{2g} \tag{2.1}$$

(a)

(b)

Fig. 2.1 Examples of open channel flows. (a) Brisbane River at Colleges Crossing, Karana Downs, Qld (Australia) on 1 September 2002 – looking upstream. (b) Rapid at Grand-Remous township on 13 July 2002, Gatineau River, Québec (Canada) – looking upstream.

where H is the total head, d is the water depth measured normal to the bed, θ is the bed slope, z_o is the bed elevation, V is the depth-averaged flow velocity and g is the gravity acceleration (Fig. 2.2). In the definition of the total head, each term is analogous to a form of energy per unit mass: $d\cos\theta$ is the pressure head, z_o is the potential head which is proportional to the

(c)

(d)

Fig. 2.1 (c) Saint-Laurent River, Montréal (Canada) on 12 July 2002 – looking downstream from Vieux-Port with Pont Jacques Cartier. (d) Escalier d'eau des jardins du Château du Touvet, 1763 (courtesy of Mr Bruno de Quinsonas) – water staircase.

potential energy per unit mass and $V^2/2g$ is the kinetic energy head. If there is no energy loss, the sum of the fluid's potential energy, kinetic energy and pressure work is a constant and the total head H is a constant. Along a streamline, the energy of the flow may be re-arranged between kinetic energy (i.e. velocity), potential energy (i.e. invert elevation) or pressure work (i.e. flow depth) but the sum of all terms must remain constant.

(e)

(f)

Fig. 2.1 (*Contd*) (e) Culvert beneath Cornwall Street, Brisbane (Australia) on 13 May 2002 during student field trip (structure no. MEL-C-X1, Chanson 1999a) – inlet view from right bank. (f) The same culvert in operation on 31 December 2001 around 6:00 a.m. near end of rainstorm – inlet view from left bank (estimated flow: 70 m³/s, flow direction from right to left).

The specific energy is defined as the total head using the invert elevation as datum. The mean specific energy equals:

$$E = H - z_0 \tag{2.2}$$

where E is the specific energy and z_0 is the bed elevation. The specific energy is analogous to the energy per unit mass, measured with the channel bottom as the datum (Fig. 2.2). The specific energy changes along a channel because of changes in bottom elevation and energy losses (e.g. friction loss). Assuming a hydrostatic pressure distribution in a rectangular channel, it is convenient to combine the specific energy definition with the continuity equation.

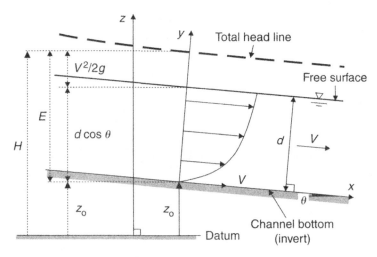

Fig. 2.2 Definition sketch of open channel flow.

The expression of the specific energy becomes:

$$E = d \cos \theta + \frac{Q^2}{2gd^2 B^2} \qquad (2.3a)$$

where Q is the total discharge and B is the free-surface width.

Remarks
1. H is also called the depth-averaged total head.
2. The term $(d \cos \theta + z_0)$ is often called the piezometric head.
3. If the velocity varies across the section, the kinetic energy term must be corrected as:

$$H = d \cos \theta + z_0 + \alpha \frac{V^2}{2g}$$

   where the kinetic energy correction coefficient α, also called the Coriolis coefficient, ranges typically between 1 and 1.05.
4. For an irregular channel, the expression of the specific energy becomes:

$$E = d \cos \theta + \frac{Q^2}{2g A^2} = d \cos \theta + \frac{V^2}{2g} \qquad (2.3b)$$

   where A is the cross-sectional area.

2.2 Fundamental principles

In fluid mechanics and hydraulics, the basic principles are the equations of continuity or conservation of mass, of momentum or conservation of momentum and conservation of energy

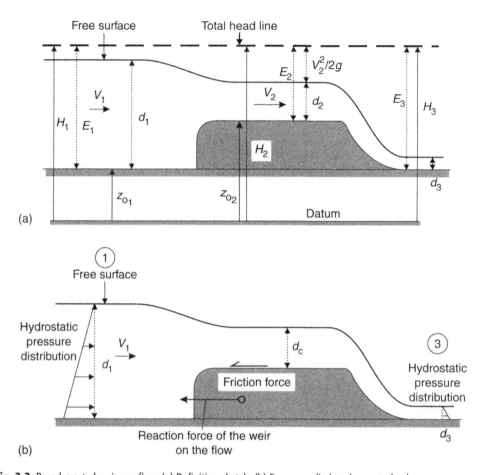

Fig. 2.3 Broad-crested weir overflow. (a) Definition sketch. (b) Forces applied to the control volume.

(Henderson 1966, Chanson 1999a). Another equation is the Bernoulli equation which may be derived from the differential form of the momentum principle. In this chapter, the simplest form of the fundamental principles is developed: i.e. for steady one-dimensional flows. Unsteady flow equations are developed in Part 3.

The law of conservation of mass, or continuity equation, states that the mass within a closed system remains constant with time. For an incompressible fluid such as water, the inflow must equal the outflow:

$$Q = V_1 A_1 = V_2 A_2 \qquad (2.4)$$

where Q is the total flow rate (i.e. volume discharge), V_1 and V_2 are the mean velocity across the cross-sections A_1 and A_2 respectively, and the subscripts 1 and 2 refer to the upstream and downstream flow cross-sections respectively.

The momentum principle states that, for a given control volume, the rate of change in momentum flux equals the sum of the forces acting on the control volume. The rate of change in momentum flux is the sum of momentum accumulation into the control volume plus the momentum flux in minus the momentum flux out. Considering the steady overflow above a broad-crested weir placed in a horizontal, rectangular channel sketched in Fig. 2.3, the forces applied to the control volume delimited by the cross-sections 1 and 3 are the pressure forces acting on sections 1 and 3, the reaction force of the weir onto the flow, the boundary friction, the weight of water of the control volume, and the reaction force of the channel bed which opposes exactly the weight of water. The momentum equation applied in the horizontal direction yields:

$$\rho Q V_3 - \rho Q V_1 = \frac{1}{2}\rho g d_1^2 B - \frac{1}{2}\rho g d_3^2 B - F_{\text{friction}} - F_{\text{weir}} \tag{2.5}$$

where ρ is the fluid density, B is the channel width, F_{friction} is the boundary friction and F_{weir} is the reaction force of the weir (Fig. 2.3).

The energy principle states that the net energy supplied to a system equals the energy that leaves the system as work is done plus the increase in energy of the system which is the sum of the potential energy, kinetic energy and internal energy. In open channel hydraulics, the energy principle is often expressed in terms of the total head:

$$H_1 = H_3 + \Delta H \tag{2.6}$$

where the total head is in metres (of water) and ΔH is the sum of the head losses between sections 1 and 3.

Remarks

1. The momentum principle is always used for hydrodynamic force calculations: e.g. force acting on a gate, flow resistance in uniform equilibrium flow. Other applications include the hydraulic jump, positive surge and bore.
2. The total momentum flux across a section equals $\rho VVA = \rho QV$. The rate of change in momentum flux is then: $(\rho QV_3 - \rho QV_1)$ in a horizontal channel, as sketched in Figure 2.3.
3. For a horizontal rectangular channel with hydrostatic distribution, the pressure force at a cross-section is:

$$\frac{1}{2}\rho g d^2 B$$

4. The reaction force F_{weir} of the weir onto the fluid equals exactly the horizontal component of the resultant of the pressure forces acting on the weir.
5. If the velocity distribution is not uniform, the momentum flux terms must be corrected by a momentum correction coefficient β, also called Boussinesq coefficient. For the above example, it yields:

$$\beta_3 \rho Q V_3 - \beta_1 \rho Q V_1 = \frac{1}{2}\rho g d_1^2 B - \frac{1}{2}\rho g d_3^2 B - F_{\text{friction}} - F_{\text{weir}}$$

6. Further discussions on the momentum principle were developed in Henderson (1966, pp. 5–9 and 66–77) and Chanson (1999a, pp. 11–17 and 51–97).
7. In open channel flows, the friction loss ΔH over a distance Δx along the flow direction is given by the Darcy equation:

$$\Delta H = f \frac{\Delta x}{D_H} \frac{V^2}{2g}$$

where f is the Darcy friction factor, V is the mean flow velocity and D_H is the hydraulic diameter or equivalent pipe diameter.

8. A broad-crested weir is a flat-crested structure with a crest length large compared to the flow thickness. When the crest is broad enough for the flow streamlines to be parallel to the crest, the pressure distribution above the crest is hydrostatic and the critical flow depth is observed on the weir crest. Broad-crested weirs are sometimes used as critical depth meters: i.e. to measure stream discharges.

Discussion: the Bernoulli equation

The Bernoulli equation derives from the momentum principle: i.e. the Navier–Stokes equation (Liggett 1993, Chanson 1999a). Considering the flow along a streamline, assuming that the gravity force is independent of the time, for a frictionless and incompressible fluid, and for a steady flow, the Navier–Stokes equation yields the differential form of the Bernoulli equation:

$$\frac{dP}{\rho} + g\,dz + V\,dV = 0 \qquad (2.7)$$

where V is the velocity, g is the gravity constant, z is the altitude, positive upwards, P is the pressure and ρ is the fluid density.

In an open channel, the integral form of Bernoulli equation is commonly written as:

$$H = z_0 + d\cos\theta + \frac{V^2}{2g} = \text{constant} \qquad (2.8a)$$

assuming a hydrostatic pressure distribution. For the smooth and short transition sketch in Fig. 2.3, the above equation may be rewritten as:

$$H_1 = H_2 = H_3 \qquad (2.8b)$$

Remarks

1. The Navier–Stokes equation was first derived by Navier in 1822 and Poisson in 1829 by an entirely different method. It was derived later in a manner similar by Barré de Saint-Venant in 1843 and Stokes in 1845.
2. Louis Navier (1785–1835) was a French engineer who primarily designed bridge but also extended Euler's equations of motion. Siméon Denis Poisson (1781–1840) was a French mathematician and scientist. He developed the theory of elasticity, a theory of electricity and a theory of magnetism. Adhémar Jean Claude Barré de Saint-Venant (1797–1886), French engineer, developed the equations of motion of a fluid particle in terms of the shear and normal forces exerted on it. George Gabriel Stokes (1819–1903), British mathematician and physicist, is known for his research in hydrodynamics and a study of elasticity.
3. The Bernoulli equation is named after the Swiss mathematician Daniel Bernoulli (1700–1782) who developed the equation in his 'Hydrodynamica, de viribus et motibus fluidorum' textbook (1st draft in 1733, first publication in 1738, Strasbourg).
4. For a one-dimensional flow, the Bernoulli equation differs the energy equation by the loss term only, although the Bernoulli principle derives from the momentum equation.

5. The Bernoulli equation is often used for smooth, short transition: e.g. sluice gate, broad-crested weir.
6. For the broad-crested weir sketched in Fig. 2.3, the Bernoulli equation may be written in terms of the specific energy:

$$E_1 = E_2 + (z_{o_2} - z_{o_1}) = E_3$$

Applications to open channel flow situations

In open channels, the application of the basic principles is a function of the flow situations. Basic flow situations are summarized in Table 2.1. These are discussed and developed in the following paragraphs. For smooth and short transitions, the continuity and Bernoulli equations form a system of two equations enabling to compute the downstream flow properties as functions of the upstream flow properties (e.g. Fig. 2.3). For a sudden transition from super- to subcritical flows, the continuity and momentum equations must be applied. The assumption of zero energy loss is untrue in a hydraulic jump. In a long prismatic channel with a constant flow rate, the flow motion reaches uniform equilibrium. That is, the flow resistance equals exactly and opposes the gravity force component in the flow direction.

2.3 Open channel hydraulics of short, frictionless transitions

Considering a smooth, short and frictionless transition, the continuity and Bernoulli equations applied to the open channel flow become:

$$V_1 A_1 = V_2 A_2 = Q \tag{2.9}$$

$$z_{o_1} + d_1 \cos \theta_1 + \frac{V_1^2}{2g} = z_{o_2} + d_2 \cos \theta_2 + \frac{V_2^2}{2g} \tag{2.10}$$

Table 2.1 Applications of the basic principles to simple flow situations

Flow situation (1)	Basic principles (2)	Remarks (3)
Steady flow Smooth, short and frictionless transition	Continuity + Bernoulli	The momentum principle may be used to calculate the forced exerted by the flow onto a structure (e.g. gate, weir)
Hydraulic jump	Continuity + momentum	The energy equation provides additionally the rate of energy dissipation
Uniform equilibrium flow	Continuity + momentum	Also called normal flow conditions
GVF	Differential form of the energy equation	Also called backwater calculations
Unsteady flow Unsteady GVF	Differential form of the continuity and momentum principles	Also called Saint-Venant equations (Part 3)

assuming a hydrostatic pressure distribution. If the discharge Q, the bed elevations, channel widths and channel slope are known, the downstream flow conditions (d_2, V_2) may be deduced from the upstream flow conditions (d_1, V_1) using the continuity and Bernoulli principles.

Application to horizontal channels

For the frictionless flow in a horizontal channel, the Bernoulli principle implies that the specific energy is constant along the channel. For a rectangular channel, it yields:

$$E = d + \frac{Q^2}{2gd^2B^2} = \text{constant} \qquad \text{Horizontal rectangular channel} \qquad (2.11)$$

That is, there is a unique relationship between specific energy E and flow depth d as shown in Fig. 2.4. In rectangular channels there is only one specific energy-flow depth curve for a given discharge per unit width Q/B. For a slow flow motion, the velocity is small, the kinetic energy term $V^2/2g$ is very small and the specific energy tends to the flow depth d (i.e. asymptote $E = d$). For a rapid flow, the velocity is large and by continuity the flow depth is small. The specific energy term tends to an infinite value when d tends to 0 (i.e. asymptote $d = 0$).

At any cross-section, the specific energy has a unique value. For a given value of specific energy and a given flow rate, there is two (meaningful) solutions, one solution (i.e. critical flow conditions) for:

$$E = \frac{3}{2} \sqrt[3]{\frac{q^2}{g}}$$

or no solution (i.e. no flow) for:

$$E < \frac{3}{2} \sqrt[3]{\frac{q^2}{g}}$$

where q is the discharge per unit width and g is the gravity acceleration. In the first case, the two possible flow depths d_1 and d_2 are called alternate depths (Fig. 2.4). The first one corresponds to a subcritical flow (i.e. $d > d_c$) and the second one to a supercritical flow ($d < d_c$), where d_c is the critical flow depth (see below). The alternate depths satisfy equation (2.11) which can be solved analytically, graphically and by trial and error.

The relationship $E = f(d)$ indicates the existence of a minimum specific energy $E_{\min}$ (Fig. 2.4). The flow conditions (d_c, V_c) such that the mean specific energy is minimum are called the critical flow conditions. For a rectangular, flat channel of constant width, the minimum specific energy $E_{\min}$ and the critical flow depth are respectively:

$$E_{\min} = \frac{3}{2}d_c \qquad (2.12)$$

$$d_c = \sqrt[3]{\frac{Q^2}{gB^2}} \qquad (2.13)$$

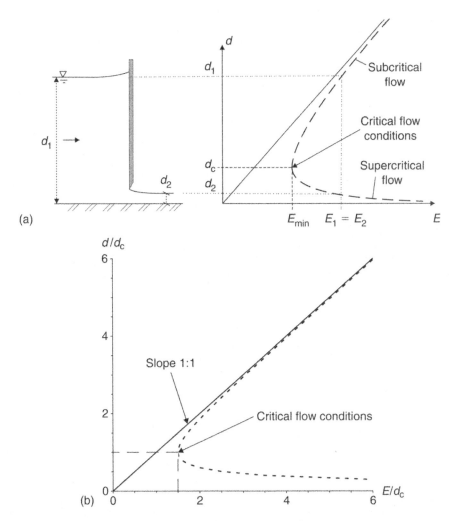

Fig. 2.4 Relationship between specific energy and flow depth. (a) Definition sketch. (b) Dimensionless specific energy curve for a flat rectangular channel.

The specific energy can be rewritten in dimensionless terms as:

$$\frac{E}{d_c} = \frac{d}{d_c} + \frac{1}{2}\left(\frac{d_c}{d}\right)^2 \tag{2.14}$$

Equation (2.14) is a unique curve, valid for any discharge. It is presented in Fig. 2.4(b).

Remarks
1. The critical flow velocity equals $V_c = \sqrt{gd_c}$ in a rectangular channel.
2. For a constant specific energy E, equation (2.11) is a cubic equation in terms of the water depth d. For $E > E_{min}$, there are three solutions (e.g. Chanson 1999a, pp. 143–145). One is negative and has no physical meaning.

3. For a channel of irregular cross-section, critical flow conditions satisfy:

$$\frac{Q^2}{g\frac{A^3}{B}} = 1$$

where A is the flow cross-sectional area and B is the free-surface width.

4. Historically, the concept of critical flow conditions was first introduced as a singularity of the backwater equation by Bélanger (1828) and later by Bazin (1865). At critical flow conditions, the GVF equation (i.e. backwater equation) cannot be solved.

5. The above definition of critical flow conditions follows the work of Bakhmeteff (1912).

Application to non-horizontal channels

Considering a short, smooth transition in a non-horizontal channel (e.g. broad-crested weir, Fig. 2.3), the upstream and downstream total heads are equal, by application of the Bernoulli equation. Hence the specific energy at the crest E_2 is the smallest in Fig. 2.3:

$$E_2 < E_1 = E_3$$

For a constant channel width B, the specific discharge ($q = Q/B$) is constant and there is only one specific energy/flow depth curve. The graphical solution (Fig. 2.4) indicates that the downstream flow depth must be smaller than the upstream one (i.e. $d_3 < d_1$) while critical flow conditions take place at the weir crest where the specific energy is minimum.

Remark

1. If the tailwater flow conditions are uncontrolled, the flow downstream of the weir is supercritical while critical flow conditions occur at the weir crest (i.e. $d_2 = d_c$) in Fig. 2.3.

Discussion

Figure 2.5 illustrates another example. The upstream flow conditions correspond to a subcritical flow. Bernoulli principle states that the upstream and downstream total heads

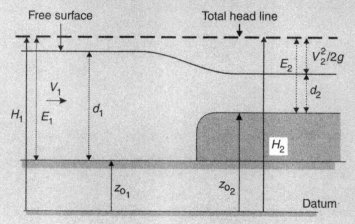

Fig. 2.5 Stepped channel transition.

must be equal. Hence the downstream specific energy is the smallest. For $B_1 = B_2$, the graphical solution (Fig. 2.4) indicates that the downstream flow depth must be smaller than the upstream one (i.e. $d_2 < d_1$) provided that $E_2 > 1.5d_c$. This flow situation is sketched in Fig. 2.5. For a supercritical upstream flow (i.e. $d_1 < d_c$), the relation would be inverted: i.e. a decrease of specific energy implies an increase in flow depth.

Note that, in Fig. 2.5, there is no flow motion for $E_1 < (z_{0_2} - z_{0_1})$. For $E_1 < (z_{0_2} - z_{0_1}) < E_2 < 1.5d_c$, the flow rate must be less than Q: i.e. choking occurs at the step.

Froude number

The Froude number is a dimensionless number proportional to the square root of the ratio of the inertial forces over the weight of fluid. For a horizontal rectangular channel, the Froude number is defined as:

$$Fr = \frac{V}{\sqrt{gd}} \qquad \text{Rectangular channel} \qquad (2.15)$$

Model studies of open channel flows and hydraulic structures are performed using a Froude similitude. That is, the Froude number must be the same for the model and the prototype.

Remarks

1. In horizontal, rectangular channels, the Froude number is unity at critical flow conditions. It may be rewritten as:

$$Fr = \left(\frac{d_c}{d}\right)^{3/2}$$

2. For a horizontal channel of irregular cross-sectional shape, the Froude number is usually defined as:

$$Fr = \frac{V}{\sqrt{g\dfrac{A}{B}}}$$

where A is the flow cross-section and B the free-surface width. With such a definition, $Fr = 1$ at critical flow conditions.

3. In open channel flows, it is strongly advised to define the Froude number such as $Fr = 1$ at critical flow conditions. That is, $Fr < 1$ for subcritical flow ($d > d_c$) and $Fr > 1$ for supercritical flow ($d < d_c$).

Discussion

The application of the Bernoulli equation is valid only within the range of assumptions: i.e. steady frictionless flow of incompressible fluid. For short and smooth transitions the energy losses are negligible and the Bernoulli equation may be applied successfully. If energy losses occur, they must be taken in account and the Bernoulli equation is no longer valid. For example, the Bernoulli equation is not valid at a hydraulic jump where turbulent energy losses are significant.

2.4 The hydraulic jump

In open channels, the transition from a rapid, supercritical flow to a slow, subcritical flow is called a hydraulic jump (Fig. 2.6). The transition occurs suddenly and it is characterized by a sudden rise of the free surface, with strong energy dissipation and mixing, large-scale turbulence, air entrainment, waves and spray (Fig. 2.6(b) and (c)).

A hydraulic jump is a marked flow discontinuity. The momentum principle is used to evaluate the basic flow properties in a hydraulic jump. Considering a horizontal, rectangular open channel of constant width B, and neglecting the shear stress at the channel bottom, the resultant of the forces acting in the x-direction are the resultant of hydrostatic pressure forces at the ends of the control volume (Fig. 2.6(a)). The continuity equation and momentum equations are:

$$V_1 d_1 B = V_2 d_2 B \tag{2.16}$$

$$\rho Q(V_2 - V_1) = \left(\frac{1}{2} \rho g d_1^2 - \frac{1}{2} \rho g d_2^2 \right) B \tag{2.17}$$

where B is the channel width and Q is the total discharge (i.e. $Q = VdB$). It yields a relationship between the upstream and downstream flow depths:

$$\frac{d_2}{d_1} = \frac{1}{2} \left(\sqrt{1 + 8Fr_1^2} - 1 \right) \tag{2.18}$$

where Fr_1 is the upstream Froude number: $Fr_1 = V_1/\sqrt{gd_1}$. The depth d_1 and d_2 are referred to as conjugate depths or sequent depths. Using equation (2.18) the momentum equation yields:

$$Fr_2 = \frac{2^{3/2} Fr_1}{\left(\sqrt{1 + 8Fr_1^2} - 1 \right)^{3/2}} \tag{2.19}$$

where Fr_2 is the downstream Froude number.

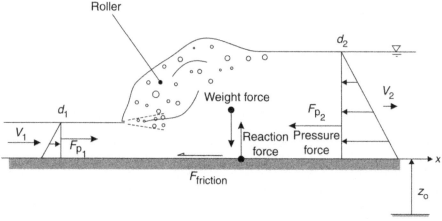

Fig. 2.6 Hydraulic jump flows. (a) Definition sketch of a hydraulic jump in a horizontal channel.

(b)

(c)

Fig. 2.6 (b) Hydraulic jump downstream of Chinchilla minimum energy loss weir spillway (Australia) on 8 November 1997 during a small overflow. (c) Hydraulic jump downstream of Awoonga dam spillway (Gladstone, Qld, Australia) in the 1970s (collection of late Prof. G.R. McKay in Apelt 1978) – Dam Stage 2: $H = 17.7\,\text{m}$, $L = 137.2\,\text{m}$, design flow: $16\,990\,\text{m}^3/\text{s}$ – the dam was heightened in 2002.

The energy equation gives the total head loss:

$$\Delta H = \frac{(d_2 - d_1)^3}{4\, d_1 d_2} \tag{2.20}$$

Remarks
1. An open channel flow can change from subcritical to supercritical in a relatively 'low-loss' manner at gates and weirs. In these cases the flow regime evolves from subcritical to supercritical with the occurrence of critical flow conditions associated with relatively small energy loss (e.g. broad-crested weir, sluice gate). The transition from supercritical to subcritical flow is, on the other hand, characterized by a strong dissipative mechanism.
2. In a hydraulic jump, the upstream flow conditions are supercritical: i.e. $Fr_1 > 1$ in equations (2.18)–(2.20).
3. Hydraulic jumps are strong energy dissipators. Hydraulic jump stilling basin are commonly designed downstream of dam spillways to dissipate the kinetic energy of the flow. For example, Fig. 2.6(b) shows a hydraulic jump downstream of a weir during a small overflow. Figure 2.6(c) presents a hydraulic jump downstream of a dam spillway during a medium flood.
4. Hydraulic jumps may be characterized by strong air entrainment (Fig. 2.6). Air is trapped at the impingement of the supercritical flow into the roller (see Chapter 17).
5. Experimental observations highlighted different types of hydraulic jumps, depending upon the upstream Froude number Fr_1 and inflow conditions (e.g. Chow 1973, p. 395; Chanson 1997, pp. 73–92; Chanson 1999a, pp. 57–64).

2.5 Open channel flow in long channels

2.5.1 Presentation

In open channel flows, flow resistance can be neglected over a short transition as a first approximation. But this assumption is invalid for long channels (Fig. 2.7). Considering a water supply canal extending over several kilometres, the boundary shear opposes the fluid motion and retards the flow. The flow resistance and gravity effects are of the same order of magnitude. In fact, the friction force equals and opposes exactly the weight force component in the flow direction at uniform equilibrium (i.e. normal flow conditions) (Fig. 2.8). The laws of flow resistance in open channels are basically the same as those in closed pipes, although, in open channel, the calculation of boundary shear stress is complicated by the existence of the free surface and the wide variety of possible cross-sectional shapes (Henderson 1966, Chanson 1999a). Another difference is the propulsive force. In open channel flows, the fluid is propelled by the weight of the flowing water resolved down a slope, whereas, in closed pipes, the flow is driven by a pressure gradient along the pipe.

(a)

(b)

Fig. 2.7 Photographs of long channels. (a) Fjord du Saguenay (more than 300 m deep) near the confluence with the Saint-Laurent River (Canada) – looking upstream on 16 July 2002 (courtesy of Mr and Mrs Chanson). (b) Petite Nation river, Outaouais region (Qué., Canada) on 14 July 2002 – looking downstream (the river is fed by massive groundwater springs).

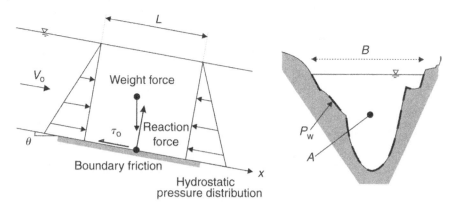

Fig. 2.8 Sketch of uniform equilibrium flow in an open channel.

For open channel flows as for pipe flows, the head loss ΔH over a distance L along the flow direction is given by the Darcy equation:

$$\Delta H = f \frac{L}{D_H} \frac{V^2}{2g} \tag{2.21}$$

where f is the Darcy–Weisbach friction factor, V is the mean flow velocity and D_H is the hydraulic diameter or equivalent pipe diameter (Fig. 2.8).

The flow regime in open channels can be either laminar or turbulent. It is commonly accepted that the flow becomes turbulent for Reynolds numbers larger than 5000–10 000 where the Reynolds number is defined as:

$$Re = \rho \frac{V D_H}{\mu} \tag{2.22}$$

where ρ is the water density and μ is the dynamic viscosity. Most open channel flows are turbulent and the friction factor may be estimated from the Colebrook–White formula:

$$\frac{1}{\sqrt{f}} = -2.0 \log_{10} \left(\frac{k_s}{3.71 D_H} + \frac{2.51}{Re \sqrt{f}} \right) \tag{2.23}$$

where k_s is the equivalent sand roughness height.

Remarks
1. The hydraulic diameter, also called equivalent pipe diameter, is defined as:

$$D_H = \frac{4A}{P_w}$$

where P_w is the wetted perimeter and A is the flow cross-sectional area (Fig. 2.8).
2. The average boundary shear stress equals:

$$\tau_o = \frac{f}{8} \rho V^2$$

3. The Colebrook–White formula is valid only in turbulent flows. In laminar flows, the Darcy friction factor equals:

$$f = \frac{64}{Re} \qquad \text{Laminar flows}$$

4. Equation (2.23) is non-linear. The Darcy–Weisbach friction factor f appears on both sides. The Colebrook–White formula must be solved by iterations using a graphical method (e.g. Moody diagram) or by trial and error.
5. Typical equivalent roughness heights are listed below:

k_s (mm)	Material
0.01–0.02	PVC (plastic)
0.3–3	Concrete
3–10	Untreated shotcrete
0.6–2	Planed wood
5–10	Rubble masonry

2.5.2 Uniform equilibrium flows

In steady open channel flows, a fundamental problem is determining the relation between the water depth, the flow velocity, the channel slope and the channel geometry. Considering a straight, prismatic channel with a constant discharge, uniform equilibrium is achieved when the flow properties (d, V) become independent of time and of position along the flow direction. The momentum equation applied in the flow direction states the exact balance between the shear forces and the gravity force component in the flow direction (Fig. 2.8). The momentum principle yields:

$$\tau_o P_w L = \rho g A L \sin\theta \qquad (2.24a)$$

where τ_o is the bottom shear stress, P_w is the wetted perimeter, L is the length of the control volume, A is the cross-sectional area and θ is the channel slope. Replacing the bottom shear stress by its expression, the momentum equation for uniform equilibrium flows becomes:

$$V_o = \sqrt{\frac{8g}{f}} \sqrt{\frac{(D_H)_o}{4} \sin\theta} \qquad (2.24b)$$

where V_o is the uniform equilibrium flow velocity and $(D_H)_o$ is the hydraulic diameter of uniform equilibrium flows.

Remarks
1. Uniform equilibrium flow conditions are also called normal flow conditions.
2. The momentum equation for steady uniform open channel flow is rewritten usually as:

$$S_f = S_o$$

where S_f is called the friction slope and S_o is the channel slope defined as:

$$S_f = -\frac{\partial H}{\partial x} = \frac{4\tau_o}{\rho g D_H}$$

$$S_o = -\frac{\partial z_o}{\partial x} = \sin\theta$$

where H is the mean total head and z_o is the bed elevation. The definitions of the friction and bottom slopes are general and applied to both uniform equilibrium and GVF.

3. The flow resistance is sometimes expressed in terms of the Chézy equation:

$$V = C_{\text{Chézy}} \sqrt{\frac{D_H}{4} \sin\theta}$$

where $C_{\text{Chézy}}$ is the Chézy coefficient (unit: $m^{1/2}/s$). The Chézy equation was first introduced as an empirical correlation. The Chézy coefficient ranges typically from $30\,m^{1/2}/s$ (small rough channel) up to $90\,m^{1/2}/s$ (large smooth channel).

4. An empirical formulation, called the Gauckler–Manning formula, was developed for turbulent flows in rough channels. The Gauckler–Manning formula is:

$$V = \frac{1}{n_{\text{Manning}}} \left(\frac{D_H}{4}\right)^{2/3} \sqrt{\sin\theta}$$

where n_{Manning} is the Gauckler–Manning coefficient (unit: $s/m^{1/3}$). The Gauckler–Manning coefficient is an empirical coefficient, found to be a characteristic of the surface roughness primarily. Typical values of n_{Manning} (in SI unit) are:

n_{Manning}	Material
0.012	Finished concrete
0.014	Unfinished concrete
0.029	Gravel
0.05	Flood plain (light brush)
0.15	Flood plain (trees)

2.5.3 GVF calculations

In most practical cases, the cross-section, depth and velocity in a channel vary in the flow direction and uniform flow conditions are not often reached. Figure 2.7 illustrates examples of longitudinal free-surface profiles in response to upstream and downstream controls. In Fig. 2.7(a), the river flow is controlled by the downstream flow conditions: i.e. the tides in the Bay and the flow conditions in the Saint-Laurent river. Figure 2.7(b) shows a supercritical flow that is controlled by the upstream flow conditions.

Considering a steady open channel flow, the differential form of the energy equation written in terms of mean total head is:

$$\frac{\partial H}{\partial x} = -f \frac{1}{D_H} \frac{V^2}{2g} \tag{2.25a}$$

where x is the distance along the channel bed, f is the Darcy friction factor, D_H is the hydraulic diameter, V is the mean flow velocity (i.e. $V = Q/A$). Equation (2.25) is called the GVF equation and it is usually rewritten as:

$$\frac{\partial H}{\partial x} = -S_f \tag{2.25b}$$

where S_f is the friction slope.

The GVF equation was first developed by J.B. Bélanger (Bélanger 1828). It is a one-dimensional model of GVF, steady flows in open channel. It may be applied to natural and artificial channels but it is important to know the limitations. GVF calculations are developed assuming a steady, non-uniform equilibrium flow, that the flow is GVF, and that the flow resistance may be estimated as in uniform equilibrium flows. That is, they do not apply to uniform equilibrium flows, nor to unsteady flows nor to rapidly varied flows (RVF, e.g. the hydraulic jump).

Remarks
1. The GVF equation is also called the backwater equation.
2. Backwater calculations may be applied to one-dimensional, steady flows. They do not apply to hydraulic jumps nor to any RVF situation. Further there is a singularity at critical flow conditions. In fact, the concept of critical flow conditions was first developed by Bélanger (1828) as the location where $d = Q^2/gA^2$ for a flat channel. It was associated with the idea of minimum specific energy by Bakhmeteff (1912). Both Bélanger and Bakhmeteff developed the concept of critical flow in relation with the singularity of the backwater equation for $d = Q^2/gA^2$ (i.e. critical flow conditions).
3. Backwater computations must start from a location where the flow conditions (d, V) are known. The results are very sensitive to the flow resistance estimate (e.g. Henderson 1966, Chanson 1999a).

Integration of the GVF equation
To solve the backwater equation, it is essential to determine correctly the boundary conditions. That is, the flow conditions upstream, downstream and along the channel reach. In practice, the boundary conditions are control devices (e.g. sluice gate, weir, reservoir) that impose flow conditions for the depth, the discharge or a relationship between the discharge and the depth, and geometric characteristics (e.g. bed elevation, channel width). Equation (2.25) is a non-linear equation which usually cannot be solved analytically, but it can be integrated numerically. One of the most common integration methods is the standard step method, developed herein and used by several numerical models.

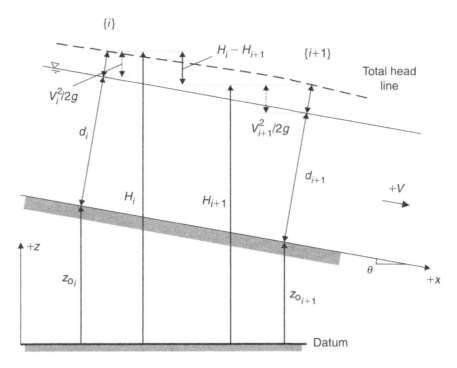

Fig. 2.9 Definition sketch of GVF calculations.

Considering the GVF sketched in Fig. 2.9, the backwater equation is integrated between two cross-sections denoted $\{i\}$ and $\{i + 1\}$. It yields:

$$\frac{H_{i+1} - H_i}{x_{i+1} - x_i} \approx -\frac{1}{2} \left(S_{f_{i+1}} + S_{f_i} \right) \tag{2.26a}$$

where the subscripts i and $i + 1$ refer to stations $\{i\}$ and $\{i + 1\}$ respectively. If the flow conditions at cross-section $\{i\}$ are known, the total head at the cross-section $\{i + 1\}$ equals:

$$H_{i+1} \approx H_i - \frac{1}{2}(S_{f_{i+1}} + S_{f_i})(x_{i+1} - x_i) \tag{2.26b}$$

2.6 Summary

Basic material	Textbook page nos.

Total head H

In a horizontal open channel, with hydrostatic pressure distribution, the depth-averaged total head H is defined as:

$$H = d + z_o + \frac{V^2}{2g}$$

where d is the water depth, z_o is the bottom elevation, V is the flow velocity and g is the gravity acceleration ($g = 9.80 \, \text{m/s}^2$ in Brisbane). Note that z_o is also called the invert elevation.

Specific energy

For a flat channel, assuming a hydrostatic pressure distribution, the specific energy is defined as:

$$E = d + \frac{V^2}{2g}$$

The specific energy is the total head, measured with the channel bottom as the datum.

Critical flow conditions

For a constant discharge Q and a given cross-section, the relationship E versus d indicates the existence of a minimum specific energy. The flow conditions (d_c, V_c) such that the mean specific energy is minimum are called the critical flow conditions. At critical flow conditions, in a rectangular channel, the specific energy equals:

$$E = E_{min} = 1.5 \, d_c$$

where d_c is the critical depth:

$$d_c = \sqrt[3]{\frac{Q^2}{gB^2}}$$

where B is the channel width. The critical velocity equals:

$$V_c = \sqrt{gd_c}$$

Bernoulli principle

Neglecting energy loss, for a steady, incompressible flow, the Bernoulli equation states:

$$H_1 = E_1 + z_{o_1} = E_2 + z_{o_2} = H_2$$

The Bernoulli principle is commonly used for short, smooth transitions when energy losses are zero (Chanson 1999a).

Hydraulic jump

A hydraulic jump is the sudden transition from a supercritical flow into a subcritical flow. The transition is characterized by a sudden rise in water level and significant energy dissipation. For a horizontal rectangular channel, the momentum principle gives the ratio of downstream to upstream depths:

$$\frac{d_2}{d_1} = \frac{1}{2}\left(\sqrt{1 + 8Fr_1^2} - 1\right)$$

where Fr_1 is the upstream Froude number.

Uniform equilibrium flow conditions

At uniform equilibrium, the momentum principle states that the gravity force component in the flow direction equals exactly the flow resistance. It yields:

$$V_o = \sqrt{\frac{8g}{f}} \sqrt{\frac{(D_H)_o}{4}} \sin\theta$$

where V_o is the uniform equilibrium flow velocity and f is the Darcy friction factor.

Basic material	Textbook page nos.
(Total head H)	24–25
(Specific energy)	30–38
(Critical flow conditions)	31–38, 44–45, 47
(Bernoulli principle)	17–18, 26–27, 33–36, 38
(Hydraulic jump)	51–67, 335–343
(Uniform equilibrium flow conditions)	71–91, 98–105

Reference: Chanson, H. (1999a). *The Hydraulics of Open Channel Flows: An Introduction.* (Butterworth-Heinemann: Oxford, UK) 512 pages. http://www.uq.edu.au/~e2hchans/reprints/errata.htm

2.7 Exercises

1. Considering a vertical sluice gate in a horizontal smooth rectangular channel, the upstream and downstream water depths are respectively 5.1 and 0.45 m. The upstream flow velocity is 0.5 m/s and the channel width is 27 m. Calculate the downstream flow depth and the force acting on the sluice.

2. Considering a broad-crested weir, draw a sketch of the weir. What is the main purpose of a broad-crested weir? A broad-crested weir is installed in a horizontal and smooth channel of rectangular cross-section. The channel width is 17 m. The bottom of the weir is 2.15 m above the channel bed. The water discharge is 24 m^3/s. Compute the depth of flow upstream of the weir, above the sill of the weir and downstream of the weir (in absence of downstream control), assuming that critical flow conditions take place at the weir crest. *Assume a frictionless flow.*

3. A hydraulic jump flow takes place in a horizontal rectangular channel. The downstream flow conditions are: $d = 5.1$ m, $q = 14$ m^2/s. Calculate the upstream flow depth and Froude number, as well as the head loss in the jump.

4. A rectangular (14 m width) concrete channel carries a discharge of 10 m^3/s. The longitudinal bed slope is 2 m/km. (a) What is the normal depth at uniform equilibrium? (b) At uniform equilibrium what is the average boundary shear stress? (c) At normal flow conditions, is the flow supercritical, supercritical or critical? Would you characterize the channel as mild, critical or steep? *For man-made channels, perform flow resistance calculations based upon the Darcy–Weisbach friction factor.*

5. At uniform equilibrium, water flows in a trapezoidal grass waterway (bottom width: 2 m, sidewall slope: 1V:3H), the flow depth is 1.1 m and the longitudinal bed slope is 21.3 m/km. Assume a Gauckler–Manning coefficient of 0.045 s/m$^{1/3}$. Calculate: (a) discharge, (b) critical depth, (c) Froude number, (d) Reynolds number and (e) Chézy coefficient. *The fluid is water at 20°C.*

6. An artificial canal carries a discharge of 12 m^3/s. The channel cross-section is rectangular (9 m bottom width, 1V:2H sideslopes). The longitudinal bed slope is 7.5 m/km. The channel bottom and sidewall consist of a mixture of sands ($k_s = 0.9$ mm). (a) What is the normal depth at uniform equilibrium? (b) At uniform equilibrium what is the average boundary shear stress and the shear velocity?

   A gauging station is set at a bridge crossing the waterway. The observed flow depth, at the gauging station, is 2.5 m. (c) Compute the flow velocity at the gauging station. (d) Calculate the Darcy friction factor (at the gauging station). (e) What is the boundary shear stress (at the gauging station)? (f) How would you describe the flow at the gauging station? (g) At the gauging station, from where is the flow controlled? Why?

PART 2
Turbulent Mixing and Dispersion in Rivers and Estuaries: An Introduction

Aichi Forest Park, Japan (photograph by H. Chanson, March 1999).

Mixing and dispersion of matter in natural rivers is of considerable importance. In some cases, the release of matter into a natural system causes significant harm to the local environment and it is essential to predict accurately the dispersion and mixing in the stream, and the extent of the impact, despite the complexity of the natural system. There are several excellent publications covering aspects of river mixing. For example, Fischer *et al.* (1979) and Rutherford (1994) provide comprehensive reviews of mixing theory, and they detail how to solve particular mixing problems. This section is an attempt to draw together the important elements from these and other important works, together with the writer's experience in lecturing open channel hydraulics, mixing in rivers and environmental fluid mechanics. The lecture material is aimed to undergraduate students who have a solid background in fluid mechanics and hydraulics. At the end of this section, the reader will have gained some basic understanding in turbulent mixing and dispersion in rivers, as well as a series of pre-design tools to estimate contaminant dispersion in river systems.

Internet resources

Internet resources of relevance include:

Biography of Hugo B. Fischer	http://www.oac.cdlib.org/dynaweb/ead/ berkeley/wrca/fischerh/
Mixing and dispersion in rivers	
Rivers seen from space	http://www.athenapub.com/rivers1.htm
Aerial photographs of rivers	ftp://geology.wisc.edu/pub/air
Mixing and dispersion in estuaries	
USACE inlets online	http://www.oceanscience.net/inletsonline/
Estuaries in South Africa	http://www.upe.ac.za/cerm/
Whirlpools	http://www.uq.edu.au/~e2hchans/whirlpl.html

3

Introduction to mixing and dispersion in natural waterways

3.1 Introduction

Dispersion of matters in natural rivers is of considerable importance. Applications include sediment and salt dispersion, injection of heated water into a cooler stream (e.g. at a cooling power plant), releases of untreated organic waste and domestic sewerage in an ecosystem, and stormwater waste disposal during floods (Figs 3.1 and 3.2). In some cases, the release of matters in a natural system may cause significant harm: e.g. smothering of seagrass and coral, release of organic or nutrient-rich wastewater into the ecosystem (e.g. from treated sewage effluent). It is therefore important to predict accurately the dispersion and mixing in the stream, and the extent of the impact despite the complexity of the natural system. For example, during an accidental release of waste that occurs in a stream, the water resource scientist needs to predict the arrival time of the contaminant cloud, the peak concentration of solute and the duration of the pollution.

In Nature, most practical applications are associated with turbulent flows, and this work will focus on turbulent dispersion and mixing in rivers and estuaries. Relevant literatures include Fischer *et al.* (1979) and Rutherford (1994). Ippen (1966), Wood *et al.* (1993) and Lewis (1997) discussed some specific aspect of mixing in the oceans (Fig. 3.3). However the topic is often poorly understood by professionals and researchers, and there is great empiricism.

Remarks
1. Hugo B. Fischer (1937–1983) was educated at the California Institute of Technology. He was a Professor of Civil Engineering at the University of California, Berkeley from 1966 until 1983.
2. Arthur Thomas Ippen (1907–1974) was Professor in Hydrodynamics and Hydraulic Engineering at MIT (USA). Educated in Germany (Technische Hochschule, Aachen), he moved to USA in 1932 and he worked at MIT from 1945 until his retirement in 1973.
3. A related topic is mixing and dispersion in the atmosphere (e.g. Csanady 1973).

(a)

(b)

Fig. 3.1 Natural river systems. (a) Shiraito-no-taki, main waterfall of the Urui River (fall height > 20 m) (Japan) on 5 June 1999. (b) Braided channel of the Fujigawa River (Japan) on 2 November 2001, looking upstream about 25 km upstream of river mouth.

(c)

Fig. 3.1 (*Contd*) (c) Gravel (sand) bar on Oyana River (Japan) on 2 November 2001 (note the large rocks).

Fig. 3.2 Sludge mixing in a sludge thickening tank, Molendinar Water Purification Plant (Gold Coast, Australia) on 4 September 2002.

Discussion

The monitoring of rivers, estuaries and marine environments is based upon key water quality parameters, including biological indicators. The effects of mixing are often ignored in assessing the overall ecosystem health of waterways, while the reliability of predictive models depends on how well these 'key parameters' best describe fundamental mechanisms such as mixing

Fig. 3.3 Plume of the Var River (top left) entering the Mediterranean Sea at Nice airport (courtesy of Prof. D.H. Peregrine). Picture taken from airplane on 30 March 2001 near full moon. The Var River may carry a lot of bauxite sediments during flood and snow melt periods.

and dispersion. Predictions of contaminant dispersion in creeks and streams are almost always based upon empirical mixing and dispersion coefficients (Chapters 7–10). These coefficients are highly sensitive to the natural system and flow conditions, and must be measured *in situ*. Experimental findings are however accurate only 'within a factor of 4' (at best!), and they can rarely be applied to another system (Fischer *et al.* 1979, Rutherford 1994). While there has been considerable research on pollutant dispersion in individual river catchments, little systematic research has been done on the turbulent mixing and dispersion in complete riverine and estuarine systems.

It is fundamental to comprehend that analytical and numerical models must be calibrated with basic field measurements, and that they must be validated with other independent field tests.

3.2 Laminar and turbulent flows

Considering a particular situation (e.g. a pipe flow, Fig. 3.4), a low-velocity flow motion is characterized by fluid particles moving along smooth paths in laminas or layers, with one layer gliding smoothly over an adjacent layer: i.e. the *laminar flow regime*.

With increasing flow velocity, there is a critical velocity above which the flow motion becomes characterized by an unpredictable behaviour, strong mixing properties and a broad spectrum of length scales: i.e. a *turbulent flow regime*. This is illustrated in Fig. 3.4 showing a modified Reynolds experiment[1] in the Hydraulics/Fluid Mechanics Laboratory at the

---

[1] In his original experiment, Osborne Reynolds used a horizontal pipe.

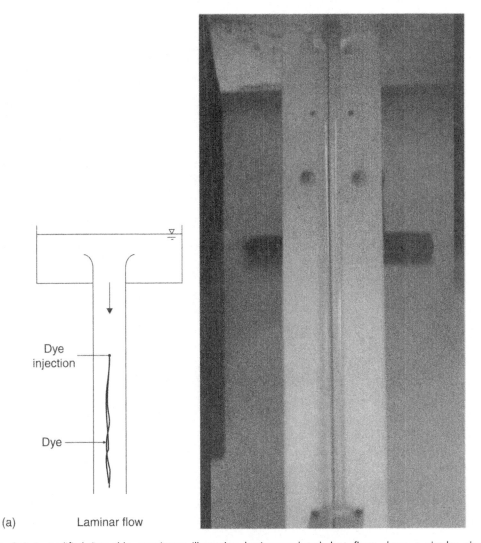

(a)   Laminar flow

Fig. 3.4 A modified Reynolds experiment illustrating laminar and turbulent flows down a circular pipe. (a) Laminar flow ($Re = 370$, $D = 0.0127\,\text{m}$ on the right photograph). Note the lamina of green dye in the middle of the pipe.

University of Queensland. In turbulent flows the fluid particles move in very irregular paths, causing an exchange of momentum from one portion of the fluid to another. Turbulent flows have great mixing potential and involve a wide range of eddy length scales. In natural streams, the flow is turbulent and strong turbulent mixing occurs.

In pipes, laminar flows are observed for Reynolds number <1000–3000, where the Reynolds number is defined as $Re = (\rho V D_H)/\mu$, where ρ and μ are the fluid density and dynamic viscosity respectively, V is the flow velocity and D_H is the hydraulic diameter or equivalent pipe diameter. Turbulent flows occur for Reynolds numbers >5000–$10\,000$ typically.

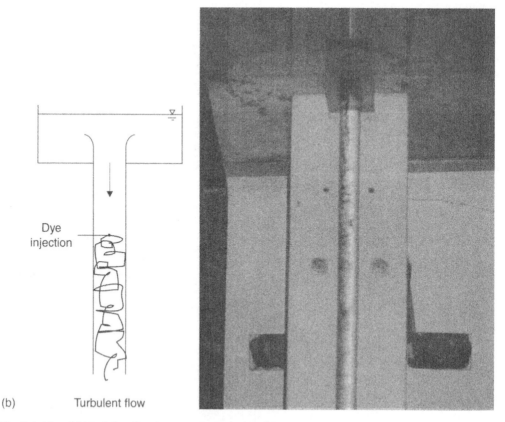

(b)   Turbulent flow

Fig. 3.4 (*Contd*) (b) Turbulent flow (*Re* = 3600, *D* = 0.0127 m on the right photograph). High-speed photograph (1/1000 shutter speed) showing the rapid dye mixing downstream of the injection point.

Remarks

1. The Reynolds number *Re* characterizes the ratio of inertial force to viscous force.
2. Osborne Reynolds (1842–1912) was a British physicist and mathematician who expressed first the Reynolds number (Reynolds 1883) and later the Reynolds stress or turbulent shear stress.
3. The hydraulic diameter is defined as:

$$D_H = 4 \frac{A}{P_w}$$

where A is the flow cross-sectional area and P_w is the wetted perimeter. D_H is also called the equivalent pipe diameter. For a circular pipe, $D_H = D$ where D is the internal pipe diameter.

Shear stress

The term *shear flow* characterizes a flow with a velocity gradient in a direction normal to the mean flow direction: e.g. a boundary layer flow along a flat plate. In a shear flow, momentum (i.e. momentum per unit volume $= \rho V$) is transferred from the region of high velocity to that of low velocity. The fluid tends to resist the shear associated with the transfer of momentum. The shear stress is proportional to the rate of transfer of momentum.

In laminar flows,[2] the Newton's law of viscosity relates the shear stress to the rate of angular deformation:

$$\tau = \mu \frac{\partial v}{\partial y} \qquad (3.1)$$

where τ is the shear stress, μ is the dynamic viscosity of the flowing fluid, v is the velocity and y is the direction normal to the flow direction. For larger shear stresses, the fluid cannot sustain the viscous shear stress and turbulence spots develop. After apparition of turbulence spots, the turbulence expands rapidly to the entire shear flow. The apparent shear stress in turbulent flow is expressed as:

$$\tau = \rho (v + v_T) \frac{\partial v}{\partial y} \qquad (3.2a)$$

where v is the kinematic viscosity (i.e. $v = \mu/\rho$) and v_T is called the 'eddy viscosity' or *momentum exchange coefficient* in turbulent flow. The momentum exchange coefficient v_T is a factor depending upon the flow motion. Practically, $v_T \gg v$ in turbulent flows and equation (3.2a) becomes:

$$\tau = \rho v_T \frac{\partial v}{\partial y} \qquad (3.2b)$$

Remarks

1. The 'eddy viscosity' concept was first introduced by the Frenchman J.V. Boussinesq (1877, 1896).
2. Joseph Valentin Boussinesq (1842–1929) was a French hydrodynamicist and Professor at the Sorbonne University (Paris). His treatise 'Essai sur la théorie des eaux courantes' (1877) remains an outstanding contribution in hydraulics literature.
3. In the mixing length theory, the momentum exchange coefficient v_T is defined as:

$$v_T = l^2 \frac{\partial v}{\partial y} \qquad (3.3)$$

where l is the mixing length.
4. The mixing length theory is a turbulence theory developed by L. Prandtl (1925). He assumed that the mixing length l is the characteristic distance travelled by a particle of fluid before its momentum is changed by the new environment.
5. Ludwig Prandtl (1875–1953) was a German physicist and aerodynamicist who introduced the concept of boundary layer (Prandtl 1904). He was a Professor at the University of Göttingen.

---

[2] That is, for a Reynolds number <1000–3000.

3.3 Basic definitions

The *mass flow rate* (or mass discharge) is the mass flux per unit time (unit: kg/s).
The *volume flow rate* (or volume discharge) is the volume flux per unit time (unit: m³/s).
The *concentration* of a contaminant C_m is defined as the mass of contaminant per unit volume (unit: kg/m³).
The *volume fraction* is defined as the volume of tracers per unit volume (unit: dimensionless).
The *dilution* is defined as the inverse of the volume fraction:

$$\text{Dilution} = \frac{1}{\text{Volume fraction}}$$

The *density* of contaminated water equals $\rho + \Delta\rho$, where ρ is the water density (unit: kg/m³). $\Delta\rho/\rho$ is typically <3%.[3] Although such changes in fluid density has little effect on fluid acceleration, they do affect buoyant discharges and the stability of density-stratified water bodies (e.g. lakes, oceans).
The *buoyancy per unit mass*, or modified gravitational acceleration, is defined as:

$$g' = g\,\frac{\Delta\rho}{\rho}$$

where g is the gravity acceleration.
Density stratification in large water bodies (lakes, oceans) is almost often stable. It is described by the vertical density profile $\rho(z)$, where z is the vertical coordinate positive upwards. Dimensional analysis shows that a relevant dimensionless number[4] is the *Richardson number* defined as:

$$Ri = \frac{g}{\rho}\,\frac{-\dfrac{\partial\rho}{\partial z}}{\left(\dfrac{\partial v}{\partial z}\right)^2}$$

where the sign minus is used for convenience to make Ri positive as the elevation z is positive upwards.
The *diffusion coefficient* (or diffusivity) is the quantity of a chemical that, in diffusing from one region to another, passes through each unit of cross-section per unit of time when the volume concentration is unity. The units of the diffusion coefficient are m²/s.

Remarks

1. The mass concentration of a substance is expressed in mass of dissolved chemical per unit volume or kg/m³. Another unit is the parts per million (ppm). The conversion is: 1 ppm = 1 mg/L.
2. The buoyancy force is a vertical force caused by pressure differences between the upper and lower surfaces of a submerged object. Calculations of the buoyancy force exerted on a submerged air bubble are developed in Appendix A (Section 3.5).

---

[3] For analyses of shallow waters (streams, estuaries, nearshore ocean waters), it is usually necessary to work with a density accuracy $\Delta\rho \sim 1 \times 10^{-4}$ kg/m³ while a density accuracy of $\Delta\rho \sim 1 \times 10^{-6}$ to 1×10^{-5} kg/m³ is required for analyses of mixing in deep lakes, reservoirs and oceans (Fischer *et al.* 1979).
[4] Used for similitude and physical modelling.

3. There are several definitions of the Richardson number. The above is sometimes called the *local gradient Richardson number*. The Richardson number was named after Lewis Fry Richardson (1881–1953), a British meteorologist who took interest in the dispersion of smoke from shell explosion during the World War I.
4. Basic fluid properties at standard atmopshere and 20°C are summarized in Table 1.1 (Chapter 1). Variations of freshwater properties with temperature are given in Appendix B (Section 3.6).

3.4 Structure of the section

This section is an attempt to draw together the important elements from important relevant works, together with the writer's experience in lecturing open channel hydraulics, mixing in rivers and environmental fluid mechanics. The material presents the basic concepts of mixing and dispersion in rivers. It is aimed to undergraduate students who have a solid background in fluid mechanics and hydraulics. At the end of the course, the reader will have gained some basic understanding in turbulent mixing and dispersion in rivers, as well as a series of pre-design tools to estimate contaminant dispersion in river systems.

A brief summary of turbulent flows follows (Chapter 4) this chapter. The next two chapters (Chapters 5 and 6) present the fundamental principle of molecular diffusion and advection. The results are extended to turbulent advective diffusion in channels: transverse mixing (Chapter 7) and longitudinal dispersion (Chapter 8). While Taylor's theory, presented in Chapter 8, is a simple prediction tool, dispersion in natural river systems is further complicated by the existence of dead zones and some possible contaminant reactions. These issues are developed in Chapter 9. Mixing and dispersion in estuarine zones are introduced in Chapter 10.

At the beginning of the book, the reader will find the list of symbols and a glossary of technical terms and names. After the conclusion, a detailed list of references is presented.

Remarks

It is worth noting the number of significant contributions by Australasian academics and scientists to the field of environmental open channel flows, and mixing and dispersion in riverine, estuarine and coastal systems. Among them, J.C. Rutherford, National Institute of Water and Atmospheric Research, Hamilton (NZ) (Rutherford 1994); Prof. Ian R. Wood, Emeritus Professor, Department of Civil Engineering, University of Canterbury, Christchurch (NZ) (Wood *et al.* 1993); Late Professor David L. Wilkinson (1944–1998), Department of Civil Engineering, University of Canterbury, Christchurch (NZ) and formerly Department of Civil Engineering, University of New South Wales, Sydney (Australia) (Wood *et al.* 1993);[5] Prof. Jorg Imberger, Department of Environmental Engineering, University of Western Australia, Perth, WA (Australia) (Fischer *et al.* 1979); Prof. Frank M. Henderson, Emeritus Professor, Department of Civil Engineering, University of Newcastle, Newcastle, NSW (Australia) (Henderson 1966); and Dr Hubert Chanson, Reader, Department of Civil Engineering, University of Queensland, Brisbane, Qld (Australia) (Chanson 1999a, 2004a).

[5] Obituary, *Journal of Hydraulic Research*, **39**(5), 565, 2001.

3.5 Appendix A – Application: buoyancy force exerted on a submerged air bubble[6]

When an air bubble is submerged in a liquid, a net upward force (i.e. buoyancy) is exerted on the bubble. Buoyancy is a vertical force caused by the pressure difference between the upper and lower surfaces of the bubble (Fig. 3A.1). To illustrate the concept of buoyancy, let us consider a diver in a swimming pool. As the pressure below him/her is larger than that immediately above, a reaction force (i.e. the buoyant force) is applied to the diver in the upward vertical direction. The buoyant force counteracts the pressure force and equals the weight of displaced liquid.

The effects of buoyancy on submerged air bubble in a liquid are often expressed in terms of the bubble rise velocity u_r of a single bubble rising steadily in a fluid at rest. The force acting on the rising bubble are the drag force $0.5C_d\rho_w u_r^2 A_{ab}$, the weight force $\rho_{air} g v_{ab}$ and the buoyant force F_b, where A_{ab} is the area of the bubble in the y-direction, C_d is the drag coefficient, g is the gravity constant, ρ_w is the water density, ρ_{air} is the air density and v_{ab} is the volume of the bubble. In the force balance, the drag force is opposed to the bubble motion direction. The buoyant force is either positive (upwards) or negative depending upon the sign of the pressure gradient $\partial P/\partial z$, z being the vertical axis positive upwards.[7] If $\partial P/\partial z$ is negative (e.g. hydrostatic pressure distribution), the buoyancy force is positive.

At equilibrium the balance of the forces yields:

$$\pm \frac{1}{2}C_d\rho_w u_r^2 A_{ab} - \rho_{air} g v_{ab} \pm F_b = 0 \qquad (3A.1)$$

where the sign $\pm$ depends upon the motion direction and the pressure gradient sign.

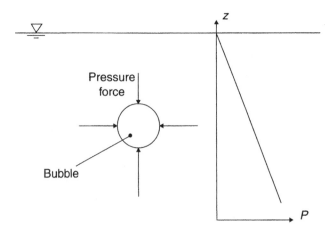

Fig. 3A.1 Sketch of pressure forces exerted on a submerged bubble.

---

[6] This section derives from Chanson (1997, Appendix C).
[7] In a hydrostatic pressure gradient, $\partial P/\partial z = -\rho_w g$.

Spherical bubble

For a spherical bubble and assuming a constant pressure gradient over the bubble height d_{ab}, the total buoyant force can be integrated over the sphere (Fig. 3A.1). It yields:

$$F_b = -\frac{\partial P}{\partial z} \frac{\pi d_{ab}^3}{6} \tag{3A.2}$$

where F_b is positive in upward direction (e.g. hydrostatic pressure gradient). The buoyant force is proportional to the pressure gradient. On Earth, F_b is proportional to the liquid density ρ_w and to the gravity acceleration g. The buoyancy is larger in denser liquids: e.g. a swimmer floats better in the water of the Dead Sea than in fresh water. In gravitationless water (e.g. waterfall) the buoyant force is zero.

The rise velocity equals:

$$u_r^2 = \pm \frac{4g d_{ab}}{3 C_d} \left(\frac{-\dfrac{\partial P}{\partial z}}{\rho_w g} - \frac{\rho_{air}}{\rho_w} \right) \tag{3A.3}$$

Bubble rise velocity in still water

For an individual air bubble rising uniformly in a fluid at rest and subjected to a hydrostatic pressure gradient, the rise velocity depends upon the value of the drag coefficient C_d which is a function of the bubble shape and velocity. A summary of experimental results is presented here.

Small air bubbles (i.e. $d_{ab} < 1$ mm) act as rigid spheres, surface tension imposing their shape. For very small bubbles (i.e. $d_{ab} < 0.1$ mm), the bubble rise velocity u_r is given by Stokes' law:

$$u_r = \frac{2}{9} \frac{g\,(\rho_w - \rho_{air})}{\mu_w} d_{ab}^2 \qquad (d_{ab} < 0.1\,\text{mm}) \tag{3A.4a}$$

where μ_w is the dynamic viscosity of water. For small rigid spherical bubbles (i.e. $0.1 < d_{ab} < 1$ mm), the rise velocity is best fitted by:

$$u_r = \frac{g \rho_w}{18 \mu_w} d_{ab}^2 \qquad (0.1 < d_{ab} < 1\,\text{mm}) \tag{3A.4b}$$

For larger bubbles (i.e. $d_{ab} > 1$ mm), Comolet (1979) showed that the bubble rise velocity can be estimated as:

$$u_r = \sqrt{\frac{2.14\sigma}{\rho_w d_{ab}} + 0.52 g d_{ab}} \qquad (d_{ab} > 1\,\text{mm}) \tag{3A.4c}$$

where σ is the surface tension between air and water.

Bubble rise velocity in a non-hydrostatic pressure gradient

Considering a bubble in a non-hydrostatic pressure distribution and neglecting the bubble weight, the rise velocity may be estimated to a first approximation as:

$$u_r = \pm(u_r)_{Hyd} \sqrt{\frac{1}{\rho(z)g}\left(\pm \frac{\partial P}{\partial z} \right)} \tag{3A.5}$$

where $(u_r)_{Hyd}$ is the bubble rise velocity in a hydrostatic pressure gradient (equation (3A.4)) and ρ is the surrounding fluid density. (Equation (3A.5) neglects the air density term.) The sign of u_r depends on the sign of $\partial P / \partial z$. For $\partial P / \partial z < 0$ (e.g. hydrostatic pressure gradient), u_r is positive.

3.6 Appendix B – Freshwater properties

Temperature (°C)	Density ρ_w (kg/m³)	Dynamic viscosity μ_w Pa s ($\times 10^{-3}$)	Surface tension σ (N/m)
(1)	(2)	(3)	(4)
0	999.9	1.792	0.0762
5	1000.0	1.519	0.0754
10	999.7	1.308	0.0748
15	999.1	1.140	0.0741
20	998.2	1.005	0.0736
25	997.1	0.894	0.0726
30	995.7	0.801	0.0718
35	994.1	0.723	0.0710
40	992.2	0.656	0.0701

3.7 Exercises

1. Considering a pipe (0.5 m diameter) discharging 0.1 L/s of sewage (1230 g/L, 0.001 Pa s), calculate the Reynolds number.
2. Considering a rectangular drain (0.04 m wide) discharging 0.8 L/s of water as an open channel flow (3 cm depth), calculate the Reynolds number and predict the flow regime.
3. Considering a plane Couette flow between a fixed wall and a conveying belt. The distance between the wall and belt is 0.05 m. The conveying belt travels at a constant speed of 0.092 m/s and the fluid is air at standard conditions. Calculate the Reynolds number, the boundary shear stress at the wall and on the centreline. If the belt is 20 m long and 0.5 m wide, calculate the shear force.
4. Rhodamine WT dye (50 ppm) is released in a 10 m³ water container. Calculate the mass of dye released.
5. Cooling water (35°C) is discharged from a power plant in a natural system (average temperature: 290 K). The outfall is discharged at the bottom of the 2.2 m deep river where the characteristic velocity is about 0.3 m/s while the free-surface velocity equals 1.05 m/s. Calculate the buoyancy per unit mass and the Richardson number.

3.8 Exercise solutions

1. $Re = 300$. Laminar flow.
2. $Re = 32 \times 10^3$. Turbulent flow.
3. $Re = 294$. Laminar flow. $\tau_o = 3.4 \times 10^{-5}$ Pa. $\tau_{CL} = 3.4 \times 10^{-5}$ Pa. $F_{shear} = 3.4 \times 10^{-4}$ N.
4. Mass = 0.5 kg.
5. $\rho = 998.77$ kg/m³. $\Delta\rho = -4.669$ kg/m³. $g' = -0.0458$ m/s². $Ri = -0.18$.

4

Turbulent shear flows

Summary

In this chapter basic turbulent flows are reviewed. Their flow properties are described and discussed.

4.1 Presentation

Turbulent shear enhances mixing and rate of spreading. The form of the velocity variation with distance may have an important effect on the degree of dispersion produced. A simple shear flow is the Couette flow (see next paragraph). For mixing applications in Nature, three types of turbulent flows are commonly encountered: (1) jets (and wakes), (2) developing boundary layer and (3) fully developed open channel flows (Fig. 4.1, Table 4.1). In a shear layer,[1] a momentum flux is transferred from the region of high velocity to that of low velocity. The shear stress characterizes the fluid resistance to the transfer of momentum.

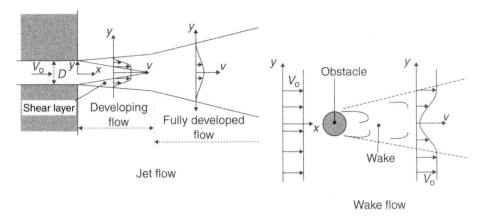

Fig. 4.1 Types of turbulent flows.

---

[1] In turbulent shear flows, the terms shear layer and mixing layer are used for the same meaning: i.e. a region of high shear associated with a velocity gradient in the direction normal to the flow.

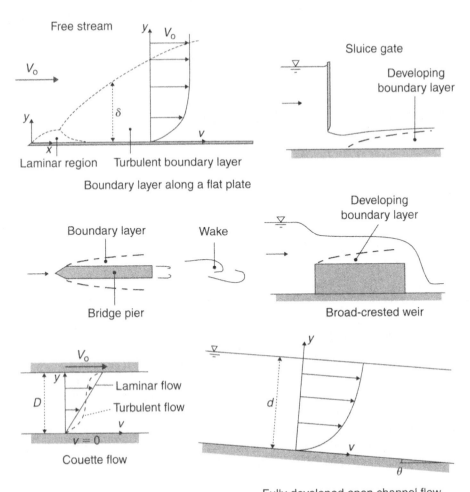

Fig. 4.1 (*Contd*).

Table 4.1 Velocity distributions in turbulent (monophase) flows

Turbulent flow situations (1)	Velocity profile (2)	Remarks (3)
Two-dimensional Couette flow	$$\frac{v}{V_*} = \frac{V_0/2}{V_*} + \frac{1}{K} \ln\left(\frac{y/D}{1 - y/D}\right)$$ $$- 0.41\left(1 - \frac{2y}{D}\right)$$ $$\frac{V_0/2}{V_*} = \frac{1}{K} \ln\left(\frac{V_*(D/2)}{v}\right) + 7.1$$	Defect law (smooth wall) K = 0.4 (von Karman constant)

(*Contd*)

Table 4.1 (*Contd*).

Turbulent flow situations (1)	Velocity profile (2)	Remarks (3)
Two-dimensional jet	$$\frac{v}{V_0} = \frac{2.67}{\sqrt{x/D}}\left(1 - \tanh^2\left(7.7\frac{y}{x}\right)\right)$$	Fully developed flow Analytical solution
Circular jet	$$\frac{v}{V_0} = \frac{5.745}{x/D}\left(\frac{1}{1 + 0.125\left(18.5\frac{y}{x}\right)^2}\right)$$	Fully developed flow Analytical solution
Wake flow	$$\frac{v}{V_0} = 1 - \frac{C}{\sqrt{x}}\exp\left(-\frac{V_0 y^2}{4\nu x}\right)$$	Two-dimensional wake in the far region: $(1 - v/V_0) \ll 1$ $C = F$ (drag on obstacle)
Boundary layer past a flat smooth horizontal plate	$$\frac{v}{V_*} = \frac{1}{K}\ln\left(\frac{V_* y}{\nu}\right) + 5.5 + \frac{\Pi}{K}Wa\left(\frac{y}{\delta}\right)$$	Turbulent zone and outer region: 30 to $70 < V_*(y/\nu)$ and $y/\delta < 1$
	$$Wa\left(\frac{y}{\delta}\right) = 2\sin^2\left(\frac{\pi}{2}\frac{y}{\delta}\right)$$	Turbulent boundary layer $K = 0.4$ (von Karman constant)
	$$\frac{\delta}{x} = 0.37\left(\frac{V_0 x}{\nu}\right)^{-1/5}$$	
Boundary layer past a rough plate	$$\frac{v}{V_*} = \frac{1}{K}\ln\left(\frac{V_* y}{\nu}\right) + 5.5 + D_2$$	Turbulent zone: $y/\delta < 0.1$–0.15 and $y/\delta < 1$; D_2 function of roughness $(D_2 \geqslant 0)$ Turbulent boundary layer
Fully developed open channel flow	$$\frac{v}{V_{\max}} = \left(\frac{y}{d}\right)^{1/N}$$	Turbulent flow; quasi-equilibrium flow conditions $K = 0.4$ (von Karman constant)
	$$N = K\sqrt{\frac{8}{f}}$$	

References: Rajaratnam (1976), Schlichting (1979), Schetz (1993), Chanson (1999a), Schlichting and Gersten (2000).

DISCUSSION

Practically it is important to understand the basic turbulent flow characteristics. Turbulent mixing is strongly affected by the location of the tracer injection. For example, faster mixing is achieved when tracer is injected in a developing shear layer of a jet, than in the jet core (paragraph 4.2). Considering an atmospheric developing boundary layer, mixing is more dynamic within the boundary shear region than in the main stream. However smoke release at higher elevations (i.e. in high velocity flow) is characterized by strong turbulent dispersion (Fig. 4.2).

Fig. 4.2 Smoke dispersion downstream of the incineration plant chimney in Tempaku-cho on 27 November 2001 at sunrise. Strong cold winds lead to rapid hot smoke dispersion.

The Couette flow

A simple shear flow is the steady flow between two parallel plates moving at different velocities and called a Couette flow (Fig. 4.1). The Couette flow is characterized by a constant shear stress distribution. In laminar flow regime, the velocity profile is linear. However, with increasing Reynolds number, turbulence develops and the velocity profile exhibits an S-shape at high Reynolds number, although the shear stress distribution remains constant between the plates and is equal to that at the wall τ_o. Turbulent Couette flows occur for $V_o D/\nu > 3 \times 10^3$, where D is the distance between the plates.

Remarks
1. Maurice Marie Alfred Couette (1858–1943) was a French scientist who experimentally measured the viscosity of fluids with a rotating viscosimeter (Couette 1890).
2. A rotating viscosimeter consists of two co-axial cylinders rotating in opposite direction. It is used to measure the viscosity of the fluid placed in the space between the cylinders. In a steady state, the torque transmitted from one cylinder to another is proportional to the fluid viscosity and relative angular velocity.
3. In a turbulent plane Couette flow, the momentum exchange coefficient may be estimated by a parabolic shape:

$$\nu_T = K V_* y \left(1 - \frac{y}{D} \right)$$

where K is the von Karman constant (K = 0.40), V_* is the shear velocity, D is the distance between plates and y is the distance normal to the plates (with $y = 0$ at one plate). The result was found to be in good agreement with experiments (Schlichting and Gersten 2000).

4.2 Jets and wakes

A jet flow is characterized by the developing flow and fully developed flow regions (Fig. 4.1). In the developing region, there is an undisturbed jet core with a velocity $V = V_o$, surrounded by developing shear layers in which momentum is transferred to the surrounding fluid. The transfer of momentum from the undisturbed jet core to the outer fluid is always associated with very-high levels of turbulence. The length of the developing flow region is about 6–12D for two-dimensional jets and 5–10D for circular jets discharging in a fluid at rest, where D is the jet thickness or jet diameter (Fig. 4.1). Further downstream, the flow becomes fully developed and the maximum velocity in a cross-section decreases with increasing distance.

In the developing flow of two-dimensional jets, the velocity in the undisturbed jet core is $V = V_o$, while the velocity profile in the developing shear layer is:

$$\frac{V}{V_o} = \frac{1}{2}\left(1 + \text{erf}\left(K_s \frac{y - y_{50}}{x}\right)\right) \tag{4.1}$$

where y_{50} is the transverse location, $V = V_o/2$, K_s is a constant inversely proportional to the rate of expansion of shear layer and erf is the error function.

In the fully developed region, the velocity distribution closely follows an analytical solution of the momentum equation:

$$\frac{V}{V_o} = \frac{2.67}{\sqrt{x/D}}\left(1 - \tanh^2\left(7.7\frac{y}{x}\right)\right) \qquad \text{Two-dimensional jet, fully developed flow} \tag{4.2a}$$

$$\frac{V}{V_o} = \frac{5.745}{x/D}\left(\frac{1}{1 + 0.125\left(18.5\frac{y}{x}\right)^2}\right) \qquad \text{Circular jet, fully developed flow} \tag{4.2b}$$

where V_o is the initial jet velocity, D is the jet thickness and jet diameter for two-dimensional and circular jets respectively. In equation (4.2b), y is the radial coordinate, sometimes denoted as r.

Discussion

The developing shear layers are characterized by very-intense turbulence, well in excess of levels observed in boundary layers and wakes. While the developing flow region is relatively short, the extent of the fully developed flow region, and the influence of the jet on the surroundings, may be felt far away.

A related flow situation is a wake behind an obstacle (Fig. 4.1). For example, the wake flow behind a bridge pier (Fig. 4.3). Schlichting (1979) developed a complete analogy between jet and wake flows in the far field.

Fig. 4.3 Wake behind bridge pier on the Mur river (Austria) in flood on 23 August 1999. Flow from right to left.

Discussion: momentum exchange coefficient in the developing flow region of a jet

In the developing flow region of a jet (e.g. Fig. 4.1 (top left) jet flow), the motion equation can be analysed as a free-shear layer. For a plane shear layer, Goertler (1942) solved the equation of motion assuming a constant eddy viscosity ν_T across the shear layer:

$$\nu_T = \frac{1}{4K_s^2} x V_o$$

where K_s is a constant which provides some information on the expansion rate of the momentum shear layer as the rate of expansion is proportional to $1/K_s$. The analytical solution of the motion equation is equation (4.1) for two-dimensional jets.

For monophase free-shear layers, K_s equals between 9 and 13.5 with a generally accepted value of 11 (Rajaratnam 1976, Schlichting 1979, Schetz 1993). However, Brattberg and Chanson (1998) showed that the value of K_s is affected by the air bubble entrainment rate in the developing region of plunging jet flows. They observed K_s to be about 5.7 for plunging jet impact velocities ranging from 3 to 8 m/s.

4.3 Boundary layer flows

A *boundary layer* is defined as the flow region next to a solid boundary where the flow field is affected by the presence of the boundary. The concept was originally introduced by Ludwig Prandtl (1904). In a boundary layer, momentum is gained from the main stream (or free stream) and contributes to the boundary layer growth. At the boundary, the velocity is zero.

A boundary layer is characterized by:

- its *thickness* δ defined in terms of 99% of the free-stream velocity:

$$\delta = y(\mathrm{v} = 0.99 V_o)$$

where y is measured perpendicular to the boundary and V_o is the free-stream velocity.

- the *displacement thickness* δ_1 defined as:

$$\delta_1 = \int_0^\delta \left(1 - \frac{V}{V_o}\right) dy$$

- the *momentum thickness* δ_2:

$$\delta_2 = \int_0^\delta \frac{V}{V_o}\left(1 - \frac{V}{V_o}\right) dy$$

- the *energy thickness* δ_3:

$$\delta_3 = \int_0^\delta \frac{V}{V_o}\left(1 - \left(\frac{V}{V_o}\right)^2\right) dy$$

Remark

Although the boundary layer thickness δ is (arbitrarily) defined in terms of 99% of the free-stream velocity, the real extent of the effects of boundary friction on the flow is probably about 1.5–2 times δ.

Velocity distribution

In a turbulent boundary layer, the flow can be divided into three regions: an inner wall region next to the wall where the turbulent stress is negligible and the viscous stress is large, an outer region where the turbulent stress is large and the viscous stress is small and an overlap region sometimes called a turbulent zone.

For a turbulent boundary layer flow along a smooth boundary with zero pressure gradient, the velocity distribution follows:

$$\frac{V}{V_*} = \frac{V_* y}{\nu} \qquad \text{Inner wall region: } \frac{V_* y}{\nu} < 5 \tag{4.3a}$$

$$\frac{V}{V_*} = \frac{1}{K}\ln\left(\frac{V_* y}{\nu}\right) + D_1 \qquad \text{Turbulent zone: } 30 \text{ to } 70 < \frac{V_* y}{\nu} \text{ and } \frac{y}{\delta} < 0.1 \text{ to } 0.15 \tag{4.3b}$$

$$\frac{V_o - V}{V_*} = -\frac{1}{K}\ln\left(\frac{y}{\delta}\right) \qquad \text{Outer region: } \frac{y}{\delta} > 0.1 \text{ to } 0.15 \tag{4.3c}$$

where V_* is the shear velocity, ν is the kinematic viscosity of the fluid, K is the von Karman constant (K = 0.40) and D_1 is a constant (D_1 = 5.5, Schlichting 1979). Equation (4.3b) is called the logarithmic profile or the '*law of the wall*'. Equation (4.3c) is called the '*velocity defect law*' or outer law.

Coles (1956) showed that equation (4.3b) can be extended to the outer region by adding a 'wake law' term to the right-hand side term:

$$\frac{V}{V_*} = \frac{1}{K} \ln\left(\frac{V_* y}{\nu}\right) + D_1 + \frac{\Pi}{K} Wa\left(\frac{y}{\delta}\right)$$

Turbulent zone and outer region: $30 \text{ to } 70 < \dfrac{V_* y}{\nu}$ (4.3d)

where Π is the wake parameter and Wa is Coles' wake function, originally estimated as (Coles 1956):

$$Wa\left(\frac{y}{\delta}\right) = 2 \sin^2\left(\frac{\pi}{2}\frac{y}{\delta}\right)$$ (4.4)

Roughness effects

Surface roughness has an important effect on the flow in the wall-dominated region (i.e. inner wall region and turbulent zone). Numerous experiments showed that, for a turbulent boundary layer along a rough plate, the 'law of the wall' follows:

$$\frac{V}{V_*} = \frac{1}{K} \ln\left(\frac{V_* y}{\nu}\right) + D_1 + D_2 \qquad \text{Turbulent zone: } \frac{y}{\delta} < 0.1 \text{ to } 0.15 \qquad (4.5)$$

where D_2 is a function of the type of the roughness height, roughness shape and spacing (e.g. Schlichting 1979, Schetz 1993). For smooth turbulent flows, D_2 equals zero.

In the turbulent zone, the roughness effect (i.e. $D_2 > 0$) implies a 'downward shift' of the velocity distribution (i.e. law of the wall). For large roughness, the so-called 'laminar sublayer' (i.e. inner region) disappears and the flow is said to be 'fully rough'.

For fully rough turbulent flows in circular pipes with uniformly distributed sand roughness, D_2, equals:

$$D_2 = 3 - \frac{1}{K} \ln\left(\frac{k_s V_*}{\nu}\right) \qquad \text{Fully rough turbulent flows in circular pipes}$$

where k_s is the equivalent sand roughness height.

Applications

In a boundary layer flow, the velocity distribution may be approximated by a power law. For a power-law velocity distribution:

$$\frac{V}{V_0} = \left(\frac{y}{\delta}\right)^{1/N}$$ (4.6)

the characteristic parameters of the boundary layers may be transformed. The displacement thickness and momentum thickness become:

$$\frac{\delta_1}{\delta} = \frac{1}{1 + N}$$ (4.7)

$$\frac{\delta_2}{\delta} = \frac{N}{(1+N)(2+N)}$$ (4.8)

where N is the velocity exponent (equation (4.6)). And the shape factor becomes:

$$\frac{\delta_1}{\delta_2} = \frac{N+2}{N}$$ (4.9)

For two-dimensional turbulent boundary layers, Schlichting (1979) indicated that separation[2] occurs for $\delta_1/\delta_2 > 1.8$–2.4. Such a condition implies separation for $N < 1.4$–2.5.

Turbulent boundary layer development along a smooth flat plate

For turbulent flows in smooth circular pipes, the Blasius resistance formula (Blasius 1913) implies that the velocity profile follows exactly a 1/7th power-law distribution. Considering a developing turbulent boundary layer on a smooth flat plate at zero incidence (and zero pressure gradient), the resistance formula deduced from the 1/7th power law of velocity distribution implies (Schlichting 1979):

$$\frac{\delta}{x} = 0.37 \left(\frac{V_o x}{\nu} \right)^{-1/5}$$ (4.10)

$$\frac{\delta_1}{x} = 0.046 \left(\frac{V_o x}{\nu} \right)^{-1/5}$$ (4.11)

$$\frac{\delta_2}{x} = 0.036 \left(\frac{V_o x}{\nu} \right)^{-1/5}$$ (4.12)

Discussion: momentum exchange coefficient in a turbulent boundary layer

In a smooth turbulent boundary layer, the mixing length l may be estimated as:

$$\frac{l}{\delta} = 0.085 \tanh \left(\frac{K}{0.085} \frac{y}{\delta} \right)$$

where tanh is the hyperbolic tangent function, K is the von Karman constant and δ is the boundary layer thickness (Schlichting and Gersten 2000, p. 557). The eddy viscosity ν_T derives then from:

$$\nu_T = l^2 \frac{\partial v}{\partial y}$$ (3.3)

---

[2] A deceleration of fluid particles leading to a reversed flow within a boundary layer is called a separation. The decelerated fluid particles are forced outwards and the boundary layer is separated from the wall.

Fig. 4.4 Mountain stream in Austria (22 August 1999). Looking upstream.

4.4 Fully developed open channel flows

Fully developed open channel flows are commonly encountered in rivers and streams (Fig. 4.4). The flow motion is primarily controlled by the gravity force and boundary friction. At uniform equilibrium, the gravity force component in the flow direction exactly equals the boundary friction force. The main differences between boundary layers and fully developed open channel flows are: (1) the absence of momentum transfer at the open channel free surface while momentum transfer does occur at the outer edge of a developing boundary layer (between the boundary layer itself and the free stream), and hence (2) the lack of boundary layer growth in fully developed open channel flows. Further air–water mass transfer may occur at the free surface, and high velocity open channel flows are characterized by significant free-surface aeration ('white waters') (Chapter 17).

The velocity distribution in fully developed turbulent open channel flows is given approximately by Prandtl's power law:

$$\frac{v}{V_{max}} = \left(\frac{y}{d}\right)^{1/N}$$ (4.13)

where y is the distance from the channel bed normal to the flow direction, d is the water depth, V_{max} is the free-surface velocity and the exponent $1/N$ varies from 1/4 down to 1/12 depending upon the boundary friction and cross-sectional shape. Chen (1990) developed a complete analysis of the velocity distribution in open channel and pipe flow with reference to flow resistance. In uniform equilibrium flows, the velocity distribution exponent is related to the flow resistance:

$$N = K\sqrt{\frac{8}{f}}$$ (4.14)

where f is the Darcy friction factor and K is the von Karman constant (K = 0.4). In engineering applications, a value $N \approx 6$ is reasonably representative of open channel flows for smooth-concrete channels, although it must be remembered that N is a function of the flow resistance (e.g. equation (4.14)).

For a wide rectangular channel, the relationship between the mean flow velocity V and the free-surface velocity V_{max} derives from the continuity equation:

$$q = Vd = \int_0^d v\, dy = \frac{N}{N+1} V_{max} d$$ (4.15)

where q is the water discharge per unit width.

Discussion: momentum exchange coefficient in fully developed open channel flows

Assuming that the mixing length equals $l = Ky$, where K is the von Karman constant (K = 0.40) and assuming a linear variation of the turbulent shear stress across the flow,[3] the eddy viscosity in a fully developed open channel flow becomes:

$$\nu_T = K V_* y \left(1 - \frac{y}{d}\right)$$

where V_* is the shear velocity. Further discussion is developed in Chapter 16.

[3] That is, $\tau(y) = \tau_0(1 - y/d)$, where τ_0 is the bed shear stress.

4.5 Mixing in turbulent shear flows

4.5.1 Presentation

In a shear flow, momentum is transferred from the region of high velocity to that of low velocity. The fluid tends to resist the shear associated with the transfer of momentum. The shear stress is proportional to the rate of transfer of momentum. In turbulent flows, the apparent shear stress flow may be expressed as:

$$\tau = \rho \nu_T \frac{\partial v}{\partial y} \tag{3.2b}$$

where ρ is the fluid density and ν_T is the eddy viscosity or *momentum exchange coefficient* in turbulent flow. The momentum exchange coefficient ν_T is a function of the flow properties.

Equation (3.2b) implies that maximum shear occurs at the location of maximum velocity gradient in the direction normal to the flow. Note that this location corresponds to an inflection point where: $\partial^2 v/\partial y^2 = 0$.

In turbulent mixing processes, mixing is related to the turbulent shear. The mixing coefficients for momentum and mass are commonly assumed the same (see paragraph 7.1, Chapter 7). For example, in open channel flows, the vertical mixing coefficient ε_v equals the momentum exchange coefficient ν_T. As a result, turbulent mixing is enhanced in regions of high momentum exchange coefficients. For example, near the singularity of a free-shear layer; at the nozzle edges of free jet; at half-depth in boundary layers and open channel flows. Practically, contaminant injection should take place in such regions of large momentum exchange coefficient to maximize the rate of turbulent mixing and dispersion.

4.5.2 Discussion: effects of contaminants on shear flows

The presence of contaminants *does interact* with turbulence (Figs 4.5 and 4.6). While turbulence enhances mixing and dispersion, contaminants in high proportion may inhibit, disturb or enhance turbulence characteristics: e.g. velocity distributions, turbulent velocity fluctuations, flow resistance.

Figure 4.5 illustrates some examples. In plunging jet flows, a large number of air bubbles may be entrapped at the plunge point and diffuse downwards. The bubble diffusion layer and the momentum shear layer do not coincide (Cummings and Chanson 1997, Brattberg and Chanson 1998). The air–water shear flow properties differ from monophase flow data (Fig. 4.5). The presence of air bubbles enhances the momentum spreading rate and the shear layer is located further outwards (i.e. away from the jet centreline).

The addition of dilute polymers and surfactants in water can be used to change the fluid properties (viscosity, surface tension) and to modify the turbulence characteristics. Experimental studies showed that very-small concentration (few ppm) of dissolved polymer substances can reduce the skin friction resistance in turbulent flows to as low as one-fourth of that in pure solvent. In pipelines and sewers, polymer additives are commonly used to reduce the skin drag, to enhance the discharge capacity: e.g. the Trans-Alaska oil pipeline, coal–water pipe flows (Fig. 4.6b). Macromolecules of polymer and air bubbles interact with the turbulent structures, inducing a modification of the flow properties as compared to clear-water flows (e.g. Gyr 1989, Bushnell and Hefner 1990, Chanson and Qiao 1994).

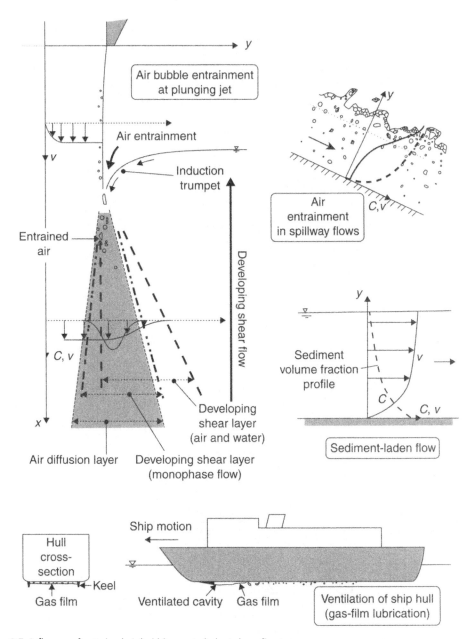

Fig. 4.5 Influence of entrained air bubbles on turbulent shear flows.

In spillway flows, large amount of air bubbles are entrained and small air bubbles next to the bottom modify the bottom shear stress, acting as macromolecules of polymer and inducing some drag reduction (Chanson 1994, Chapter 17). For example, at the downstream end of Karun dam spillway, flow velocities were observed to be in excess of 40 m/s that is nearly 30% greater than predicted. The unexpected velocity increase was the result of some drag reduction caused by free-surface aeration.

(a)

(b)

Fig. 4.6 Photographs of turbulent shear flows with high 'contaminant' fraction. (a) Sediment-laden flow downstream of Mount St Helens (USA) with large (timber log) debris at a bridge crossing (courtesy of the US Army Corps of Engineers, Portland District). (b) Trans-Alaska pipeline along the Richardson Highway just north of Paxson, AK in September 2000 (courtesy of Steve STAPP). Completed in 1978, the pipeline (1.2 m diameter) is about 1300 km long and carries about 32 000 m^3 of oil per day. Note the heat exchangers used to prevent thawing of the permafrost and the zig-zag configuration to allow for expansion or contraction of the pipe because of temperature changes. The design also allows for pipeline movement caused by a 8.5 earthquake. Drag reduction agent (DRA) is injected into the oil to reduce the energy loss due to turbulence in the oil.

(c)

Fig. 4.6 (*Contd*) (c) Air bubble entrainment at Shasta dam (USA) for low discharge on 6 August 1999 (courtesy of Daniel Stephens).

In laboratory and river flows, suspended sediment is observed sometimes to increase the flow velocity and to decrease the friction factor. Historical cases include observations of suspended silt flood flows in the Nile, Indus and Mississippi rivers. Chanson (1994, 1997) showed conclusively the cases of drag reduction with suspended-sediment flows.

Similar cases of drag reduction include ventilation downstream of hydrofoils for high-speed boats (40–80 knots) and gas-film lubrication of ship hull[4] (Fig. 4.5).

4.6 Exercises

1. A circular jet (0.5 m diameter) discharges 0.6 m³/s of water in the middle of a lake. Assuming that the flow is driven by the initial jet momentum, calculate the flow velocity: (a) 15 m downstream on the jet centreline and (b) 20 m downstream and 8 m away from the centreline (i.e. $x = 20$ m, $r = 8$ m).
2. During a storm, the wind blows over a sandy beach (0.5 mm sand particle). The wind boundary layer is about 100 m high at the beach and the free-stream velocity at the outer edge of the wind boundary layer is 35 m/s. Calculate the shear velocity, the displacement and momentum thickness.
3. For the above problem (wind storm), calculate the virtual origin of the turbulent boundary layer.

---

[4]Controlled air injection along ship hull results in the reduction of skin friction drag.

4. The free-surface velocity in a river is 0.65 m/s and the water depth is 0.95 m. Assuming a Darcy friction factor $f = 0.03$, calculate the water discharge per unit width.
5. Considering a stream flow in a wide rectangular channel, the discharge per unit width is 0.41 m²/s and the water depth is 1.9 m. Plot the velocity profile and momentum exchange coefficient profile. Calculate the velocity and momentum exchange coefficient at 0.6 m above the bed. *Calculate the shear velocity as Chanson (1999a, pp. 74). Assume f = 0.041.*

4.7 Exercise solutions

1. $V_0 = 3.056$ m/s: (a) $v = 0.58$ m/s and (b) $v = 0.056$ m/s.
2. First we must compute the bed shear stress or the shear velocity.
   The Reynolds number $V_0\delta/\nu$ of the boundary layer is 2.3×10^8. The flow is turbulent. As the sand diameter is very small compared to the boundary layer thickness, the boundary layer flow is assumed smooth turbulent. In a turbulent boundary layer along a smooth boundary, the mean bed shear stress equals:

$$\tau_0 = 0.0225\rho V_0^2 \left(\frac{\nu}{V_0\delta}\right)^{1/4}$$

   where ρ is the fluid density, ν is the kinematic viscosity, δ is the boundary layer thickness and V_0 is the free-stream velocity (at the outer edge of the boundary layer) (Schlichting 1979, p. 637).
   For the beach, it yields: $\tau_0 = 0.27$ Pa and $V_* = 0.47$ m/s.
   Assuming a 1/7th power-law velocity distribution (i.e. smooth turbulent flow), the displacement and momentum thicknesses equal: $\delta_1 = 12.5$ m, $\delta_2 = 9.7$ m.

Notes
The displacement thickness characterizes the displacement of the main flow due to slowing down of the fluid particles in the boundary layer. The momentum thickness characterizes the displacement of free-stream momentum transport.

3. Virtual origin: ~43 km upstream.
4. $q = 0.54$ m²/s.
5. Using equation (4.14), $N = 5.6$, $V_{max} = 0.2518$ m/s, $v(y = 0.6\,m) = 0.205$ m/s and $V_* = 0.0154$ m/s − $\nu_T(y = 0.6\,m) = 0.0025$ m²/s.

5

Diffusion: basic theory

Summary
In this chapter the basic equation of molecular diffusion and simple applications are developed.

5.1 Basic equations

The basic diffusion of matter, also called molecular diffusion, is described by Fick's law, first stated by Fick (1855). Fick's law states that the transfer rate of mass across an interface normal to the x-direction and in a quiescent fluid varies directly as the coefficient of molecular diffusion D_m and the negative gradient of solute concentration. For a one-dimensional process:

$$\dot{m} = -D_m \frac{\partial C_m}{\partial x} \tag{5.1}$$

where $\dot{m}$ is the solute mass flux and C_m is the mass concentration of matter in liquid. The coefficient of proportionality D_m is called the *molecular diffusion coefficient*. Equation (5.1) implies a mass flux from a region of high mass concentration to one of smaller concentration. An example is the transfer of atmospheric gases at the free surface of a water body. Dissolution of oxygen from the atmosphere to the water yields some re-oxygenation.

The continuity equation (i.e. conservation of mass) for the contaminant states that spatial rate of change of mass flow rate per unit area equals minus the time rate of change of mass:

$$\frac{\partial \dot{m}}{\partial x} + \frac{\partial C_m}{\partial t} = 0 \tag{5.2}$$

Replacing into equation (5.1), it yields:

$$\frac{\partial C_m}{\partial t} = D_m \frac{\partial^2 C_m}{\partial x^2} \tag{5.3a}$$

For diffusion in a three-dimensional system, the combination of equations (5.1) and (5.2) gives:

$$\frac{\partial C_m}{\partial t} = D_m \left(\frac{\partial^2 C_m}{\partial x^2} + \frac{\partial^2 C_m}{\partial y^2} + \frac{\partial^2 C_m}{\partial z^2} \right) \tag{5.3b}$$

Equations (5.3a) and (5.3b) are called the *diffusion equations*. It may be solved analytically for a number of basic boundary conditions. Mathematical solutions of the diffusion equation (and heat equation) were addressed in two classical references (Crank 1956, Carslaw and Jaeger 1959). Since equations (5.3a) and (5.3b) are linear, the *theory of superposition* may be used to build up solutions with more complex problems and boundary conditions: e.g. spreading of mass caused by two successive slugs.

Discussion: theory of superposition

If the functions ϕ_1 and ϕ_2 are solutions of the diffusion equation subject to the respective boundary conditions $B_1(\phi_1)$ and $B_2(\phi_2)$, any linear combination of these solutions, $(a\phi_1 + b\phi_2)$, satisfies the diffusion equation and the boundary conditions $aB_1(\phi_1) + bB_2(\phi_2)$. This is the principle of superposition for homogeneous differential equations.

Figure 5.1(a) and (b) illustrates a simple example. Figure 5.1(a) shows the solution of diffusion equation for the sudden injection of mass slug at the origin. By adding an uniform velocity (current), the solution is simply the superposition of Fig. 5.1(a) plus the advection of the centre of mass (Fig. 5.1(b)).

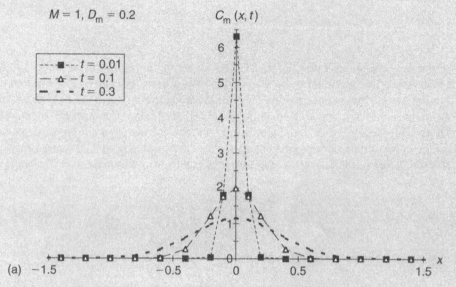

Fig. 5.1 (a) Application of the theory of superposition. (a) Diffusion downstream a sudden mass slug injection. Gaussian distribution solutions of equation (5.4) for $M = 1$ and $D_m = 0.2$.

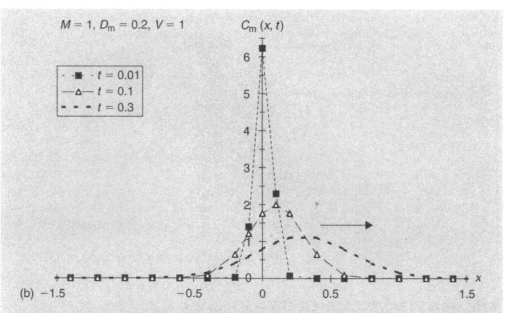

Fig. 5.1 (b) Advection downstream a sudden mass slug injection for $M = 1$, $D_m = 0.2$, $V = 1$ – equation (6.5).

Notes

1. Adolf Eugen Fick was a 19th Century German physiologist who applied Fourier's (1822) law of heat flow to molecular diffusion process.
2. Typical values of molecular diffusion coefficients for solutes in water are in the range 5×10^{-10} to $2 \times 10^{-9} \, \mathrm{m^2/s}$. D_m is a property of the fluid. For a given solvent (i.e. fluid), solute (i.e. tracer), concentration and temperature, D_m is a constant.
3. Turbulent enhances mixing drastically. Turbulent diffusion may be described also by equations (5.1)–(5.3) in which the molecular diffusion D_m is replaced by a turbulent diffusion coefficient that is a function of the flow conditions (Chapter 7).
4. The theory of superposition may be applied to the diffusion equations (5.3a) and (5.3b) because it is linear.

5.2 Applications

5.2.1 Initial mass slug

Initial mass slug introduced at t = 0 and x = 0

A simple example is the one-dimensional spreading of a mass M of contaminant introduced suddenly at $t = 0$ at the origin ($x = 0$) in an infinite (one-dimensional) medium with zero contaminant concentration. The fluid is at rest everywhere (i.e. $V = 0$). The fundamental solution of the diffusion equation (5.3a) is:

$$C_m(x,t) = \frac{M}{\sqrt{4\pi D_m t}} \exp\left(-\frac{x^2}{4 D_m t}\right) \qquad \text{for } t > 0 \qquad (5.4)$$

Equation (5.4) is called a *Gaussian distribution* or random distribution. The mean equals zero and the standard deviation σ equals $\sqrt{2D_m t}$. In the particular case of $M = 1$, it is known as the normal distribution. Equation (5.4) is plotted in Fig. 5.1(a). The curve has a bell shape.

DISCUSSION

The Gaussian distribution is given by:

$$C_m = C_{max} \exp\left(-\frac{1}{2}\left(\frac{x-m}{\sigma}\right)^2\right)$$

where m is the mean and σ is the standard deviation.

For a Gaussian distribution of tracers, the standard deviation σ may be used as a characteristic length scale of spreading. Ninety-five per cent of the total mass is spread between $(m - 2\sigma)$ and $(m + 2\sigma)$, where m is the mean. Hence an adequate estimate of the width of a dispersing cloud is about 4σ (Fischer *et al.* 1979, p. 41).

For an initial mass slug, the length of contaminant cloud at a time t is $4\sigma = 4\sqrt{2D_m t}$.

Initial mass slug introduced at t = 0 and x = x₀

Considering an initial mass slug introduced at $t = 0$ and $x = x_o$, the analytical solution of the diffusion equations (5.3a) and (5.3b) is:

$$C_m(x,t) = \frac{M}{\sqrt{4\pi D_m t}} \exp\left(-\frac{(x-x_o)^2}{4D_m t}\right) \qquad \text{for } t > 0 \qquad (5.5)$$

Two initial mass slug introduced at t = 0

Considering two separate slugs (mass M_1 and M_2) introduced at $t = 0$, $x = x_1$ and $x = x_2$ respectively (Fig. 5.2), the solution of the diffusion equation is:

$$C_m(x,t) = \frac{M_1}{\sqrt{4\pi D_m t}} \exp\left(-\frac{(x-x_1)^2}{4D_m t}\right) + \frac{M_2}{\sqrt{4\pi D_m t}} \exp\left(-\frac{(x-x_2)^2}{4D_m t}\right) \qquad \text{for } t > 0 \quad (5.6)$$

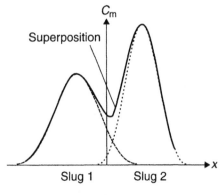

Fig. 5.2 Application of the theory of superposition: diffusion downstream a sudden injection of two mass slugs.

The solution is based upon the assumption that the mass slugs diffuse independently because of the fundamental premise that the motion of individual particles is independent of the concentration of other particles (Fischer *et al.* 1979, p. 42).

5.2.2 Initial step function $C_m(x, 0)$

Considering a sudden increase (i.e. step) in mass concentration at $t = 0$, the boundary conditions are:

$$C_m(x, t < 0) = 0 \quad \text{everywhere for } t < 0$$

$$C_m(x, 0) = 0 \quad x < 0$$

$$C_m(x, 0) = C_o \quad \text{for } x > 0$$

The solution of the diffusion equation may be resolved as a particular case of superposition integral. It yields:

$$C_m(x, t) = \frac{C_o}{2}\left(1 + \text{erf}\left(\frac{x}{\sqrt{4D_m t}}\right)\right) \quad \text{for } t > 0 \qquad (5.7)$$

where the *error function* erf is defined as:

$$\text{erf}(u) = \frac{2}{\sqrt{\pi}} \int_0^u \exp(-\tau^2)\, d\tau$$

Equation (5.7) is shown in Fig. 5.3. Details of the error function erf are given in Appendix A (Section 5.3).

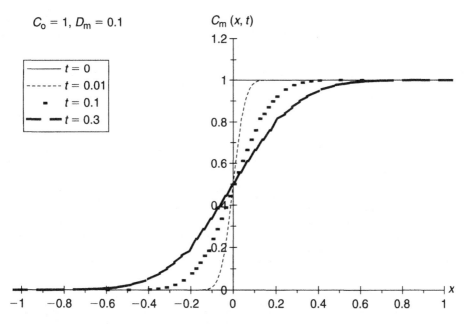

$C_o = 1, D_m = 0.1$

$C_m(x, t)$

Legend:
—— $t = 0$
- - - - $t = 0.01$
■ $t = 0.1$
— ■ — $t = 0.3$

Fig. 5.3 Contaminant diffusion for an initial step distribution, solutions of equation (5.7) for $C_o = 1$ and $D_m = 0.1$.

> **Note**
> At the origin, the mass concentration becomes a constant for $t > 0$: $C_m(x = 0, t > 0) = C_o/2$.

5.2.3 Sudden increase in mass concentration at the origin

The concentration is initially zero everywhere. At the initial time $t = 0$, the concentration is suddenly raised to C_o at the origin $x = 0$ and held constant: $C_m (0, t \geq 0) = C_o$. The analytical solution of the diffusion equations (5.3a) and (5.3b) is:

$$C_m(x,t) = C_o\left(1 - \mathrm{erf}\left(\frac{x}{\sqrt{4D_m t}}\right)\right) \qquad \text{for } x > 0 \qquad (5.8)$$

Equation (5.8) is that of an advancing front (Fig. 5.4). At the limit $t = +\infty$, $C_m = C_o$ everywhere.

The result may be extended, using the theory of superposition, when the mass concentration at the origin C_o varies with time. The solution of the diffusion equation is:

$$C_m(x,t) = \int_{-\infty}^{t} \frac{\partial C_o(\tau)}{\partial \tau}\left(1 - \mathrm{erf}\left(\frac{x}{\sqrt{4D_m(t-\tau)}}\right)\right)d\tau \qquad \text{for } x > 0 \qquad (5.9)$$

DISCUSSION
The step function (Section 5.2.2) is a limiting of the sudden increase in mass concentration at the origin with constant mass concentration at the origin. In Section 5.2.2, the mass concentration at the origin was $C_m (x = 0, t > 0) = C_o/2$.

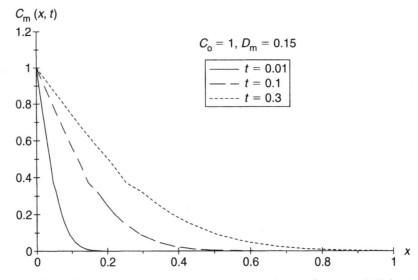

Fig. 5.4 Spread of a sudden concentration increase at the origin, solutions of equation (5.8) for $C_o = 1$ and $D_m = 0.15$.

5.2.4 Effects of solid boundaries

When the spreading (e.g. of a mass slug) is restricted by a solid boundary, the principle of superposition and the *method of images* may be used. The spreading pattern resulting from a combination of two mass slugs of equal strength includes a line of zero concentration gradient midway between them (Fig. 5.5). Since the mass flux is zero according to Fick's law (equation (5.1)), it can be considered as a boundary wall[1] without affecting the other half of the diffusion pattern.

A simple example is the spreading of a mass slug introduced at $x = 0$ and $t = 0$, with a wall at $x = -L$ (Fig. 5.5). At the wall there is no transport through the boundary. That is, the concentration gradient must be zero at the wall:

$$\dot{m}(x = -L, t) = -D_m \frac{\partial C_m}{\partial x} = 0 \qquad \text{Boundary condition at the wall: } x = -L$$

In order to ensure no mass transport at the wall, a *mirror image* of mass slug, with mass M injected at $x = -2L$, is superposed to the real mass slug of mass M injected at $x = 0$. The flow due to the mirror image of the mass slug is superposed onto that due to the mass slug itself. It yields:

$$C_m(x,t) = \frac{M}{\sqrt{4\pi D_m t}} \left(\exp\left(-\frac{x^2}{4D_m t}\right) + \exp\left(-\frac{(x+2L)^2}{4D_m t}\right) \right) \qquad \text{for } t > 0 \quad (5.10)$$

Equation (5.10) is the solution of the superposition of two mass slugs of equal mass injected at $x = -2L$ and $x = 0$. It is also the solution of a mass slug injected at $x = 0$ with a solid boundary at $x = -L$.

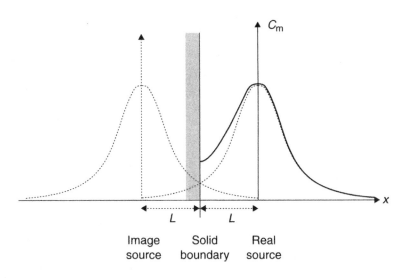

Fig. 5.5 Spread of a sudden concentration increase at the origin with one boundary.

---

[1] There is no mass flux through a wall and any solid boundary.

Problems involving straight or circular boundaries can be solved by the method of images. Considering a mass slug injected at the origin in between two solid walls located at $x = -L$ and $x = +L$, the solution of the problem is:

$$C_{\mathrm{m}}(x,t) = \frac{M}{\sqrt{4\pi D_{\mathrm{m}}t}} \sum_{i=-\infty}^{+\infty} \exp\left(-\frac{(x+2iL)^2}{4D_{\mathrm{m}}t}\right) \quad \text{for } t > 0 \quad (5.11)$$

The solution is obtained by adding an infinity of mass slug source on both positive and negative axis (Fischer *et al.* 1979, pp. 47–48).

Note

The method of images is a tool by which straight solid boundaries are treated as symmetry lines. The problem is solved analytically by combining the method of images with the theory of superposition.

In Fig. 5.5, the real slug is located at $x = 0$. The solid boundary is located at $x = -L$. Hence the image slug (or mirror slug) must be located at $x = -2L$ to verify zero mass flux at $x = -L$. (Remember: there is no transport through the boundary.)

5.3 Appendix A – Mathematical aids

Differential operators
Gradient:

$$\overrightarrow{\mathrm{grad}}\ \Phi(x,y,z) = \nabla\Phi(x,y,z) = \vec{\mathbf{i}}\frac{\partial\Phi}{\partial x} + \vec{\mathbf{j}}\frac{\partial\Phi}{\partial y} + \vec{\mathbf{k}}\frac{\partial\Phi}{\partial z} \quad \text{Cartesian coordinate}$$

Divergence:

$$\mathrm{div}\ \vec{F}(x,y,z) = \nabla\vec{F}(x,y,z) = \frac{\partial F_x}{\partial x} + \frac{\partial F_y}{\partial y} + \frac{\partial F_z}{\partial z}$$

Curl:

$$\overrightarrow{\mathrm{curl}}\ \vec{F}(x,y,z) = \nabla\wedge\vec{F}(x,y,z) = \vec{\mathbf{i}}\left(\frac{\partial F_z}{\partial y} - \frac{\partial F_y}{\partial z}\right) + \vec{\mathbf{j}}\left(\frac{\partial F_x}{\partial z} - \frac{\partial F_z}{\partial x}\right) + \vec{\mathbf{k}}\left(\frac{\partial F_y}{\partial x} - \frac{\partial F_x}{\partial y}\right)$$

Laplacian operator:

$$\Delta\Phi(x,y,z) = \nabla\times\nabla\Phi(x,y,z) = \mathrm{div}\ \overrightarrow{\mathrm{grad}}\ \Phi(x,y,z) = \frac{\partial^2\Phi}{\partial x^2} + \frac{\partial^2\Phi}{\partial y^2} + \frac{\partial^2\Phi}{\partial z^2}$$
$$\text{Laplacian of scalar}$$

$$\Delta\vec{F}(x,y,z) = \nabla\times\nabla\vec{F}(x,y,z) = \vec{\mathbf{i}}\,\Delta F_x + \vec{\mathbf{j}}\,\Delta F_y + \vec{\mathbf{k}}\,\Delta F_z \quad \text{Laplacian of vector}$$

Table 5A.1 Values of the error function erf

u	erf(u)	u	erf(u)
0	0	1	0.8427
0.1	0.1129	1.2	0.9103
0.2	0.2227	1.4	0.9523
0.3	0.3286	1.6	0.9763
0.4	0.4284	1.8	0.9891
0.5	0.5205	2	0.9953
0.6	0.6309	2.5	0.9996
0.7	0.6778	3	0.99998
0.8	0.7421	$+\infty$	1
0.9	0.7969		

Error function

The Gaussian error function, or function erf, is defined as:

$$\text{erf}(u) = \frac{2}{\sqrt{\pi}} \int_0^u \exp(-t^2)\,dt$$

Tabulated values are given in Table 5A.1. Basic properties of the function are:

$$\text{erf}(0) = 0$$

$$\text{erf}(+\infty) = 1$$

$$\text{erf}(-u) = -\text{erf}(u)$$

$$\text{erf}(u) = \frac{1}{\sqrt{\pi}}\left(u - \frac{u^3}{3 \times 1!} + \frac{u^5}{5 \times 2!} - \frac{u^7}{7 \times 3!} + \cdots\right)$$

$$\text{erf}(u) \approx 1 - \frac{\exp(-u^2)}{\sqrt{\pi}u}\left(1 - \frac{1}{2u^2} - \frac{1 \times 3}{\left(2x^2\right)^2} - \frac{1 \times 3 \times 5}{\left(2x^2\right)^3} + \cdots\right)$$

where $n! = 1 \times 2 \times 3 \times \cdots \times n$.

The complementary Gaussian error function erfc is defined as:

$$\text{erfc}(u) = 1 - \text{erf}(u) = \frac{2}{\sqrt{\pi}} \int_u^{+\infty} \exp(-t^2)\,dt$$

Note

In first approximation, the function erf(u) may be correlated by:

$$\text{erf}(u) \approx u(1.375511 - 0.61044u + 0.088439u^2) \qquad 0 \le u < 2$$

$$\text{erf}(u) \approx \tanh(1.198787u) \qquad -\infty < u < +\infty$$

with a normalized correlation coefficient of 0.99952 and 0.9992 respectively. In many applications, the above correlations are not accurate enough, and Table 5A.1 should be used.

Notation

x, y, z	Cartesian coordinates
r, θ, z	polar coordinates
$\partial/\partial x$	partial differentiation with respect to the x-coordinate
$\partial/\partial y, \partial/\partial z$	partial differential (Cartesian coordinate)
$\partial/\partial r, \partial/\partial \theta$	partial differential (polar coordinate)
$\partial/\partial t$	partial differential with respect to time t
D/Dt	absolute derivative
$N!$	N-factorial: $N! = 1 \times 2 \times 3 \times 4 \times \cdots \times (N-1) \times N$

Constants

e	constant such as $\text{Ln}(e) = 1$: $e = 2.718\,281\,828\,459\,045\,235\,360\,287$
π	$\pi = 3.141\,592\,653\,589\,793\,238\,462\,643$
$\sqrt{2}$	$\sqrt{2} = 1.414\,213\,562\,373\,095\,048\,8$
$\sqrt{3}$	$\sqrt{3} = 1.732\,050\,807\,568\,877\,293\,5$

Mathematical bibliography

Beyer, W.H. (1982). *CRC Standard Mathematical Tables*. (CRC Press Inc.: Boca Raton, Florida, USA).
Korn, G.A. and Korn, T.M. (1961). *Mathematical Handbook for Scientist and Engineers*. (McGraw-Hill Book Comp.: New York, USA).
Spiegel, M.R. (1968). *Mathematical Handbook of Formulas and Tables*. (McGraw-Hill Inc.: New York, USA).

5.4 Exercises

1. A 3.1 kg mass of dye is injected in the centre of large pipe. In the absence of flow and assuming molecular diffusion only, calculate the time at which the mass concentration equals 0.1 g/L at the injection point. *Assume $D_m = 0.89 \times 10^{-2}\,m^2/s$.*
2. Considering a one-dimensional semi-infinite reservoir bounded at one end by a solid boundary (e.g. a narrow dam reservoir), a 5 kg mass slug of contaminant ($D_m = 1.1 \times 10^{-2}\,m^2/s$) is injected 12 m from the straight boundary (e.g. concrete dam wall). Calculate the tracer concentration at the boundary 5 min after injection. Estimate the maximum tracer concentration at the boundary and the time (after injection) at which it occurs.
3. A 10 km long pipeline is full of fresh water. At one end of the pipeline, a contaminant is injected in such a fashion that the contaminant concentration is kept constant and equals 0.14 g/L. Assuming $D_m = 1.4 \times 10^{-3}\,m^2/s$ and an infinitely long pipe, calculate the time at which the pollutant concentration exceeds 0.007 and 0.01 g/L at 4.2 km from the injection point.

5.5 Exercise solutions

1. $t = 8500\,s$ (2 h 21 min).
2. (a) $C_m = 2.8 \times 10^{-5}\,kg/m^3$ and (b) $C_m = 0.2\,kg/m^3$ and $t = 6480\,s$ (1.8 h).
3. (a) $t = 18\,900$ days (0.007 g/L) and (b) $t = 22\,000$ days (0.01 g/L).

6

Advective diffusion

Summary
The basic equation of molecular advective diffusion is presented and simple applications are shown.

6.1 Basic equations

The previous section was developed assuming molecular diffusion with no transport and zero velocity. That is, the fluid was assumed stationary, mass transport occurring by diffusion only. *Advection* is the transport by an imposed current; a movement of a mass of fluid which enhances change in temperature or in other physical or chemical properties of fluid. In this section, we shall assume that transport by advection and diffusion are *separate additive processes*. That is, the diffusion takes place within the moving fluid as in a stationary fluid.

The total mass transport rate equals:

$$\dot{m} = VC_m + \left(-D_m \frac{\partial C_m}{\partial x}\right) \tag{6.1}$$

where V is the fluid velocity. In equation (6.1), the first term (VC_m) is the *advective flux* while the second term is the *diffusive flux*.

For a one-dimensional flow, the *advective diffusion equation* is:

$$\frac{\partial C_m}{\partial t} + V \frac{\partial C_m}{\partial x} = D_m \frac{\partial^2 C_m}{\partial x^2} \tag{6.2a}$$

For a three-dimensional flow, it is:

$$\frac{\partial C_m}{\partial t} + V_x \frac{\partial C_m}{\partial x} + V_y \frac{\partial C_m}{\partial y} + V_z \frac{\partial C_m}{\partial z} = D_m \left(\frac{\partial^2 C_m}{\partial x^2} + \frac{\partial^2 C_m}{\partial y^2} + \frac{\partial^2 C_m}{\partial z^2} \right) \tag{6.2b}$$

DISCUSSION
The advective diffusion theory is based upon the key assumption that diffusion and advection are two separate additive processes (Fischer *et al.* 1979, p. 50). Hence the theory of superposition is applicable (Chapter 5).

6.2 Basic applications

6.2.1 Advective diffusion of a sharp front

A simple one-dimensional application is a fluid moving in the x-direction at the velocity V, when the gradients in the y-direction are small (i.e. $\partial/\partial y \ll \partial/\partial x$). The two-dimensional advective diffusion equation (6.2b) yields:

$$\frac{\partial C_m}{\partial t} + V \frac{\partial C_m}{\partial x} = D_m \frac{\partial^2 C_m}{\partial x^2} \qquad (6.3)$$

Considering a sharp front at $t = 0$, the boundary conditions are:

$$C_m(x,0) = 0 \qquad \text{for } x > 0$$
$$C_m(x,0) = C \qquad \text{for } x < 0$$

This is the case of a pipe filled with one fluid and being displaced by another fluid moving at a velocity V (Fig. 6.1). Practical applications include the cleaning of sewers with freshwater and the steady injection of antibiotics in a blood vessel.

The solution of the advective diffusion equation is:

$$C_m(x,t) = \frac{C_o}{2}\left(1 - \text{erf}\left(\frac{x - Vt}{\sqrt{4D_m t}}\right)\right) \qquad \text{for } t > 0 \qquad (6.4)$$

Remark
Practical applications include an 'idealized' pipe filled with one fluid and being displaced by another, e.g. the cleaning of sewers with freshwater, the pumping of warm petroleum in an oil pipeline filled with cold fluid and the injection of antibiotics in blood vessels.

6.2.2 Initial mass slug introduced at t = 0 and x = 0

Another simple example is the sudden injection of a mass slug (mass M) at the origin at $t = 0$. The solution of the problem is similar to the diffusion of an instantaneous mass slug injection in a fluid at rest (i.e. equation (5.4)). When the velocity is non-zero, molecular diffusion takes place around the location of the centroid which is advected at a velocity V. At the time t, the location of centroid is $X = Vt$.

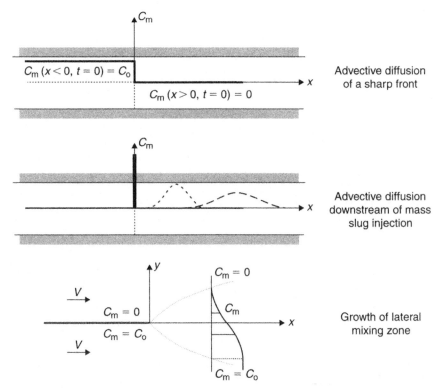

Fig. 6.1 Basic applications of advective diffusion.

For a one-dimensional flow, the solution of the advective diffusion equation is:

$$C_m(x,t) = \frac{M}{\sqrt{4\pi D_m t}} \exp\left(-\frac{(x - Vt)^2}{4D_m t}\right) \quad \text{for } t > 0 \qquad (6.5)$$

Equation (6.5) is shown in Figure 5.1(b).

6.2.3 Transverse mixing of two streams with different concentrations

Considering the transverse mixing of two streams of different concentrations flowing side by side with the same velocity (Fig. 6.1), the diffusive transport in the x-direction is smaller than the advective transport. For the steady flow, the advective diffusion equation yields:

$$V\frac{\partial C_m}{\partial x} = D_m \frac{\partial^2 C_m}{\partial y^2} \qquad (6.6)$$

The boundary conditions are:

$$C_m(0, y) = 0 \quad \text{for } y > 0$$
$$C_m(0, y) = C_o \quad \text{for } y < 0$$

The solution of equation (6.6) is:

$$C_m(x,t) = \frac{C_o}{2}\left(1 - \text{erf}\left(\frac{y}{\sqrt{4D_m\dfrac{x}{V}}}\right)\right) \qquad \text{for } x > 0 \tag{6.7}$$

> **Remark**
> A practical application is the confluence of two rivers (Figs 6.2, 7.2(a) and (b)). Figure 6.2 shows a false colour thermal infrared image of the confluence of two shallow streams. One stream is 5.2°C cooler than the other and the transverse mixing of heat is clearly visible downstream of the confluence. Figure 7.2(a) and (b) presents the confluence of two streams with different sediment concentrations (i.e. murkiness).

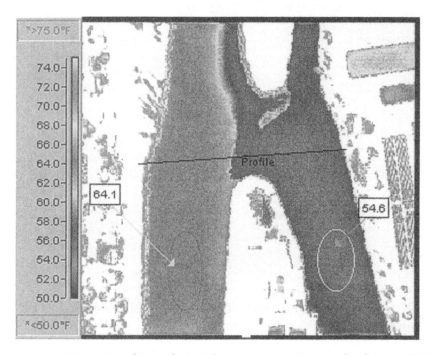

Fig. 6.2 Transverse mixing at the confluence of two shallow water streams (courtesy of Prof. Steven J. Wright). False colour thermal infrared image. Flow from the bottom to the top. The temperature of the stream on right is 17.8°C (64.1°F) while the temperature of the other stream is 12.6°C (54.6°F).

6.2.4 Sudden mass contamination in a river

A more complex case is the sudden contamination of a stream with a steady concentration C_o introduced at the origin at $t = 0$ and for $t > 0$. The boundary conditions are:

$$C_m(0,t) = C_o \qquad \text{for } 0 < t < +\infty$$
$$C_m(x,0) = 0 \qquad \text{for } 0 < x < +\infty$$

The advective diffusion equation is:

$$\frac{\partial C_m}{\partial t} + V \frac{\partial C_m}{\partial x} = D_m \frac{\partial^2 C_m}{\partial x^2} \tag{6.8}$$

The solution of equation (6.8) is:

$$C_m(x,t) = \frac{C_0}{2}\left[1 - \text{erf}\left(\frac{x - Vt}{\sqrt{4D_m t}}\right) + \exp\left(\frac{Vx}{D_m}\right)\left(1 - \text{erf}\left(\frac{x + Vt}{\sqrt{4D_m t}}\right)\right)\right] \quad \text{for } 0 < t < +\infty \tag{6.9}$$

Note that the final solution ($t = +\infty$) is the contamination of the entire river, i.e. $C_m = C_0$ everywhere.

6.3 Two- and three-dimensional applications

Consider a point source at the origin discharging mass at a rate $\dot{M}$ in a fluid moving at a velocity V in the x-direction. The problem is a steady case. The solution of the advective diffusion equation for the three-dimensional mass concentration is:

$$C_m(x,y,z) = \frac{\dot{M}}{4\pi D_m x} \exp\left(-\frac{(y^2 + z^2)}{4D_m \dfrac{x}{V}}\right) \qquad x \gg \frac{2D_m}{V} \tag{6.10}$$

Note that the solution is valid for $x \gg 2D_m/V$. Practically the relevant time scale is very small.

For a point source in a two-dimensional plane $\{x, y\}$, the solution of the advective diffusion equation is:

$$C_m(x,y) = \frac{\dot{M}}{V\sqrt{4\pi D_m \dfrac{x}{V}}} \exp\left(-\frac{y^2}{4D_m \dfrac{x}{V}}\right) \qquad x \gg \frac{2D_m}{V} \tag{6.11}$$

where $\dot{M}$ is the mass discharge per unit width.[1] Note again that the solution is valid for $x \gg 2D_m/V$.

Remarks
1. For a point injection in a three-dimensional medium, the cloud of contaminant is often called improperly a 'diffusion cone'. The term is inappropriate because it is an advective diffusion process and the tracer cloud is not a true cone although it has an axi-symmetrical shape.
2. The above analytical solutions (equations (6.10) and (6.11)) are used to analyse transverse mixing of pollutant discharge from a pipe in a river (Chapter 7).

---

[1] In a three-dimensional problem, $\dot{M}$ would be the mass flow rate of the line source per unit width (i.e. $\dot{M} = \dot{M}/W$).

6.4 Exercises

1. An initial mass slug (mass $M = 1$) is introduced suddenly at the origin at $t = 0$. Assuming $D_m = 0.2$ and $V = 1$, (1) calculate the maximum mass concentration at $t = 0.3$ and (2) calculate the mass concentration for $x = 0.07$ and $t = 0.3$.
2. A pipeline is initially filled with clear water. At $t = 0$, contaminated waters ($C_o = 55$ ppm, $D_m = 2 \times 10^{-9} m^2/s$) are flushed into the pipeline at one end and the average flow velocity is 0.95 m/s. Estimate the width of the interface (defined between 5% and 95% of the initial concentration C_o) 50 km downstream.
3. A one-dimensional stream ($V = 0.35$ m/s) is suddenly contaminated with a steady concentration ($C_o = 185$ ppm, $D_m = 1.8 \times 10^{-1} m^2/s$) introduced at $t = 0$. Estimate the time at which the tracer concentration will be 20 ppm at 12 m from the injection.

Note: In the above exercises, large values of diffusion coefficients were used for simplicity of calculations and more meaningful results. Such values are not representative of solutes in water.

6.5 Exercise solutions

1. The solution of the problem is similar to the diffusion of an instantaneous mass slug injection in a fluid at rest (i.e. equation (5.4)). When the velocity is non-zero, molecular diffusion takes place around the location of the centroid which is advected at a velocity V. At the time t, the location of centroid is $X = Vt$.

    For a one-dimensional flow, the advective diffusion equation is:

$$\frac{\partial C_m}{\partial t} + V \frac{\partial C_m}{\partial x} = D_m \frac{\partial^2 C_m}{\partial x^2} \tag{6.2a}$$

    The solution is:

$$C_m(x,t) = \frac{M}{\sqrt{4\pi D_m t}} \exp\left(-\frac{(x - Vt)^2}{4 D_m t}\right) \qquad \text{for } t > 0 \tag{6.5}$$

    For $t = 0.3$, $C_{max} = 1.1$ at $x = 0.3$. For $t = 0.3$ and $x = 0.07$, $C_m = 0.15$.

2. For a sharp front:

$$C_m(x,t) = \frac{C_o}{2}\left(1 - \text{erf}\left(\frac{x - Vt}{\sqrt{4 D_m t}}\right)\right)$$

    The distribution is symmetrical. For $t = 50\,000/0.95$ s, $x(C_m = 0.05C_o) = 50\,000 + 0.024$ m and $x(C_m = 0.95C_o) = 50\,000 - 0.024$ m. Hence the interface width equals 4.8 cm.

3. $t \sim 2$ s.

7

Turbulent dispersion and mixing: 1. Vertical and transverse mixing

Summary

In this chapter, the basic equation of advective diffusion are applied to vertical and transverse mixing in natural river systems. Basic applications are developed.

7.1 Introduction

Natural rivers are characterized by turbulent flows. Turbulence is generated by boundary friction and it increases significantly the rate of mixing. While Chapters 5 and 6 consider molecular diffusion and advection, the developments may be extended to turbulent dispersion and mixing with few changes (Fischer *et al.* 1979, Lewis 1997). The molecular diffusion coefficient is replaced basically by *mixing coefficients*: i.e. the vertical and transverse mixing coefficients ε_v and ε_t respectively. The turbulent advection and spread of particles in the longitudinal direction is described by a *dispersion coefficient K* (Chapter 8).

When a tracer is injected into a homogeneous stream flow (Fig. 7.1), the advective transport may be divided into three stages from upstream to downstream:

(a) mixing near the outlet driven by initial momentum and buoyancy,
(b) transverse mixing of the effluent by turbulent transport,
(c) longitudinal shear flow dispersion.

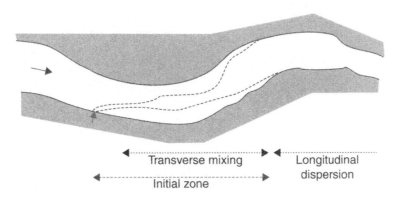

Fig. 7.1 Mixing and dispersion of an effluent in a river.

Fig. 7.2 Examples of mixing in natural systems. (a) Confluence of Colorado River and Little Colorado River in the Grand Canyon of Arizona on 12 April 1966 (courtesy of Dr Lou Maher) – note the light (reddish) colour of the Little Colorado River; the Colorado River has left its red mud behind the Glen Canyon Dam 60 miles upstream. (b) Confluence of the Colorado and Dolorus Rivers, 20 miles North-East of Moab, Utah on 13 April 1966 (courtesy of Dr Lou Maher). (c) Pimpama Creek, Redlands, Queensland on 14 October 1999 (courtesy of the Waterways Scientific Services, Queensland Environment protection Agency) – looking upstream – note freshwater outfall near left bank during construction of a weir to reduce saltwater intrusion in the upper reach.

In natural streams, turbulence contributes to the enhancement of mass spreading (Fig. 7.2). Turbulent flows are also characterized by unsteady velocities and pressures at any point, leading to unsteady mass concentration.

Discussion

Turbulent mixing and dispersion are assumed to take place in a similar fashion as molecular advective diffusion. Analytical solutions developed in Chapters 5 and 6 may be used by replacing the molecular diffusion coefficient D_m by the appropriate turbulent process coefficient:

	Molecular diffusion coefficient	Turbulent mixing/ dispersion coefficient
Vertical mixing	D_m	ε_v
Transverse mixing	D_m	ε_t
Longitudinal dispersion	D_m	K

Notes

1. Usually the term *mixing* refers to lateral spreading (transverse and vertical) caused by turbulence. The term *dispersion*, or shear dispersion, characterized longitudinal spreading caused by turbulent shear.
2. While the molecular diffusion coefficient D_m is a property of the fluid, the turbulent mixing and dispersion coefficients are properties of the flow rather than of the fluid. Further they are much larger than D_m.

7.2 Flow resistance in open channel flows

In natural stream flows, friction losses and flow resistance are always significant. The *boundary shear stress* equals:

$$\tau_o = \frac{f}{8} \rho V^2 \tag{7.1}$$

where f is the Darcy–Weisbach friction factor, ρ is the density of the flowing fluid and V is the mean flow velocity. The *shear velocity* V_* is defined as: $V_* = \sqrt{\tau_o/\rho}$. In equation (7.1), the Darcy friction factor is a function of the Reynolds number VD_H/ν and relative roughness k_s/D_H, where k_s is the equivalent roughness height and D_H is the hydraulic diameter[1] (Appendix A, Section 7.6).

---

[1] The hydraulic diameter, also called equivalent pipe diameter, is defined as four times the cross-sectional area divided by the wetted perimeter.

For steady flows, the continuity equation states that: $Q = VA$ where Q is the flow rate and A is the flow cross-sectional area. At uniform equilibrium (i.e. normal flow conditions), the momentum principle states that the boundary shear force equals exactly the weight force component in the flow direction. The uniform equilibrium flow velocity (i.e. normal velocity) equals:

$$V = \sqrt{\frac{8g}{f}} \sqrt{\frac{D_H}{4} \sin \theta}$$

(7.2)

and the average shear velocity becomes:

$$V_* = \sqrt{g \frac{D_H}{4} \sin \theta} \qquad \text{Uniform equilibrium flow} \qquad (7.3)$$

where θ is the bed slope.

In natural systems, estimates of mixing and dispersion coefficients rely heavily upon accurate estimate of the hydraulic properties of the river flow. Each student must know how to calculate basic hydraulic properties before embarking into mixing and dispersion estimates.

Discussion: flow resistance estimate in natural streams

Flow resistance calculations in open channels must be performed in term of the Darcy friction factor. In turbulent flows, however, the choice of the boundary equivalent roughness height is important. Hydraulic handbooks (e.g. Idelchik 1969, 1986) provide a selection of appropriate roughness heights for standard materials. Main limitations of the Darcy equation for turbulent flows include:

- the friction factor may be estimated for relative roughness $k_s/D_H < 0.05$ and
- classical correlations were validated for uniform-size roughness and regular roughness patterns.

In simple words, the Darcy flow resistance equation is accurate in man-made channels with well-defined roughness height. But flow resistance cannot be accurately predicted for complex roughness patterns: e.g. vegetation, composite cross-section, flood plain roughness (trees, houses, cars), shallow waters over large roughness, braided channels, meandering channel beds (Figs 3.1 and 7.2). See Chanson (1999a, pp. 85–91) for further discussion.

7.3 Vertical and transverse (lateral) mixing in turbulent river flows

In natural rivers, the flow is highly turbulent (Figs 3.1, 7.2 and 7.3). Fluid particles fluctuate randomly while the stream is advected with a time-averaged velocity V. Figure 7.3 shows instantaneous velocity measurements in a subtropical creek, in terms of the velocity modulus and direction. Note the data scatter. The random process may be modelled by a 'random walk' model (Appendix B, Section 7.7).

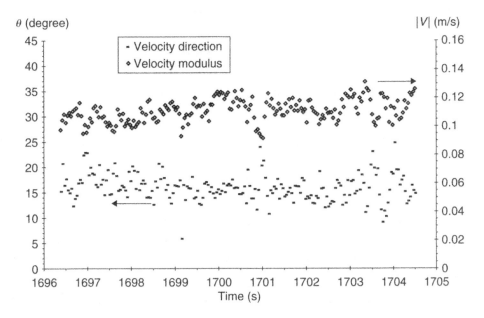

Fig. 7.3 Turbulent velocity fluctuations in a natural creek – instantaneous velocity modulus and direction in Eprapah Creek, Queensland at 0.5 m beneath the free surface (after Chanson *et al.* 2003) – measurements conducted with an acoustic Doppler velocity meter recording 25 data per second.

Mixing is related to the turbulent shear: i.e. vertical mixing is induced by bottom friction and transverse mixing is generated by bank roughness (Fig. 7.4). The mixing coefficients for momentum and mass are assumed the same.[2] That is, the local mixing coefficient ε equals the momentum exchange coefficient ν_T. This approximation is nearly always assumed in river mixing and sediment suspension studies.

In fully developed open channel flows, the average *vertical mixing coefficient* (unit: m²/s) is about:

$$\varepsilon_v = 0.067 \, dV_* \qquad (7.4)$$

The result is valid for a wide range of flows (e.g. Rutherford 1994, pp. 59–60).

For fully developed open channel flows in straight rectangular channels, experimental observations (Fischer *et al.* 1979) suggest that the *transverse mixing coefficient* (unit: m²/s) is about:

$$\varepsilon_t = 0.15 \, dV_* \qquad \text{Straight rectangular channels} \qquad (7.5)$$

A more recent review proposed: $\varepsilon_t / dV_* = 0.13$ (Rutherford 1994, p. 102).

In natural rivers, bends and sidewall irregularities enhance transverse mixing. Secondary currents[3] cause tracers to move in opposite directions at different depths and this behaviour

---

[2] This assumption is not strictly correct. Chanson (1997) discussed specifically the limitations of this assumption for air bubble entrainment in turbulent shear flows. See also Chapter 4, paragraph 4.5.2.

[3] Secondary currents are perpendicular to the main current and they are caused by Reynolds stresses in any non-circular conduits. In natural rivers, they are significant at bends, and between a flood plain and the main channel.

greatly increases the rate of transverse mixing. For slowly meandering channels with moderate sidewall irregularities, a practical approximation is:

$$\varepsilon_t = 0.6\,dV_* \qquad \text{Slowly meandering channels with moderate sidewall irregularities} \quad (7.6)$$

Larger transverse mixing coefficients are observed in sharply meandering rivers as shown in Figure 7.5 (see review in Fischer *et al.* 1979, pp. 109–112; Rutherford 1994, pp. 112–113).

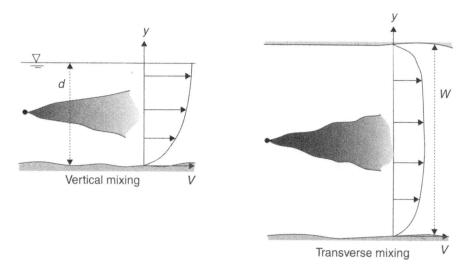

Fig. 7.4 Mixing in open channel flows.

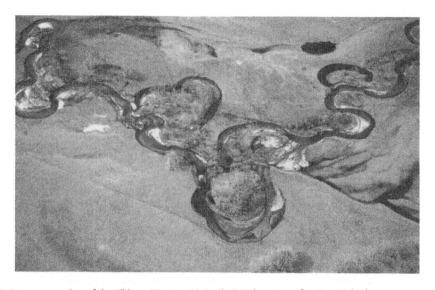

Fig. 7.5 Stream meanders of the Elkhorn River on 16 April 1966 (courtesy of Dr Lou Maher).

Notes
1. In the middle of a stream, it is convenient to neglect the effect of depth variations on tracer concentration. That is, most calculations are conducted assuming a constant depth. Although untrue this approximation is relatively robust.
2. Assuming that the mixing length equals: $l = 0.4(d + z)$, the momentum exchange coefficient in fully developed open channel flow becomes: $\nu_T(z) = KV_*(d + z)(-z/d)$, where z is the vertical positive upwards and $z = 0$ at the free surface (Fig. 7.6) (Chapter 4, paragraph 4.4). Integrating over the flow depth, the depth-average eddy viscosity equals:

$$\nu_T = \frac{1}{d}\int_{z=-d}^{z=0} \nu_T(z)\,dz = \frac{K}{6}V_*d$$

For $K = 0.40$, the result yields equation (7.4) and ε_v equals the depth-averaged momentum exchange coefficient ν_T.
3. The analysis of field data in meandering channels by Rutherford (1994, pp. 111–112) gave:

$$0.3 < \frac{\varepsilon_t}{dV_*} < 0.9 \qquad \text{Meandering channels}$$

for 31 field data measured on 16 streams across the world. Deng *et al.* (2002) re-analysed 70 field data measured in 30 streams in USA; 70% of the data set satisfied:

$$0.18 < \frac{\varepsilon_t}{dV_*} < 0.9 \qquad \text{Meandering channels}$$

4. For strongly curved channels, an analysis of field data by Rutherford (1994, pp. 112–113) suggested that:

$$1 < \frac{\varepsilon_t}{dV_*} < 3 \qquad \text{Curved channels}$$

although larger values may be experienced in sharp bends.

Boxall *et al.* (2003) presented detailed measurements the effect of channel curvature on transverse mixing coefficients.

Remarks: role of turbulence and secondary currents in open channels
In natural waterways, the flow motion is characterized by unpredictable behaviour, strong mixing properties and broad spectrum of length scales. But the turbulent velocity components are not independently random: they are correlated with each other in space and time (Nezu and Nakagawa 1993). There is some coherence. Coherent structures may be classified into two categories: (1) bursting phenomena and (2) large-scale vertical motion. The former is generated in the fluid layers next to the boundary where the flow consists of high-speed and low-speed streaks with regular spanwise spacing. In fully developed open channel flows, strong bursting event may induce scars and boil marks at the free surface.

Secondary currents are generated by boundary shear stress, and non-homogeneity and anisotropy of turbulence. They are evidenced by presence of circulation superposed to the longitudinal fluid advection, and the flow streamlines often exhibit a spiral form. In straight prismatic open channels, secondary currents are affected primarily by sidewall effects, free-surface effects and bed roughness (Henderson 1996, p. 88; Nezu and Nakagawa 1993, pp. 85, 91). The free-surface damps fluctuations normal to it. Secondary currents are responsible for large-size eddies which exhibit a coherent behaviour because they are generated by some interactions with the mean flow. They retain their structure while they are advected downstream over significant distances. In contrast small-scale eddies are nearly isotropic and behave randomly.

In meandering channels, secondary currents are enhanced by the centrifugal force, and their velocity may be about 20–30% of mainstream velocity. A dominant feature is a helicoidal flow pattern which induces typically scour at the outer bend and deposition at the inner bank, yielding a quasi-triangular flow cross-section (Rozovskii 1957, Blanckaert and Graf 2001b).

Discussion

Vertical mixing is the result of turbulence generated by bed friction. In natural streams, the time scale of vertical mixing (d^2/ε_v) is nearly two orders of magnitude smaller than that of transverse mixing (W^2/ε_t). Most rivers are much wider than deep (i.e. $W/d > 10$ typically) while the vertical mixing coefficient is about 10 times smaller than the transverse mixing coefficient. Often vertical and transverse mixing may be considered separately because contaminants are well mixed over the depth long before they are well mixed across the channel.

Considering a natural stream with $V = 1$ m/s, $d = 2$ m, $W = 30$ m and $V_* = 0.05$ m/s, the vertical mixing time scale (d^2/ε_v) is about 600 s (10 min). The transverse mixing time scale (W^2/ε_t) is about 15 000 s (4.2 h).

Importantly: Diffusion coefficient estimates (equations (7.4)–(7.6)) were developed for gradually varied flows and uniform equilibrium flows. They do not apply to rapidly varied flow conditions. For example, hydraulic jumps are known for their very-strong mixing properties (Henderson 1966, Chanson 1999a). Experimental observations of vertical and transverse mixing coefficients in hydraulic jumps are presented and discussed in Appendix C (Section 7.8).

Remarks

Two practical rules of thumb are:

- contaminants become well mixed vertically within a longitudinal distance of 50 times the water depth and
- tracers becomes well mixed across the channel about 100–300 channel widths downstream of a point source near mid-stream (Rutherford 1994).

7.4 Turbulent mixing applications

Usually, vertical mixing is rapid in natural systems while transverse mixing is a much longer process because the channel width is much greater than the water depth. In many practical problems, it can be assumed that the contaminant is nearly uniformly distributed over the vertical and the problem becomes a two-dimensional spread from a line source.

7.4.1 Transverse mixing downstream of a continuous point source

Considering a point source at the origin discharging mass at a rate $\dot{M}$ over the depth d in an infinitely wide rectangular channel, the basic advection equation for a two-dimensional steady flow becomes:

$$V \frac{\partial C_m}{\partial x} = \varepsilon_t \left(\frac{\partial^2 C_m}{\partial x^2} + \frac{\partial^2 C_m}{\partial y^2} \right)$$

where x is the direction in the flow direction and y is the transverse direction. The solution of the dispersion equation is:

$$C_m = \frac{\dot{M}}{Vd \sqrt{4\pi\varepsilon_t \frac{x}{V}}} \exp \left(-\frac{y^2}{4\varepsilon_t \frac{x}{V}} \right) \qquad \text{Infinitely wide channel} \qquad (7.7)$$

where ε_t is the transverse mixing coefficient, x is the coordinate in the flow direction (i.e. following the mean flow path and river meanders), and y is the transverse direction (Fig. 7.6). Note the unit of $\dot{M}$ in kg/s/m. Equation (7.7) may be compared with equation (6.11).

For a channel of finite width W, the effluent source being located at $y = y_o$, the principle of superposition and method of images (Chapter 5) give the downstream concentration distribution:

$$\frac{C_m}{C_o} = \frac{1}{\sqrt{4\pi x'}} \sum_{i=-\infty}^{+\infty} \left(\exp \left(-\frac{(y' - 2i - y_o')^2}{4x'} \right) + \exp \left(-\frac{(y' - 2i + y_o')^2}{4x'} \right) \right) \qquad (7.8)$$

where y is the transverse direction ($y = 0$ on the right bank) (Fig. 7.6) and C_o, x' and y' are dimensionless parameters defined as:

$$C_o = \frac{\dot{M}}{VdW}$$

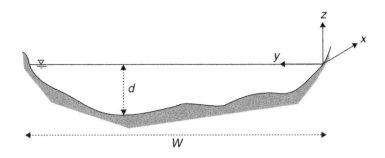

Fig. 7.6 Definition sketch of a natural river system.

$$x' = \frac{x \varepsilon_t}{V W^2}$$

$$y' = \frac{y}{W}$$

and the unit of $\dot{M}$ is kg/s.

7.4.2 Transverse mixing downstream of a mass slug injection

Considering a mass of contaminant M injected at $t = 0$ at the origin over the depth d in an infinitely wide rectangular channel, the basic advection equation for a two-dimensional unsteady flow is:

$$\frac{\partial C_m}{\partial t} + V \frac{\partial C_m}{\partial x} = \varepsilon_t \left(\frac{\partial^2 C_m}{\partial x^2} + \frac{\partial^2 C_m}{\partial y^2} \right)$$

The solution of the advective diffusion equation is:

$$C_m = \frac{\frac{M}{d}}{4 \pi t \varepsilon_t} \exp \left(-\frac{(x - Vt)^2 + y^2}{4 \varepsilon_t t} \right) \qquad \text{Infinitely wide channel} \qquad (7.9)$$

where y is the transverse direction (Fig. 7.6) and ε_t is the transverse mixing coefficient. Note that M is in kg, the mass injection is assumed to be instantaneous and the channel to be infinitely wide.

Remarks
1. Equation (7.9) assumes a velocity vector: $V_x = V$ and $V_y = 0$. That is, x is the longitudinal coordinate along the main flow direction.
2. Turbulent diffusion takes place around the centroid which is advected at a velocity V. At the time t, the location of centroid is $x = Vt$.

7.4.3 Complete transverse mixing

Practically a reasonable distance for complete transverse mixing from a centreline discharge is:

$$L \sim 0.1 \frac{V W^2}{\varepsilon_t} \qquad \text{Complete mixing of centreline discharge} \qquad (7.10)$$

where *complete mixing* is defined as the concentration is within 5% of its mean value everywhere in the cross-section.

If the effluent is discharged at the side of the channel (e.g. Fig. 7.2(c)), the width over which the mixing take place is twice that for a centreline injection. Solid boundaries, such as the channel bank, do lead to zero concentration gradients because there is zero mass flux. A reasonable distance for complete mixing from a side discharge is:

$$L \sim 0.4 \frac{VW^2}{\varepsilon_t} \qquad \text{Complete mixing of side discharge} \qquad (7.11)$$

The latter may be rewritten as:

$$x' = \frac{L\varepsilon_t}{VW^2} \sim 0.4 \qquad \text{Complete mixing of side discharge} \qquad (7.11)$$

Notes

Rutherford (1994) proposed a slightly different definition for which 'fully mixed' is defined as the ratio of minimum to maximum concentration across the channel is 98%. With this definition, it yields:

$$L \sim 0.134 \frac{VW^2}{\varepsilon_t} \qquad \text{Centreline discharge}$$

$$L \sim 0.536 \frac{VW^2}{\varepsilon_t} \qquad \text{Side discharge}$$

7.5 Discussion

7.5.1 Initial mixing

If the effluent discharge has enough momentum or buoyancy, the initial discharge forms a plume over some fraction of the cross-section. Further turbulent mixing occurs and it may be computed by a superposition of line sources (see Chapter 5, equation (5.9)). It yields:

$$C_m = \int_0^1 \frac{C_o(y_o')}{\sqrt{4\pi x'}} \sum_{i=-\infty}^{+\infty} \left(\exp\left(-\frac{(y' - 2i - y_o')^2}{4x'} \right) + \exp\left(-\frac{(y' - 2i + y_o')^2}{4x'} \right) \right) dy_o' \qquad (7.12)$$

where $C_o(y_o')$ is the distributed source characterizing the initial mixing plume.

7.5.2 Applications

1. An industry releases $5\,\text{Mm}^3$/day of effluents containing $250\,\text{ppm}$ of a chemical near the centre of a very wide, slowly meandering stream. The creek is $8\,\text{m}$ deep, the mean velocity is $0.5\,\text{m/s}$ and the shear velocity is $0.1\,\text{m/s}$. Assuming that the effluent is completely mixed over the vertical, determine the width of the plume and the maximum concentration $500\,\text{m}$ downstream of the discharge point.

Solution

The rate of input is:

$$\dot{M} = 5 \times 10^6\,\text{m}^3/\text{day} \times 250\,\text{ppm} = 14.5\,\text{kg/s}$$

The transverse mixing coefficient is about:

$$\varepsilon_t = 0.48\,\text{m}^2/\text{s} \quad (\text{equation (7.6)})$$

The width of the plume may be approximated by

$$4\sigma = 4\sqrt{2\varepsilon_t\,x/V} = 124\,\text{m}$$

The maximum chemical concentration is: $C_m = \dot{M}/Vd\sqrt{4\pi\varepsilon_t\,x/V} = 0.047\,\text{kg/m}^3$ (or 47 ppm) (equation (7.7)). It occurs on the stream centreline.

2. A plant discharges a chemical at the side of a straight rectangular concrete channel. The channel is 50 m wide. The water flows at uniform equilibrium. The water depth equals 2.5 m, the velocity is 1.05 m/s and the bed slope is 0.0001. What is the channel length required for complete mixing? (Complete mixing is defined as to mean that the concentration of the chemical varies no more than 5% over the cross-section.)

Solution

The shear velocity equals:

$$V_* = \sqrt{gD_H/4\sin\theta} \approx \sqrt{9.80 \times 2.5 \times 0.0001} = 0.0495\ \text{m/s}$$

The transverse mixing coefficient is about:

$$\varepsilon_t = 0.0186\,\text{m}^2/\text{s} \quad (\text{equation (7.5)})$$

As the effluent discharges at the side, the length for complete mixing equals:

$$L \sim 0.4VW^2/\varepsilon_t = (0.4 \times 1.05 \times 50^2)/0.0186 = 5645\ \text{m}$$

3. Considering two streams which flow together at a smooth junction, we will assume that the density is nearly the same and that mixing is caused by turbulence only. For example, two sources supply a treatment plant and the flow is blended before processing. Each source delivers 15 m³/s. At the junction the water flows down a single rectangular channel (14 m wide) with a slope of 0.0008. The channel is concrete lined. Assuming a straight channel, what is the channel length required to provide complete mixing? *Assume that uniform equilibrium flow conditions are achieved.*

Solution

Firstly the hydraulic characteristics of the concrete channel must be calculated using the momentum principle. The process is iterative (Henderson 1966, Chanson 1999a). Assuming $k_s = 1$ mm (smooth concrete), it yields: $V = 1.99$ m/s, $d = 1.07$ m, $V_* = 0.09$ m/s.

Secondly we assume that one stream is contaminated with a concentration C_o and the other is not ($C_m = 0$). After complete mixing, the concentration will be $0.5C_o$.

An upper limit of the length for complete mixing derives assuming that all the contamination is discharged at the wall: $L = 0.4VW^2/\varepsilon_t$. For a straight channel, it is about $108\,000\,\text{m}$. The actual distance will be $<10.8\,\text{km}$ because the contaminants are discharged uniformly over half of the channel width.

The exact solution may be achieved by applying a method of superposition based upon equation (5.7) or equation (7.12). (Fischer *et al.* (1979, pp. 118–119) developed the complete analytical solution.) The final result is about $8.5\,\text{km}$ for complete mixing (i.e. the concentration is 5% of the mean everywhere).

7.6 Appendix A – Friction factor calculations

The laws of flow resistance in open channels are essentially the same as those in closed pipes (Henderson 1966, Chanson 1999a, 2004b). In laminar flows, the Darcy friction factor may be estimated as:

$$f = \frac{64}{Re} \qquad Re < 2 \times 10^3 \qquad (7\text{A}.1)$$

where Re is the Reynolds number defined as: $Re = \rho V D_H/\mu$ and D_H is the hydraulic diameter.[4]

In turbulent flows, the Darcy friction factor may be calculated from the Colebrook–White formula:

$$\frac{1}{\sqrt{f}} = -2.0 \log_{10}\left(\frac{k_s}{3.71\,D_H} + \frac{2.51}{Re\sqrt{f}}\right) \qquad Re > 1 \times 10^4 \qquad (7\text{A}.2)$$

where k_s is the equivalent sand roughness height. Note that the friction factor f appears on both side of equation (7A.2) which must be solved by iterations. The friction factor may be initialized by a less accurate expression called Altsul's formula:

$$f = 0.1\left(1.46\frac{k_s}{D_H} + \frac{100}{Re}\right)^{1/4} \qquad Re > 1 \times 10^4 \qquad (7\text{A}.3)$$

7.7 Appendix B – Random walk model

In turbulent flows, the chaotic motion of the fluid and contaminant particles may be modelled by the 'random walk'. The random walk model assumes that each particle is advected by the main current and displaced longitudinally, vertically and laterally by turbulent eddies. With time, the particles become separated further and further by the random (chaotic) motion. Each particle represents a mass unit and the number of particles in a given volume is a measure of concentration.

---

[4] The hydraulic diameter equals four times the flow cross-sectional area divided by the wetted perimeter. It is also called the equivalent pipe diameter. Indeed, for a circular pipe flow, the hydraulic diameter equals exactly the pipe diameter.

Random walk models simulate discrete mixing at a series of time steps δt. It is assumed that turbulence is homogeneous and stationary: i.e. each step is uncorrelated to the previous one. For a long diffusion time T, the location of the centroid is $X = V_x T$ and the mean square displacement σ_x^2 of particles in the x-direction is:

$$\sigma_x^2 = 2 D_x T \qquad \text{Long diffusion time} \qquad (7B.1)$$

where D_x is the turbulent mixing coefficient in the x-direction. The probability that a particle is between x and $(x + \delta x)$ is given by:

$$\text{Probability}(x,T) = \frac{1}{\sqrt{2\pi}\sigma_x} \exp\left(-\frac{(x-X)^2}{2\sigma_x^2}\right) \qquad \text{Long diffusion time} \qquad (7B.2)$$

During a time step δt, a particle is advected in the x-direction by the ambient current V_x and moved by the turbulent displacement. The net change in position δx equals:

$$\delta x = V_x \, \delta t + \text{RAND} \sqrt{2D_x \delta t} \qquad (7B.3)$$

where RAND is a random number between -0.5 and $+0.5$. For long diffusion times and a large number of particles, the random walk model yields a standard normal (Gaussian) distribution. It may be extended to the y- and z-directions in a three-dimensional flow. Figure 7B.1 presents a typical result for turbulent mixing of two particles, injected at the origin in a two-dimensional plane.

The random walk model can deal with boundaries, such as sea surface, bed and banks. When the random step taken by a particle reach a boundary, it may be allowed to reflect off the boundary. Lewis (1997, pp. 149–150) illustrated the concept and presented one application.

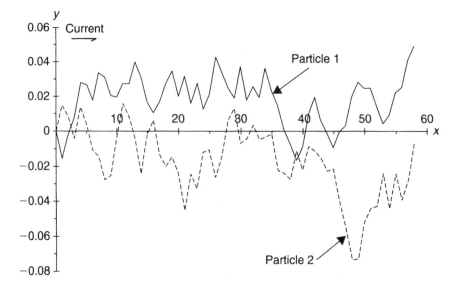

Fig. 7B.1 Random walk of two particles suddenly injected at the origin in a two-dimensional plane for $V_x = 1$, $V_y = 0$, $D_x = D_y = 1$ and $\delta t = 1$.

Application to turbulent dispersion

According to Langevin's model of turbulent dispersion (Pope 2000, pp. 501–502), the Lagrangian velocity auto-correlation function in turbulent flows[5] is:

$$R_{xx}(\delta_t) = -\exp\left(-\frac{\delta_t}{T_L}\right)$$

where T_L is the Lagrangian time scale which is related to the dominant eddy size l and a typical turbulent velocity u' by $T_L \approx l/u'$, and δ_t is the spacing between the two velocity points.

The variance of the momentum spread must then satisfy:

$$\sigma_x^2 = 2u'^2 T_L\left(t - T_L\left(1 - \exp\left(-\frac{t}{T_L}\right)\right)\right)$$

where t is the time, u'^2 is the velocity variance for each particle. Initially, for $t \ll T_L$, the velocity are highly correlated and the spread grows linearly with time: $\sigma_x \approx u't$. For very large times ($T_L \ll t$), it yields:

$$\sigma_x \approx \sqrt{2u'^2 T_L t}$$

The latter is similar to equation (7B.1):

$$\sigma_x^2 \approx 2u'^2 T_L t = 2u'lt = 2D_x t \qquad (T_L \ll t)$$

assuming that $D_x = u'l$.

Remarks

1. Paul Langevin (1879–1946) was a French physicist, specialist in magnetism, ultrasonics and relativity. He was a member of the French pioneering team of atomic researchers, which included Pierre and Marie Curie.

   In 1905, Albert Einstein identified Brownian motion as due to imbalances in the forces on a particle resulting from molecular impacts from the liquid. Shortly thereafter, Paul Langevin formulated a theory in which the minute fluctuations in the position of the particle were due explicitly to a random force. His approach had great utility in describing molecular fluctuations in other systems, including non-equilibrium thermodynamics.

2. A Lagrangian method is the study of a process in a system of coordinates moving with an individual particle. For example, study of ocean currents with buoys. A different method is the Eulerian method.

7.8 Appendix C – Turbulent mixing in hydraulic jumps and bores

Most diffusion coefficient estimates (equations (7.4)–(7.6)) were developed for gradually varied flows and uniform equilibrium flows. They do not apply to rapidly varied flow

---

[5]At a given time.

conditions: e.g. hydraulic jumps, positive surges, tidal bores. Experimental observations of mixing coefficients in hydraulic jumps and bores are summarized in Table 7C.1.

In laboratory hydraulic jumps, the vertical diffusion coefficient of entrained air bubbles was observed to be:

$$\frac{\varepsilon_v}{V_1 d_1} \approx 4.5 \times 10^{-2} \qquad 5.0 < Fr_1 < 8.5$$

where d_1 is the upstream water depth, V_1 is the upstream flow velocity and $Fr_1 = V_1/\sqrt{gd_1}$ (Chanson and Brattberg 2000). For dye and salt injection at the jump toe, complete vertical and transverse mixing was very rapid implying a transverse mixing coefficient estimate:[6]

$$\frac{\varepsilon_t}{V_1 d_1} \approx 0.14 \qquad 5.9 < Fr_1 < 7.7$$

In the Ord River, transverse sediment diffusivity ε_t was estimated to be about $0.71\,\text{m}^2/\text{s}$ (Table 7C.1). For comparison, measured transverse diffusivities were about 0.014–$0.02\,\text{m}^2/\text{s}$ in the Severn River that has a similar water depth and possibly smaller width (Elliott *et al.*'s work, in Lewis 1997).

Table 7C.1 Experimental observations of lateral mixing in hydraulic jumps and bores

Experimental data	Fr_1 (m/s)	d_1 (m)	$\dfrac{\varepsilon_v}{V_1 d_1}$	$\dfrac{\varepsilon_t}{V_1 d_1}$	Remarks
(1)	(2)	(3)	(4)[a]	(5)[a]	
Chanson and	5.01	0.0158	1.5×10^{-2}	–	$W = 0.25\,\text{m}$. Air bubble
Brattberg (2000)	5.67	0.0158	6.2×10^{-2}	–	entrainment at jump toe
	6.05	0.017	6.1×10^{-2}	–	
	6.32	0.014	5.0×10^{-2}	–	
	8.03	0.0158	5.2×10^{-2}	–	
	8.11	0.0158	3.0×10^{-2}	–	
	8.48	0.014	4.5×10^{-2}	–	
Bhargava and	7.30	0.0070	–	0.222	$W = 0.3\,\text{m}$. Dye and
Ojha (1990)	7.40	0.0072	–	0.227	salt injection at jump
	6.37	0.0091	–	0.212	toe on centreline
	7.62	0.0109	–	0.123	
	8.11	0.0119	–	0.110	
	6.14	0.0150	–	0.105	
	5.90	0.0167	–	0.102	
	6.94	0.0162	–	0.083	
	6.42	0.0180	–	0.081	
Wolanski *et al.* (2001)	1.2 to 1.3	–	–	$\varepsilon_t = 0.71\,\text{m}^2/\text{s}^a$	Undular tidal bore of the Ord river (East branch) on 31 August 1999. $W = 390\,\text{m}$ (at mean still water level)

Notes: [a]Data re-analysis by the writer.

[6]Re-analysis of Bhargava and Ojha's (1990) data.

Remarks
1. A positive surge results from a sudden change in flow that increases the depth.
2. When the surge is of tidal origin, it is usually termed a bore. The difference of name does not mean a difference in principle (Henderson 1966, Chanson 1999a).

7.9 Exercises

1. Considering a wide rectangular channel, the water depth is 3.2 m and the flow rate is 2.8 m^2/s. Assuming a gravel bed (k_s = 12 mm), calculate the boundary shear stress and the shear velocity.
2. Water flows at uniform equilibrium in a 4.5 m wide, concrete-lined rectangular channel. The observed water depth is 0.85 m and the bed slope is 0.0011. Calculate the flow rate, the shear velocity and the vertical mixing coefficient.
3. A natural stream discharges 4.9 m^3/s at equilibrium down a sandy bed slope (k_s = 2 mm, slope: 0.0004). The channel width is 27 m. (a) Calculate the water depth and the shear velocity. (b) Estimate the vertical and transverse mixing coefficient.
4. A sewage plant releases 250 m^3/day of effluents containing 195 ppm of a chemical near the right bank of a wide, slowly meandering stream. The creek is 2.4 m deep and the bed slope is 0.000 11. Calculate the flow rate and the shear velocity. Assuming that the effluent is completely mixed over the vertical, determine the width of the plume, the maximum concentration 1200 m downstream of the discharge point and its location. *Assume uniform equilibrium flow conditions in the river (k$_s$ = 5 mm).*
5. A barrel of arsenic (2.1 tonnes) falls accidentally from a trailer crossing a 1 km long wide river. The river flows at 0.12 m/s and the water depth is 4.6 m (it is assumed that the container falls on the river centreline). Calculate the contaminant concentration 500 m downstream of the injection point: (a) on the centreline, 1 h after the accident, (b) on the centreline, 1 h 20 min after the accident and (c) 100 m from the centreline, 1 h 20 min after the accident. *Assume V$_*$ = 0.0062 m/s. 1 tonne = 1 metric ton = 1000 kg.*
6. Generate a random walk model using a spreadsheet (Appendix B). Assuming V = 1.2, D_x = 0.8 and δt = 1, study the dispersion of 100 particles in a one-dimensional flow. At a time T = 90, calculate the probability distribution function of the tracer concentration. (Perform the random walk computation using δt = 1. Plot the PDF using bins of Δx = 0.2.)
7. A mass slug of chemical (15 kg) is released accidentally in a natural stream on the channel centreline when a palette plunges into the creek at a crossing. The natural river has the following channel characteristics during the event: water depth: 0.80 m, width: 72 m, bed slope: 0.0003, short grass: k_s = 3 mm. *Assume uniform equilibrium flow conditions in a rectangular channel and assume a slowly meandering stream.*

    Calculate basic mixing and dispersion coefficients of the natural riverine system, and the length of the initial zone.

7.10 Exercise solutions

1. V_* = 0.043 m/s.
2. Q = 6.14 m^3/s, V_* = 0.082 m/s, ε_v = 0.0046 m^2/s.

3. (a) $d = 0.29\,\text{m}$, $V_* = 0.033\,\text{m/s}$.
   (b) $\varepsilon_v = 6.4 \times 10^{-4}\,\text{m}^2/\text{s}$, $\varepsilon_t = 1.4 \times 10^{-3}\,\text{m}^2/\text{s}$ (straight channel) or $\varepsilon_t = 5.7 \times 10^{-3}\,\text{m}^2/\text{s}$
   (slowly meandering channel).
4. $q = 2.6\,\text{m}^2/\text{s}$, $V = 1.09\,\text{m/s}$, $V_* = 0.050\,\text{m/s}$, $\varepsilon_v = 8.05 \times 10^{-3}\,\text{m}^2/\text{s}$, $\varepsilon_t = 7.21 \times 10^{-2}\,\text{m}^2/\text{s}$
   (slowly meandering channel).

   The half-plume width is calculated based upon the method of images, whereby the bank is a line of symmetry. Using equation (7.7), the virtual mass discharge of sewer is: $2 \times 250\,\text{m}^3/\text{day} \times 195\,\text{ppm} = 0.001\,13\,\text{kg/s}$.

   Maximum contaminant concentration occurs on the virtual centreline, that is on the right bank where the effluent is released: $C_{\max}(x = 1200\,\text{m}) = 1.37 \times 10^{-5}\,\text{kg/m}^3$.

   The width of the plume (i.e. measured from the right bank) is half of the width of the virtual plume released on the centreline: i.e. $2\sigma = 2\sqrt{2\varepsilon_t x / V} = 252\,\text{m}$.
5. Using equation (7.9):
   (a) $C_m = 4.2 \times 10^{-9}\,\text{kg/m}^3$ at $x = 500\,\text{m}$, $y = 0$ and $t = 1\,\text{h}$.
   (b) $C_m = 1 \times 10^{-8}\,\text{kg/m}^3$ at $x = 500\,\text{m}$, $y = 0$ and $t = 1\,\text{h}\,20\,\text{min}$.
   (c) $C_m = 6.2 \times 10^{-22}\,\text{kg/m}^3$ at $x = 500\,\text{m}$, $y = 100\,\text{m}$ and $t = 1\,\text{day}$.
6. Using a spreadsheet, set one row per particle. Apply equation (7B.3) to 100 discrete particles. For $T = 90$, perform an histogram analysis. Note that the location of centroid is $X = 90 \times 1.2 = 108$.

   For a Gaussian probability distribution function, the square of the standard deviation equals:

   $$\sigma_x^2 = 2D_x T = 2 \times 0.8 \times 90 = 144 \qquad \text{equation (7B.1)}$$

   and the normal (Gaussian) probability distribution function satisfies:

   $$\text{Probability}(x, T) \times \delta x = \frac{1}{\sqrt{2\pi 144}} \exp\left(-\frac{(x - 108)^2}{2 \times 144} \right) \qquad \text{equation (7B.2)}$$

7. At uniform equilibrium flow: $d = 0.792\,\text{m}$, $V = 0.98\,\text{m/s}$, $V_* = 0.048\,\text{m/s}$.
   $\varepsilon_v = 0.003\,\text{m}^2/\text{s}$, $\varepsilon_t = 0.023\,\text{m}^2/\text{s}$ (slowly meandering channel), $K = 1405\,\text{m}^2/\text{s}$ (natural river).
   Length of initial zone: $22\,\text{km}$ (centreline discharge).

8

Turbulent dispersion and mixing: 2. Longitudinal dispersion

Summary

In this chapter, Taylor's dispersion theory is introduced. Then the dispersion equation is applied to natural river systems. Basic applications are discussed.

8.1 Introduction

Shear flow dispersion is the longitudinal stretching of matter caused by velocity shear. For example, in a circular pipe, the velocity is larger on the centreline than near the wall. As a result, the longitudinal spread of matter is faster on the centreline, leading to greater separation of the effluent particles than by molecular diffusion. The rate of separation is caused by the difference in advective velocity. Turbulent shear dispersion is also called longitudinal dispersion. The basic theory of shear dispersion is derived from Taylor's (1953, 1954) work.

Considering a tracer released from an instantaneous line source, the contaminant travels more slowly near the channel banks than in the centreline (Fig. 8.1). As a result, the initial line source is stretched longitudinally and spreading occurs both along and across the channel as illustrated in Fig. 8.1. The rate of longitudinal dispersion reflects the balance between velocity shear which acts to spread tracer along the channel and transverse mixing which promotes uniform concentrations across the channel and hence counteracts the effects of velocity shear (Rutherford 1994, pp. 178–179).

Discussion

Taylor's theory is based upon the assumptions that turbulent mixing processes occur in a similar fashion as molecular diffusion, and it is valid only away from the injection point: i.e. downstream of the initial zone (see paragraph 8.2 and Fig. 7.1). Further most solutions of the dispersion equation imply that there are constant mixing conditions, uniform velocity distribution and idealized reflection at the channel boundaries. This monograph is no exception.

> *Longitudinal dispersion and transverse mixing are strongly interrelated.* Longitudinal dispersion is a delicate balance between velocity shear spreading tracer longitudinally versus transverse mixing promoting uniform transverse concentration. It will be shown that the longitudinal dispersion coefficient K is inversely proportional to the transverse mixing coefficient ε_t: i.e. $K \propto 1/\varepsilon_t$.
>
> ### Note
> Sir Geoffrey Ingram Taylor (1886–1975) was a British fluid dynamicist professor based in Cambridge. He established the basic developments of shear dispersion (Taylor 1953, 1954).

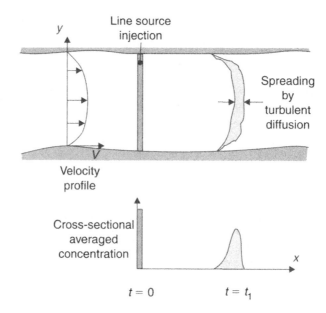

Fig. 8.1 Sketch of longitudinal dispersion of a line source injection by vertical and transverse velocity shear in a river.

8.2 One-dimensional turbulent dispersion

For one-dimensional flows, the *dispersion equation* is:

$$\frac{\partial C_{\mathrm{m}}}{\partial t} + V \frac{\partial C_{\mathrm{m}}}{\partial x} = K \frac{\partial^2 C_{\mathrm{m}}}{\partial x^2} \qquad (8.1)$$

where C_{m} and V are the *mean* concentration and velocity respectively, and K is the *dispersion coefficient* (unit: m²/s). Equation (8.1) implies that turbulent mixing and longitudinal dispersion take place in a similar fashion as molecular advective diffusion. Further most solutions assume constant mixing conditions (K constant), uniform velocity distribution (V constant) and idealized reflection of mass at channel boundaries.[1] Note that equation (8.1) is valid only away from the injection point.[2] That is, downstream of the initial zone.

The dispersion coefficient K is a function of the flow conditions (e.g. velocity profile). For turbulent flows in circular pipes, K equals:

$$K = 5.05\,DV_* \qquad\qquad \text{Turbulent pipe flow} \qquad\qquad (8.2)$$

where V_* is the shear velocity and D is the pipe diameter.

For turbulent open channel flows, the dispersion coefficient is:

$$K = 5.93\,dV_* \qquad\qquad \text{Turbulent open channel flow} \qquad (8.3)$$

where d is the flow depth. The result is based upon the logarithmic velocity distribution (Chapter 4, paragraph 4.3).

Equation (8.2) was first developed by Taylor (1954) while equation (8.3) was developed by Elder (1959). In any case the dispersion coefficient is proportional to a characteristic length scale times the shear velocity, V_*, which is a measure of shear stress and velocity gradient near the boundary (Chapter 7, paragraph 7.2). In the next section, however, it is shown that equations (8.2) and (8.3) are not appropriate for natural river systems.

8.3 Longitudinal dispersion in natural streams

8.3.1 Basic equation

After a contaminant is well mixed across the entire cross-section, the final stage of the mixing process is the reduction of longitudinal gradients by dispersion. However, if the effluent discharge is a constant, there is no need to be concerned by dispersion. Longitudinal dispersion is important in applications characterized by non-constant effluent releases: e.g. accidental spill a quantity of pollutant, daily variations of sewage effluents released by a water treatment plant.

The advection–dispersion equation for one-dimensional flow is:

$$\frac{\partial C_m}{\partial t} + V\frac{\partial C_m}{\partial x} = K\frac{\partial^2 C_m}{\partial x^2} \qquad\qquad (8.1)$$

Considering the instantaneous slug release (mass M, side discharge) in a one-dimensional flow, the solution of the dispersion equation (8.1) is a Gaussian distribution:

$$C_m = \frac{M}{A\sqrt{4\pi Kt}}\exp\left(-\frac{(x-Vt)^2}{4Kt}\right) \qquad x' > 0.4 \qquad\qquad (8.4)$$

where A is the channel cross-section, t is the time of travel (or elapsed time since tracer release) and V is the flow velocity.

---

[1] See Chapter 9 for further discussion on dead zones.

[2] For $t \gg 0.4\frac{D^2}{K}$, where D is the characteristic length scale (e.g. pipe diameter) and K is the dispersion coefficient.

Discussion

When applying Taylor's dispersion analysis to natural rivers, one must understand that the advective dispersion equation (8.1) is a *one-dimensional* analysis and *it does not apply to the initial zone* (Fig. 8.1). The initial zone is characterized by vertical and transverse mixing. Basically the dispersion equation (8.1) does not apply for $x' = x\varepsilon_t/(VW^2) < 0.4$ with a side discharge (Chapter 7, paragraph 7.4.3).

For $0.4 < x' < 1$, the dispersion equation is valid and the longitudinal concentration distribution is skewed, decaying towards a Gaussian distribution. For $x' > 1$, the solution of the dispersion equation is a Gaussian distribution (equation (8.4)). In both the cases, the *virtual origin* of the chemical cloud is approximately (Fischer *et al.* p. 136, Fig. 5.14):

$$x' = \frac{x\varepsilon_t}{VW^2} = 0.07 \qquad \text{Virtual origin} \qquad (8.5)$$

In the case of a slug of contaminants (mass M) released suddenly at the origin into a natural stream, the longitudinal length of the cloud after the initial mixing may be estimated as four times the standard deviation $\sigma = \sqrt{2Kt}$. That is:

$$\text{Length of cloud} \sim 4\sigma = 4\sqrt{2K\frac{x - 0.07(VW^2/\varepsilon_t)}{V}} \qquad x' > 0.4^{(3)} \qquad (8.6)$$

The peak concentration within the dispersed cloud equals:

$$C_{\max} = \frac{M}{A\sqrt{4\pi K\dfrac{x}{V}}} \qquad x' > 0.4 \qquad (8.7)$$

Comments

1. Many natural streams have bends, sandbars, sidepools and other natural changes, and every irregularity in the channel contributes to longitudinal dispersion. Some channels may be so irregular that no reasonable approximation of dispersion is possible: e.g. a mountain stream consisting of pools and riffles (see Chapter 9, paragraph 9.2).
2. Sometimes the virtual origin of the tracer cloud may be neglected for calculations performed far downstream ($x' \gg 0.07$). This is often the case, like in the 'frozen cloud' approximation method (paragraph 8.4).

8.3.2 Dispersion coefficient in natural rivers

In natural rivers, the coefficient of dispersion was found to be about: $K/(dV_*) \sim 9\text{--}7500$ (Fischer *et al.* 1979, pp. 125–127). Such values are well in excess of Elder's (1959) estimate (equation (8.3)). Natural rivers are characterized by transverse variations of the velocity which enhance dispersion. Figure 8.2 shows a typical cross-sectional velocity distribution in a stream.

---

[3] For a side discharge.

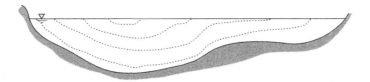

Fig. 8.2 Cross-sectional velocity distribution in a natural stream: lines of constant velocity magnitude.

Experimental measurements suggest that the coefficient of dispersion in real rivers may be estimated as:

$$K = 0.011 \frac{V^2 W^2}{d V_*} \tag{8.8}$$

The above equation agrees with observations *within a factor of 4.*

Discussion

It is uppermost important to understand that *longitudinal dispersion and transverse mixing are closely related.* This is illustrated by combining equation (7.5) (or equation (7.6)) with equation (8.8) which yields:

$$K \propto \frac{1}{\varepsilon_t}$$

Rutherford (1994) and Deng *et al.* (2001) developed this important issue in more details.

Further estimates of dispersion coefficient (e.g. equation (8.8)) were developed for quasi-uniform equilibrium flow conditions. They may be applied to gradually varied flows but they should *not* be used in rapidly varied flows (e.g. hydraulic jumps).

Notes

1. Rutherford (1994, pp. 193–204) discussed specifically the uncertainty of the empirical formulae for the dispersion coefficient. He further reviewed a large number of field data and he proposed a range:

$$2 < \frac{K}{W V_*} < 50$$

   In practice, the dispersion coefficient must be measured with field tests.
2. Most studies assume a constant dispersion coefficient K, but this is not always correct. Hunt (1999) re-analysed dispersion data for mountain streams. He showed an increase in the dispersion coefficient with distance downstream, caused by relatively large dispersion from velocity shear near the leading and trailing edges of the tracer cloud:

$$K \propto V x \qquad \text{Mountain streams}$$

   where the coefficient of proportionality must be estimated experimentally.

3. Deng *et al.* (2001) re-analysed a number of field observations for straight, natural rivers. They proposed an estimate for the longitudinal dispersion coefficient:

$$K = \frac{0.15}{8E_t} \frac{W^{5/3}V^2}{d^{2/3}V_*} \qquad \text{Straight rivers}$$

where E_t is basically a dimensionless transverse mixing coefficient:

$$E_t = 0.145 + \frac{1}{3520} \frac{V}{V_*} \left(\frac{W}{d}\right)^{1.38}$$

Applications

Application No. 1

Considering the dispersion of a slug of chemical injected at the side of a stream, the river flows at uniform equilibrium. The channel is 50 m wide, it may be assumed rectangular and the river bed is made of small gravels ($k_s = 5$ mm). The water depth is 2.4 m and the bed slope is 0.0003.

(a) Estimate the longitudinal dispersion coefficient.
(b) Calculate the channel length required to provide complete transverse mixing. (Note that Taylor's dispersion analysis does not apply in the initial zone characterized by transverse mixing.)
(c) Estimate the peak concentration that will be observed 38 000 m downstream of the injection point, and the length of the chemical cloud at the time when the peak passes that point. (Assume a contaminant mass of 10 kg.)

Solution

Firstly the hydraulic characteristics of the river flow must be calculated using the momentum principle. The process is iterative (Henderson 1966, Chanson 1999a). For $k_s = 5$ mm, it yields: $V = 1.73$ m/s, $Q = 207$ m³/s and $V_* = 0.080$ m/s.

(a) The longitudinal dispersion coefficient equals:

$$K = 0.011 \times \frac{1.73^2 \times 50^2}{2.4 \times 0.080} = 429 \text{ m}^2/\text{s} \qquad \text{(equation (8.8))}$$

(b) A reasonable distance for complete mixing from a side discharge is: $L \sim 0.4VW^2/\varepsilon_t$, or $x' = 0.4$. In the absence of further information, the transverse mixing coefficient is estimated as for a slowly meandering stream with some wall roughness: $\varepsilon_t = 0.6dV_* = 0.6 \times 2.4 \times 0.080 = 0.1152$ m²/s (equation (7.6)). It yields:

$$L = 0.4 \times 1.73 \times 50^2/0.1152 = 15\,000 \text{ m}$$

(c) At the observation station where $x = 38\,000$ m, $x' = x\varepsilon_t/(VW^2) = 38\,000 \times 0.1152/(1.73 \times 50^2) = 1.01$. We may assume the longitudinal concentration distribution to be Gaussian. The length of the real cloud may be estimated by assuming that the cloud started with zero variance at $x' = 0.07$ (i.e. $x = 2600$ m).

The variance of the cloud at the observation station is basically a linear function of time for $x' > 0.4$. It yields: $\sigma^2 = 2Kt = 2K(38\,000 - 2600)/1.73 = 17.6 \times 10^6\,\mathrm{m}^2$.

For a Gaussian distribution, the length of the cloud may be approximately estimated as 4σ:

$$\text{Length of cloud: } 4\sigma = 16\,800\,\mathrm{m} \qquad \text{Equation (8.6)}$$

The peak concentration is:

$$C_\mathrm{m} = \frac{M}{A\sqrt{4\pi K \dfrac{x}{V}}} = \frac{10}{50 \times 2.4 \times \sqrt{4 \times 3.14 \times 429 \times \dfrac{38\,000}{1.73}}} = 7.7 \times 10^{-6}\,\mathrm{kg/m}^3$$

Equation (8.7)

Application No. 2

Rhodamine WT dye is released as a slug into the Bremer River to estimate the longitudinal dispersion coefficient. The dye cloud is monitored at four locations. At the first two, the cloud was poorly mixed and not amenable to a simple analysis. At the third location (located 3800 m downstream of the release point), a dye concentration $C_\mathrm{m} = 0.0004\,\mathrm{kg/m}^3$ was measured 45 min after the initial release. The peak dye concentration $C_{\max} = 0.0025\,\mathrm{kg/m}^3$ passed the fourth measuring station, located 2300 m downstream of the release point, 39 min after the initial release.

Calculate the longitudinal dispersion coefficient.

Solution

Calculations are conducted between the third and fourth measurement locations, assuming that the dye is well mixed and that there is no decay of rhodamine (source/sink). All the calculations derive from:

$$C_\mathrm{m} = \frac{M}{A\sqrt{4\pi Kt}} \exp\left(-\frac{(x - Vt)^2}{4Kt}\right) \qquad (8.4)$$

At the fourth measuring station, maximum concentration is obtained for $(x - Vt) = 0$. That is, the average velocity of the flow is: $V = 2300/(39 \times 60) = 0.98\,\mathrm{m/s}$.

At the third measuring station, equation (8.4) yields:

$$0.0004 = \frac{M}{A\sqrt{4\pi K \times 2700}} \exp\left(-\frac{(3800 - 0.98 \times 2700)^2}{4K \times 2700}\right)$$

At the fourth measuring station, equation (8.4) yields:

$$0.0025 = \frac{M}{A\sqrt{4\pi K \times 2340}}$$

It yields: $K = 70\,\mathrm{m}^2/\mathrm{s}$.

Note that the peak concentration for $t = 45$ min satisfies:

$$\frac{M}{A\sqrt{4\pi K \times 2700}} = 0.0025\sqrt{\frac{39}{45}} = 0.00233\,\mathrm{kg/m}^3$$

8.4 Approximate models for longitudinal dispersion

8.4.1 The 'frozen cloud' approximation

In many applications, advection dominates dispersion (i.e. $x \gg K/V$) and maximum tracer concentration occurs at $t \approx x/V$. The concept of *frozen cloud* assumes that no longitudinal dispersion occurs during the time taken for the tracer to pass a sample site. Although not strictly correct in natural streams, the frozen cloud approximation is commonly made when analysing longitudinal dispersion data.

With that approximation, hypothetical concentration versus time prediction *at a fixed point* x may be estimated as:

$$C_m(t) = C_{max} \exp\left(-\frac{(x - Vt)^2}{4K\dfrac{x}{V}}\right) \qquad x' > 1 \tag{8.9}$$

where x is fixed (constant) and C_{max} is estimated from equation (8.7). Equation (8.9) is derived from equation (8.4) for the frozen cloud approximation.

The prediction of concentration versus distance at a given time t may be estimated as:

$$C_m(x) = \frac{M}{A\sqrt{4\pi K\left(2t - \dfrac{x}{V}\right)}} \exp\left(-\frac{(x - Vt)^2}{4K\left(2t - \dfrac{x}{V}\right)}\right) \tag{8.10}$$

where t is fixed (constant). Note that equation (8.10) derives from equation (8.4).

> **DISCUSSION**
>
> Let us consider a tracer cloud such that the centroid passes a location $x = x_1$ at $t = T_1$. The frozen cloud assumption implies: $x_1 = VT_1$. Equation (8.4) expresses the relationship: $C_m = C_m(x, t)$. Under the frozen cloud approximation, it may be transformed to yield:
>
> $$C_m(t) = C_m(x_1 + V(T_1 - t); T_1) = \frac{M}{A\sqrt{4\pi KT_1}} \exp\left(-\frac{(x_1 - Vt)^2}{4KT_1}\right)$$
>
> Concentration versus time at a fixed location x_1 (equation (8.9))
>
> $$C_m(x) = C_m\left(x_1; T_1 + \frac{x_1 - x}{V}\right) = \frac{M}{A\sqrt{4\pi K\left(2T_1 - \dfrac{x}{V}\right)}} \exp\left(-\frac{(x - x_1)^2}{4K\left(2T_1 - \dfrac{x}{V}\right)}\right)$$
>
> Concentration versus distance at a fixed time T_1 (equation (8.10))
>
> Rutherford (1994, pp. 209–215) discussed the limitations of the frozen cloud assumption. In any case, equations (8.9) and (8.10) may apply only downstream of the initial region ($x' \gg 0.07$), assuming constant dispersion coefficient K and uniform velocity V.

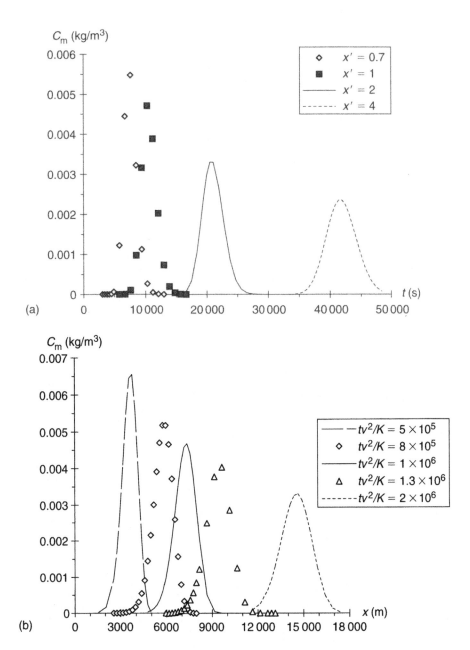

Fig. 8.3 Solution of the dispersion equation (8.1) using the 'frozen cloud' approximation. (a) Concentration versus time measured at a fixed point (equation (8.9)): $M = 100$, $d = 2$, $W = 25$, $V = 0.5$, $V_* = 0.05$. (b) Concentration versus distance measured at a fixed time (equation (8.10)): $M = 100$, $d = 2$, $W = 25$, $V = 0.5$, $V_* = 0.05$.

In Fig. 8.3, both cases are considered. Figure 8.3(a) shows the tracer concentration versus time at a fixed location x. Maximum concentration occurs for $t = x/V$. Figure 8.3(b) presents the concentration versus distance at a given time t. Maximum concentration takes place at $x = Vt$.

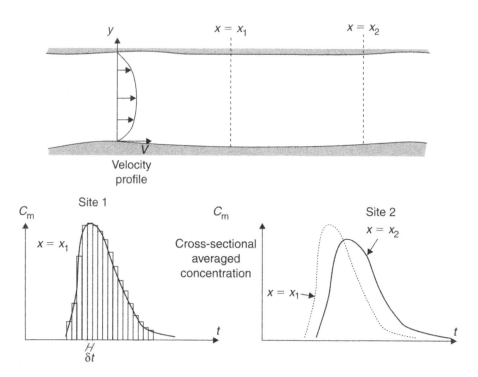

Fig. 8.4 Application of the 'frozen cloud' approximation: predicting the concentration versus time distribution at a downstream location.

Application

If the concentration versus time distribution (i.e. temporal concentration profile) is known at Site 1, the concentration versus time at a downstream Site 2 may be predicted using the frozen cloud approximation and an extension of equation (8.9). The mean velocity in the reach between Sites 1 and 2 is estimated as:

$$V = \frac{x_2 - x_1}{T_2 - T_1}$$

where T_1 and T_2 are the mean times of passage (i.e. time of passage of centroid) at Sites 1 and 2, respectively.

The problem is solved by applying the method of superposition for a succession of instantaneous mass slugs, injected at $x = x_1$, and applying the frozen cloud approximation. Each slug has a mass $\delta M = C_m Q \delta t$, where C_m is the time-averaged concentration at $x = x_1$ during the time interval δt and Q is the flow rate (Fig. 8.4).

Such a prediction method is valid only if *both* Sites 1 and 2 are downstream of the injection point and outside of the initial zone (Fig. 7.1), the entire concentration versus time profile is recorded at Site 1, and tracer loss is negligible between Sites 1 and 2. The values of V and K, deduced from the method, are *averages* between Sites 1 and 2.

8.4.2 Discussion: the Hayami solution

When temporal concentration profiles are measured at two different locations (e.g. Applications, paragraph 8.4.1), the frozen cloud approximation may be applied, provided that the measurement sites are outside of the initial zone, that the entire concentration versus time distributions are recorded and that the tracer loss is negligible between the two sites. The measurement outcomes are reach-averaged values of V and K.

The dispersion equation (8.1) has another solution in addition to equation (8.4). It is called the Hayami solution (Barnett 1983):

$$C_m = \frac{Mx}{AVt\sqrt{4\pi Kt}} \exp\left(-\frac{(x - Vt)^2}{4Kt}\right) \qquad x' > 0.4 \qquad (8.11)$$

Equation (8.11) can be used to predict the downstream propagation of a tracer cloud without needing to invoke the frozen cloud approximation.

DISCUSSION
Equation (8.11) is simply equation (8.4) multiplied by $x/(Vt)$. It satisfies the boundary condition that $C_m(x = 0, t \neq 0) = 0$, for all values but $t = 0$ when an instantaneous slug of mass M is introduced at the origin. Rutherford (1994, pp. 214–215) presented a comparison between the Taylor and Hayami solutions (i.e. equations (8.4) and (8.11)).

8.5 Design applications

8.5.1 Application No. 1

Investigations are undertaken for a scheme to abstract drinking water from the Logan River to supply Cedar Grove. One important design consideration is the time that it would take for a pollutant accidentally discharged into the river (e.g. following a rail/road tanker accident) to reach the intake site and the length of the time that concentrations would be above some critical value. To assist the design of off-river storage, calculations are based upon the following assumptions.

$Q = 250\,\text{m}^3/\text{s}$ Average width: 45 m   Average bed slope: 0.0009   Gravel bed: $k_s = 25\,\text{mm}$

Consider the case of road bridge 35 km upstream of the abstraction site and the instantaneous injection of 250 kg of a chemical which is likely to cause water treatment problems at concentrations exceeding $5\,\text{mg/m}^3$.

Solution
Hydraulic calculations at uniform equilibrium flow conditions yield:

$$d = 2.15\,\text{m} \quad V_* = 0.132\,\text{m/s}$$

The mean travel time is:

$$t_{\text{mean}} = 13565\,\text{s}\ (3.8\,\text{h}) \qquad (\text{equation } (8.4))$$

Transverse mixing coefficient:

$$\varepsilon_t = 0.17\,\text{m}^2/\text{s} \qquad (\text{equation } (8.6))$$

Longitudinal dispersion coefficient:

$$K = 523\,\text{m}^2/\text{s} \qquad (\text{equation } (8.8))$$

Dimensionless distance:

$$x' = 1.14$$

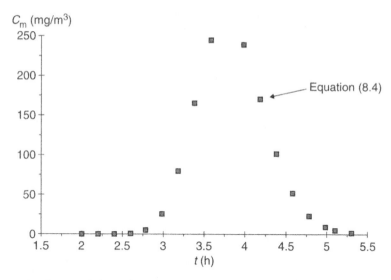

Fig. 8.5 Cross-sectional averaged chemical concentrations predicted at the proposed intake ($x = 35\,\text{km}$).

Note that, although the true longitudinal concentration distribution is skewed for $x' < 1$, the shape of the cloud is approaching the Gaussian distribution for $x' > 0.7$. For $x' > 1$, the frozen cloud approximation may be used (Fig. 8.3).

Length of complete mixing:

$$L = 12.3\,\text{km} \qquad \text{assuming side discharge (Section 7.4.2)}$$

Peak concentration at proposed intake:

$$C_{max} = 2.74 \times 10^{-4}\,\text{kg/m}^3\ (274\,\text{mg/m}^3) \qquad \text{(equation (8.7))}$$

The peak concentration is above the acceptable limit by a factor of 55.

Applying equation (8.9), the concentration at the intake reaches $5\,\text{mg/m}^3$ about 2.8 h after injection. It will drop below this limit about 5.1 h after injection (Fig. 8.5).

8.5.2 Application No. 2

Norman Creek, in Southern Brisbane, has the following channel characteristics during a flood event:

   Water depth: 1.16 m   Width: 55 m   Bed slope: 0.002   Short grass: $k_s = 3\,\text{mm}$

A mass slug of chemical (7.5 kg) is released accidentally in the natural stream on the channel centreline (when a truck plunge into the creek at a crossing):

(a) Calculate the hydraulic characteristics of the stream in flood.
(b) Predict the dispersion coefficient and the length of the initial zone.

A measurement station is located 25 km downstream of the injection point.

(c) Calculate the maximum tracer concentration at that station and the arrival time.

(d) The detection limit of the chemical is $0.2\,\mathrm{mg/m^3}$. Calculate the length of time during which the chemical will be detectable at the measurement station.

Assume uniform equilibrium flow conditions in a rectangular channel. Neglect vertical mixing and assume a slowly meandering stream.

Solution

Uniform equilibrium flow calculations yield:

$$V = 3.13\,\mathrm{m/s} \quad V_* = 0.148\,\mathrm{m/s}$$

The dispersion coefficient equals:

$$K = 1890\,\mathrm{m^2/s} \qquad \text{(equation (8.8))}$$

The length of the initial zone equals: $9170\,\mathrm{m}$ (Chapter 7, paragraph 7.4.3).

Note that the measurement station is located outside of the initial zone. Hence the dispersion theory is valid.

The maximum tracer concentration at the measurement station equals: $8.5 \times 10^{-6}\,\mathrm{kg/m^3}$ (equation (8.7)) and it is observed $8000\,\mathrm{s}$ ($2\,\mathrm{h}\ 13\,\mathrm{min}$) after injection.

The length of the time during which the chemical is detectable of the measurement station equals: $7810\,\mathrm{s}$. That is, between $53\,\mathrm{min}$ and $3\,\mathrm{h}$ after injection (using a frozen cloud approximation).

DISCUSSION

Note the near-critical flow conditions ($Fr = 0.9$) which are likely to be characterized by free-surface undulations. The undular flow pattern may induce scour beneath wave troughs (Kennedy 1963, Chanson 2000), while secondary currents may enhance mixing and dispersion.

8.6 Exercises

1. A natural stream has the following channel characteristics: flow rate: $18.5\,\mathrm{m^3/s}$, water depth: $0.786\,\mathrm{m}$, width: $15\,\mathrm{m}$, bed slope: 0.001, gravel: $k_s = 5\,\mathrm{mm}$. *Assume uniform equilibrium flow conditions in a rectangular channel. Neglect vertical mixing and assume a slowly meandering stream.*

A barrel of dye ($2.2\,\mathrm{kg}$) is suddenly released in the natural stream from the channel bank (i.e. side). A measurement station is located $18\,\mathrm{km}$ downstream of the injection point. (a) Calculate the maximum mass concentration in the cross-section located at $500\,\mathrm{m}$ downstream of the injection point. (b) Calculate the maximum tracer concentration at that station and the arrival time. (c) The detection limit of the chemical is $0.3\,\mathrm{mg/m^3}$.

Calculate the length of time during which the chemical will be detectable at the measurement station.

2. The flood plain of Oxley Creek, in Brisbane, has the following channel characteristics during a flood event: water depth: 1.16 m, width: 55 m, bed slope: 0.0002, short grass: $k_s = 0.003$ m. *Assume uniform equilibrium flow conditions in a rectangular channel. Neglect vertical mixing. Assume a slowly meandering stream.*

During a test, a barrel of dye (7.5 kg) is released in the natural stream on the channel centreline (from a bridge). A measurement station is located 15 km downstream of the injection point. (a) Calculate the basic hydraulic, mixing and dispersion parameters. (b) Calculate the maximum tracer concentration at that station and the arrival time. (c) The detection limit of the chemical is 0.2 mg/m³. Calculate the length of time during which the chemical will be detectable at the measurement station.

3. Dye profiles were measured at in a natural stream. The observed water depth was about 0.45 m and the average channel breadth is 6.9 m. (The initial injection was quasi-instantaneous.) The table below shows cross-sectional averaged dye concentrations at two measurement sites located respectively 14 and 18.2 km downstream of the injection point. (a) Preliminary calculations suggested that the site locations were downstream of the initial zone. Is this correct? (b) Using the frozen cloud approximation, estimate the flow velocity and longitudinal dispersion coefficient between the two measurement sites.

Assume a conservative tracer. Divide the concentration versus time distribution at Site 1 into a series of four small mass slugs $\delta M(t)$ injected during a time interval $\delta t = 1$ h. The channel bed is made of small gravels ($k_s = 4$ mm).

Location (1)	Time (h) (2)	C_m (mg/m³) (3)	Location (4)	Time (h) (5)	C_m (mg/m³) (6)
Site 1	6	0.03	Site 2	8.75	0.08
Site 1	6.25	0.11	Site 2	9	0.14
Site 1	6.5	0.25	Site 2	9.25	0.24
Site 1	6.75	0.43	Site 2	9.5	0.34
Site 1	7	0.41	Site 2	9.75	0.33
Site 1	7.25	0.32	Site 2	10	0.29
Site 1	7.5	0.25	Site 2	10.25	0.25
Site 1	7.75	0.19	Site 2	10.5	0.18
Site 1	8	0.11	Site 2	10.75	0.13
Site 1	8.25	0.05	Site 2	11	0.08
Site 1	8.5	0.02	Site 2	11.25	0.06
Site 1	8.75	0.01	Site 2	11.5	0.05

4. During a flood, the river flow characteristics are: flow rate: 9.5 m³/s, width: 14 m, bed slope: 0.00015, coarse sand: $k_s = 5$ mm. A barrel of dye (2.8 kg) is suddenly released in the natural stream from the channel bank (i.e. side). A measurement station is located 12 km downstream of the injection point. *Assume uniform equilibrium flow conditions in a rectangular channel. Neglect vertical mixing and assume a slowly meandering stream.*

Calculate the maximum tracer concentration at that gauging station and the arrival time. The detection limit of the chemical is 0.2 mg/m³. Calculate the length of time during which the chemical will be detectable at the measurement station. Calculate the relationship of dye concentration versus time. *On the graph, add relevant comments to highlight the duration of dye detection at the measurement station.*

5. A chemical is accidentally released in a natural stream between 2:30 a.m. and 3:00 a.m. at a rate of 12 kg/h for the first 15 min and later 8 kg/h for the next 15 min. Calculate the period during which concentration exceeds 0.1 mg/m^3 as well as the maximum concentration, at a site located 11 km downstream of the source. Plot the chemical concentration versus time on the same graph. The river characteristics are: $Q = 2.3$ m^3/s, width: 9.5 m, bed slope: 0.0003, small gravel bed: $k_s = 15$ mm.

Present all your results on a graph. *Add relevant comments to highlight your results. Use a graph with the horizontal axis being the clock time (00:00 equals midnight) Approximate the chemical release by two mass slugs released at 2:42:30 a.m. and 2:57:30 a.m. and apply the method of superposition.*

8.7 Exercise solutions

1. (a) The maximum mass concentration in the cross-section located at 500 m downstream of the injection point equals: 3.5×10^{-2} kg/m^3 using the principle of superposition and method of images.

(b and c) The calculations must start with the following estimates: transverse mixing coefficient, dispersion coefficient, initial length, maximum tracer concentration at the measurement station and length of the time during which the chemical is detectable at the measurement station.

$\varepsilon_t = 0.039$ m^2/s

$K = 9.8$ m^2/s

$C_{\max} = 5.1 \times 10^{-5}$ kg/m^3

Duration: 1 h 40 min

2. $V = \sqrt{\dfrac{8g}{f}} \sqrt{\dfrac{D_H}{4}} \sin\theta$ $\qquad\qquad$ $V = 0.99$ m/s

$\tau_o = \dfrac{f}{8} \rho V^2$ $\qquad\qquad$ $\tau_o = 2.2$ Pa

$V_* = \sqrt{\dfrac{\tau_o}{\rho}}$ $\qquad\qquad$ $V_* = 0.047$ m/s

$\varepsilon_v = 0.067 \, dV_*$

$\varepsilon_t \approx 0.6 \, dV_*$ Slowly meandering channels $\qquad$ $\varepsilon_t = 0.033$ m^2/s

$K = 0.011 \dfrac{V^2 W^2}{dV_*}$ $\qquad\qquad$ $K = 600$ m^2/s

$L \sim 0.1 \dfrac{V W^2}{\varepsilon_t}$ Centreline discharge $\qquad$ $L = 9.1$ km

$x = 15$ km (dispersion zone)

$T = \dfrac{x}{V}$ (time of passage of centroid) $\qquad$ $T = 4$ h 13 min

$$C_m = \frac{M}{A\sqrt{4\pi K \dfrac{x}{V}}} \quad \text{(peak)} \qquad\qquad C_{max} = 1.1 \times 10^{-5} \, \text{kg/m}^3$$

$C_m > 0.2 \, \text{mg/m}^3$ for $2990 \, \text{s}$ (50 min) $< t < 27400 \, \text{s}$ (7 h 40 min).

3. (a) First the temporal moment must be computed at Sites 1 and 2:

	Time of passage of centroid (s)	Velocity (m/s)
Site 1	25 580	–
Site 2	34 200	0.415

For that flow velocity, the shear velocity is: $V_* = 0.024 \, \text{m/s}$. Hence:

$$\varepsilon_v = 0.0007 \, \text{m}^2/\text{s}$$
$$\varepsilon_t = 0.0063 \, \text{m}^2/\text{s}$$

For a side discharge, the length of the injection zone is: 1.26 km. The measurement Sites 1 and 2 are hence in the dispersion zone and the frozen cloud approximation may be applied.

Secondly, the concentration versus time distribution at Site 1 is divided into a series of four small mass slugs $\delta M(t)$ injected during a time interval δt such as:

$$C_m(t) = \frac{\delta M(t)}{A V \delta t}$$

where A is the flow cross-sectional area.

Time (s) (1)	$\delta M/A$ (kg/m^2) (2)
22 838	1.4×10^{-4}
25 313	4.3×10^{-4}
27 788	2.0×10^{-4}
30 263	3.0×10^{-5}

Note that the mass slugs must be selected to satisfy the conservation of mass for the contaminants.

Then the concentration versus time distribution at Site 2 is predicted by superposition of all small mass slugs injected at Site 1.

Note that both Sites 1 and 2 are far downstream of the initial zone. Hence the longitudinal dispersion of small mass slugs between Sites 1 and 2 is not affected by the initial zone.
The principle of superposition is used for the 'frozen cloud' approximation.

Using the 'frozen cloud' approximation, the solution of the longitudinal dispersion equation is:

$$C_m(t) = \frac{\delta M}{A\sqrt{4\pi K T_1}}\exp\left(-\frac{(x_1 - Vt)^2}{4 K T_1}\right)$$ Concentration versus time at fixed x_1

(equation (8.9))

where $x_1 = 18\,200 - 14\,000 = 4200\,\text{m}$, $T_1 = 32\,950 - 22\,838 = 10\,112\,\text{s}$, $V = 0.415\,\text{m/s}$.

The concentration at Site 2 is the superposition of the four mass slugs injected at Site 1. The dispersion coefficient is deduced from the best fit of the data with the prediction. Using four mass slugs, best fit is achieved for $15 < K < 20\,\text{m}^2/\text{s}$.
  Note however that the result does not agree well with equation (8.8).

4. At uniform equilibrium: $d = 0.99\,\text{m}$, $V = 0.69\,\text{m/s}$, $V_* = 0.036\,\text{m/s}$, $\varepsilon_t = 0.021\,\text{m}^2/\text{s}$ (slowly meandering channel) and $K = 28.8\,\text{m}^2/\text{s}$ (natural river).

Length of initial zone: 2.5 km (side discharge)

$x = 12\,\text{km}$ (dispersion zone).

$T = \dfrac{x}{V}$ (time of passage of centroid) $T = 17\,500\,\text{s}$

$$C_{max} = \frac{M}{A\sqrt{4\pi K \dfrac{x}{V}}} \quad \text{(peak)} \qquad C_{max} = 8.04 \times 10^{-5}\,\text{kg/m}^3$$

$C_m > 0.2\,\text{mg/m}^3$ for $2990\,\text{s} < t < 27\,400\,\text{s}$
The relationship between dye concentration and time since injection is shown in Fig. 8.6.

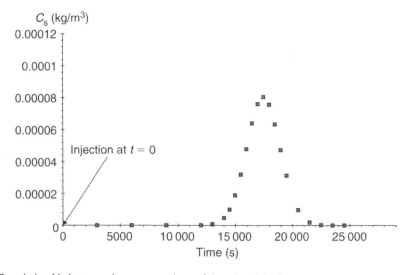

Fig. 8.6 The relationship between dye concentration and time since injection.

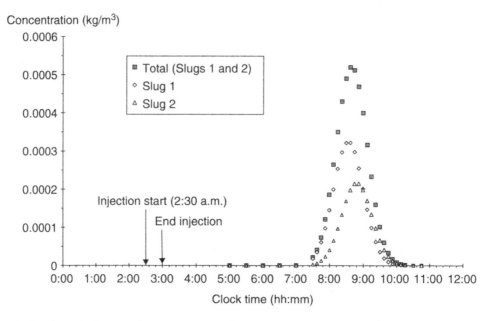

Fig. 8.7 The relationship between dye concentration and time since injection. The horizontal axis is the clock time in seconds.

5. At uniform equilibrium: $d = 0.46\,\text{m}$, $V = 0.52\,\text{m/s}$, $V_* = 0.035\,\text{m/s}$, $\varepsilon_t = 0.0098\,\text{m}^2/\text{s}$ (slowly meandering channel) and $K = 6.6\,\text{m}^2/\text{s}$ (natural river).
Length of initial zone: 1.9 km (side discharge).

   The principle of superposition is applied for two mass slugs and applying the frozen cloud approximation.

Slug	Time of injection	Mass (kg)
	2:42:30 a.m.	3
2	2:57:30 a.m.	2

The relationship between dye concentration and time since injection is shown in Fig. 8.7. Note that the horizontal axis is the clock time in seconds (e.g. 04:00 a.m. = 14 400 s).

9

Turbulent dispersion in natural systems

Summary

Turbulent dispersion in natural systems is discussed in this chapter. The role of dead zones is detailed, while the dispersion and transport of reactive contaminant are developed latter.

9.1 Introduction

Taylor's dispersion theory (Chapter 8) is a one-dimensional analysis of longitudinal contaminant dispersion. It assumes that the turbulence is homogeneous across the river and that there is no reaction (e.g. substance decay). In natural river systems, both assumptions are untrue. First it is unlikely that natural rivers behave as one-dimensional flows and that turbulence is homogeneous in an irregular channel (e.g. Figs 9.1 and 9.3). Second water is a powerful solvent, which may react with contaminants, with the river bed and with the atmosphere.

Definitions

Longitudinal dispersion of contaminants in a natural system may be characterized by the longitudinal and temporal moments. The longitudinal moments of cross-sectional averaged contaminant concentration give the following basic spatial characteristics:

$$M(t) = \int_{x=-\infty}^{x=+\infty} C_{\mathrm{m}}(x,t)\,\mathrm{d}x \qquad \text{Mass of contaminant} \qquad (9.1)$$

$$X(t) = \frac{1}{M(t)} \int_{x=-\infty}^{x=+\infty} x C_{\mathrm{m}}(x,t)\,\mathrm{d}x \qquad \text{Location of centroid} \qquad (9.2)$$

$$\sigma_x^2(t) = \frac{1}{M(t)} \int_{x=-\infty}^{x=+\infty} x^2 C_{\mathrm{m}}(x,t)\,\mathrm{d}x \qquad \text{Variance of contaminant profile} \qquad (9.3)$$

(a)   (b)

(c)

Fig. 9.1 Natural streams. (a) Brisbane River (QLD, Australia) at Colleges Crossing, Karana Downs on 7 April 2002, looking upstream. (b) Moggill Creek, Brisbane (Australia) at Rafting Ground Reserve on 20 June 2002, looking downstream at low tide (near the confluence with the Brisbane River). (c) Oxley Creek, Brisbane (Australia) in August 1999, looking downstream. On the far left, note the abutment ruins of Brookbent Road bridge. The bridge was destroyed during a flood event in 1996.

(d)

Fig. 9.1 (*Contd*) (d) Norman Creek (Brisbane, Australia) during a flood on 31 December 2001, looking upstream, with a hydraulic jump in foreground.

$$Sk(t) = \frac{1}{M(t)} \int_{x=-\infty}^{x=+\infty} x^3 C_m(x,t) \, dx \qquad \text{Skewness of contaminant profile} \qquad (9.4)$$

The basic temporal moments of cross-sectional averaged contaminant concentration give the following temporal characteristics:

$$T(x) = \frac{\int_{t=-\infty}^{t=+\infty} t C_m(x,t) \, dt}{\int_{t=-\infty}^{t=+\infty} C_m(x,t) \, dt} \qquad \text{Time of passage of centroid at the location } x \qquad (9.5)$$

$$\sigma_t^2(x) = \frac{\int_{t=-\infty}^{t=+\infty} t^2 C_m(x,t) \, dt}{\int_{t=-\infty}^{t=+\infty} C_m(x,t) \, dt} \qquad \text{Temporal variance at the location } x \qquad (9.6)$$

Spatial and temporal moments may be calculated from field measurements. They are used to select an appropriate dispersion model, and its relevant parameters (e.g. dispersion coefficient).

Notes

1. The longitudinal moments of cross-sectional averaged contaminant concentration are also called spatial moments. The ith spatial moment is defined as:

$$m_i(t) = \int_{x=-\infty}^{x=+\infty} x^i C_m(x,t)\, dx$$

2. The ith temporal moment is defined as:

$$\tau_i(x) = \int_{t=-\infty}^{t=+\infty} t^i C_m(x,t)\, dt$$

3. The spatial variance has the unit m^2 while the temporal variance has the unit s^2. Note that σ_t^2 increases with time and that is proportional to t^2.
4. The spatial and temporal variances satisfy:

$$V^2 \sigma_t^2(X) > \sigma_x^2(T)$$

where T is the time of passage of centroid at the location X (Rutherford 1994, pp. 188–189). In the longitudinal dispersion region ($x' > 0.4$, Fig. 7.1), the term $(V^2\sigma_t^2(X) - \sigma_x^2(T))$ becomes a constant. For large times t, the following approximation holds:

$$V^2\sigma_t^2 \approx \sigma_x^2 \qquad \text{for large times } t$$

9.2 Longitudinal dispersion in natural rivers with dead zones

9.2.1 Introduction

In natural rivers, there are regions of secondary currents and flow recirculations. Recirculation and stagnant waters may be associated with irregularities of the river bed and banks. They are known as peripheral dead zones, and they can trap and release some water and tracer volumes. In natural channels, dead zones may be found along the banks and at the bed (Fig. 9.2(a)). Examples of bed dead zones include large obstacles, trees, wooden debris, large rocks and bed forms (Fig. 9.3). Figure 9.3 illustrates examples of streams with dead zones, predominantly along the banks. Lateral dead zones may be caused by riparian vegetation, by groynes for river bank stabilization, by submerged trees in flood plains, and by houses and cars in flooded townships (Fig. 9.3(a) and (c)). Figure 9.3(d) and (e) presents artificial dead zones, introduced to assist in river habitat restoration.

Dead zones are thought to explain long tails of tracer observed in natural rivers. The existence of dead zones implies that the turbulence is not homogeneous across the river, and that the time taken for contaminant particles to sample the entire flow is significantly enhanced (i.e. the length of the initial zone is increased).

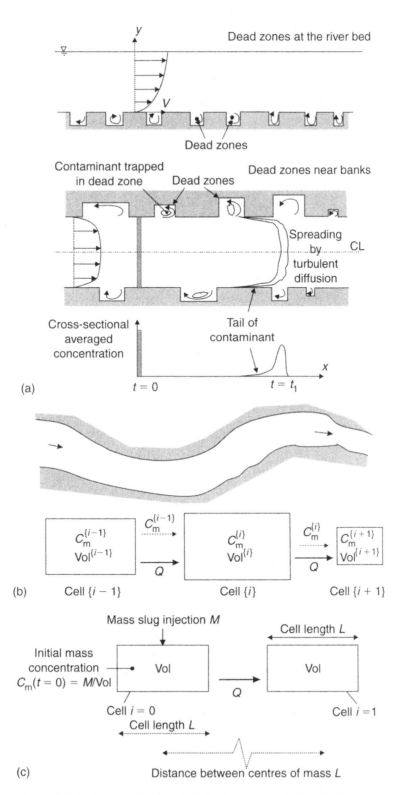

Fig. 9.2 River systems with dead zones. (a) Sketch of idealized dead zones. (b) Longitudinal model in a river with dead zones. (c) Aggregated dead zone (ADZ) model: two-cell model.

(a)

(b)

Fig. 9.3 Examples of dead zones in streams and rivers. (a) Mur River (Graz, Austria) in flood on 21 August 1999, looking downstream. Note the submerged trees on the left bank (top left) which created recirculating flows. (b) Tochi-Shiro River, Japan on 1 November 2001. Note the massive bed load material (size > 4 m) which induces large dead zones.

(c)

(d)

(e)

Fig. 9.3 (*Contd*) (c) Small dead zone during the flood of the Seine River at Caudebec-en-Caux, France on 12 March 2001 (courtesy of Ms Nathalie Lemiere, Sequana-Normandie). View from the right bank. The photograph was taken in the afternoon at high tide. Spring tides (coefficient: 113 and tidal range: 8.5 m on 12/3/2001) induced significant back-water effect. (d) River restoration test section, looking upstream on 30 March 1999, ARRC River habitat research centre, Gifu, Japan. The rectangular pond (i.e. dead zone) was designed to facilitate habitat restoration (see also Fig. 9.3(e)). (e) River restoration test section, looking downstream on 30 March 1999, ARRC River habitat research centre. This is a different type of river restoration system, compared to Fig. 9.3(d).

9.2.2 Basic equation

A river reach is modelled by a sequence of sub-reaches, or cells, which flows into each other. A simplified dead zone model may be developed assuming that:

1. each cell is well mixed,
2. the outflow concentration equals the cell concentration (Fig. 9.2(b)),
3. the inflow concentration equals the outflow concentration of the upstream cell, but with a time delay Δt.

Considering a non-conservative contaminant (decay rate k), the mass balance equation in a cell $\{i\}$ is:

$$\mathrm{Vol}^{\{i\}} \frac{\partial C_{\mathrm{m}}^{\{i\}}(t)}{\partial t} = Q C_{\mathrm{m}}^{\{i-1\}}(t - \Delta t) - (Q + k\,\mathrm{Vol}^{\{i\}}) C_{\mathrm{m}}^{\{i\}}(t) \tag{9.7}$$

where Vol is the cell volume, Q is the river discharge, $C_{\mathrm{m}}^{\{i\}}$ is the average contaminant concentration in cell $\{i\}$ and k is the first-order decay rate coefficient (Section 9.3).

Remarks
1. There are basically three models of longitudinal dispersion with dead zone: the transient storage model, the cells in series (CIS) model and the ADZ model (Rutherford 1994, pp. 218–229). The above development is the basic equation of the ADZ model first introduced by Beer and Young (1983). The present form derives from the work of Wallis *et al.* (1989). Rutherford recommended the ADZ model, although he warned of the lack of physical meaning of the ADZ model coefficients.
2. The dispersion and transport of reactive contaminants are discussed in Sections 9.3 and 9.4, where values of the decay rate constant are introduced.

9.2.3 Analytical solutions (instantaneous mass slug injection)

Considering the instantaneous injection of a mass slug M at $t = 0$ in the first sub-reach ($i = 0$), the solution of equation (9.7) for that cell (i.e. $i = 0$) is:

$$C_{\mathrm{m}}(t) = \frac{M}{\mathrm{Vol}} \exp\left(-\gamma(t - \Delta t)\right) \qquad \text{for } t > \Delta t \tag{9.8}$$

where $\gamma = k + 1/\Delta T$ and $\Delta T = \mathrm{Vol}/Q$ is the cell residence time. Note that $C_m(t = 0) = M/\mathrm{Vol}$: i.e. an instantaneous dilution.

For a two-cell system (Fig. 9.2(c)), the solution of equation (9.7) for the second cell ($i = 1$) is:

$$C_m(t) = \gamma(t - \Delta t)\frac{M}{\mathrm{Vol}}\exp(-\gamma(t - \Delta t)) \qquad \text{Two-cell system} \qquad (9.9a)$$

Equation (9.9a) is the analytical solution for the second cell.

Considering an instantaneous mass slug injection of mass M at $t = 0$ in the initial sub-reach ($i = 0$) of a river system made of $(n + 1)$ identical sub-reaches, placed in series, the contaminant concentration in the last cell (i.e. $i = n$) is:

$$C_m(t) = \frac{\gamma^n (t - \Delta t)^n}{n!}\frac{M}{\mathrm{Vol}}\exp(-\gamma(t - \Delta t)) \qquad (n + 1)\text{-cell system} \qquad (9.9b)$$

where $n!$ is the n-factorial: $n! = 1 \times 2 \times 3 \times \cdots \times n$.

Equation (9.9) is plotted in Fig. 9.4 for a conservative contaminant (i.e. $k = 0$) and a decaying substance (i.e. $k = 0.003$), and for two values of n. Note that the number of cells is directly related to the distance downstream. With increasing value of n, the time of passage of centroid increases and the peak concentration decreases. Qualitatively this trend is consistent with Taylor's dispersion model for the sudden injection of a mass slug of contaminant (e.g. Fig. 8.3). For small values of n (e.g. $n = 5$), the concentration versus time distribution is skewed, while it becomes more symmetrical for larger values of n (e.g. $n = 20$).

Remarks

1. For an instantaneous mass slug injection of mass M at $t = 0$ in two-cell river model, the time of passage of the centroid in the second cell is:

$$T = \Delta T + \Delta t$$

2. For an instantaneous mass slug injection of mass M at $t = 0$ in a river system made of $(n + 1)$ identical cells placed in series, the time of passage of the centroid T in the last cell ($i = n + 1$) equals:

$$T = n(\Delta T + \Delta t)$$

3. In practice, the ADZ 'dead zone' model is best applied to predict the longitudinal dispersion between two sites located both downstream of an injection point. The concentration versus time distribution at the upstream site is divided as a succession of small mass slugs as illustrated in Fig. 8.4. The final solution is obtained by applying the method of superposition.

4. Remember that, for an instantaneous mass slug injection in a two-cell model, the mass of contaminant M is instantaneously diluted into the initial cell volume ($i = 0$) (Fig. 9.2(c)). Then it decays with time in the initial cell according to:

$$C_m(t) = \frac{M}{\mathrm{Vol}}\exp(-\gamma(t - \Delta t)) \qquad \text{Initial cell } (i = 0) \qquad (9.8)$$

5. The ADZ 'dead zone' model is not a longitudinal (Taylor's) dispersion model. It is a one-dimensional box model, based upon a stirred tank reaction model (Beer and Young 1983). It may be applied to both the initial and dispersion zones; it is more general than the 'frozen cloud' approximation model.

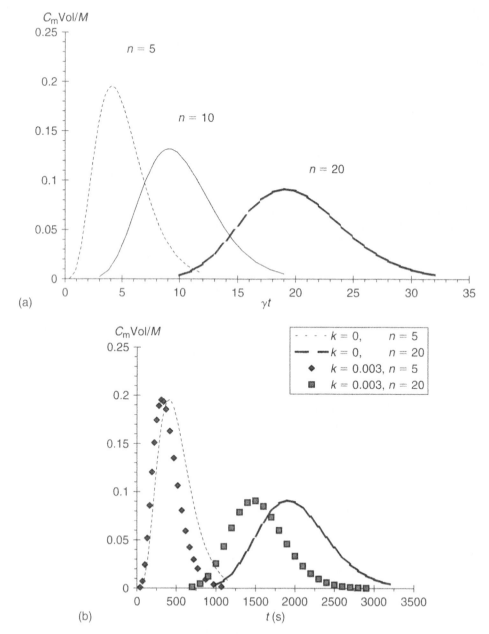

Fig. 9.4 Dimensionless contaminant concentrations using the dead cell model. (a) Conservative contaminant ($k = 0$), $\Delta T = 100\,\text{s}$, $\Delta t = 10\,\text{s}$. (b) Conservative and decaying contaminants ($k = 0$ and 0.003), $\Delta T = 100\,\text{s}$, $\Delta t = 10\,\text{s}$.

Discussion: application to natural systems

For a natural system, each sub-reach has unique characteristics and equation (9.7) must be integrated numerically. It requires however some prediction of the values of the cell volume Vol and cell residence time ΔT, and of the pure time delay Δt. Wallis *et al.* (1989) discussed practical considerations.

But Rutherford (1994, pp. 26–227) emphasized the difficulties to estimate accurately the parameters Vol, ΔT and Δt, while using meaningful estimates.

9.3 Dispersion and transport of reactive contaminants

9.3.1 Basic equation

For non-conservative substances, such as biochemical oxygen demand (BOD) in a sewage effluent or as heat in a power station, the dispersion equation may be extended by adding a reaction term. The one-dimensional mass transport equation becomes:

$$\frac{\partial C_m}{\partial t} + V\frac{\partial C_m}{\partial x} = K\frac{\partial^2 C_m}{\partial x^2} - k\,C_m \qquad (9.10)$$

where k is the decay rate and K is the longitudinal dispersion coefficient.

Equation (9.10) is an extension of the dispersion equation (8.1), by introducing a single first-order reaction term $(-kC_m)$ where $k > 0$ implies a contaminant decay.

Remarks

1. Many substances may be considered to undergo, in first approximation, a first-order decay. In a stagnant water body, the rate of decay is given by:

$$\frac{\partial C_m}{\partial t} = -k\,C_m$$

The integration yields:

$$C_m = C_{max}\exp(-k\,t)$$

The result implies that the time required for decay of the contaminant to a factor e^{-1} is $1/k$. In flowing waters, the fluid is advected to a distance V/k during the time $1/k$.
2. The unit of the decay rate constant k is s^{-1}.
3. The coefficient of reaction k depends upon the contaminant itself and on the temperature. For organic substances in treated and untreated wastewater, k is typically between 0.05 and 0.30 days^{-1} (i.e. $0.6 \times 10^{-6} < k < 3.5 \times 10^{-6}\,s^{-1}$).
4. The one-dimensional dispersion analysis (i.e. equations (8.1) and (9.10)) does not apply in the *initial zone* (Fig. 7.1). Equation (9.10) assumes further a constant rate of decay k, as well as a constant dispersion coefficient K and uniform velocity V in the reach.

9.3.2 Applications

Sudden mass slug contamination in a river

Considering an instantaneous slug release (mass M) in a one-dimensional flow, the solution of dispersion equation (9.10) is:

$$C_m = \frac{M}{A\sqrt{4\pi Kt}}\exp\left(-\frac{(x-Vt)^2}{4Kt} - k\,t\right) \qquad (9.11)$$

where A is the flow cross-sectional area.

In the absence of reaction ($k = 0$), equation (9.11) becomes equation (8.4). Equation (9.11) is plotted in Fig. 9.5 for a reactive contaminant (thick curves) and in the absence of reaction (thin curves). Figure 9.5 shows that the mass of contaminant is advected downstream with the average velocity V. Note that the total mass of the reactive substance does not conserve itself. There is a weak continuous decay in the total mass of the substance according to:

$$\int_{x=-\infty}^{x=+\infty} C_m\,dx = \exp(-k\,t) \qquad (9.12)$$

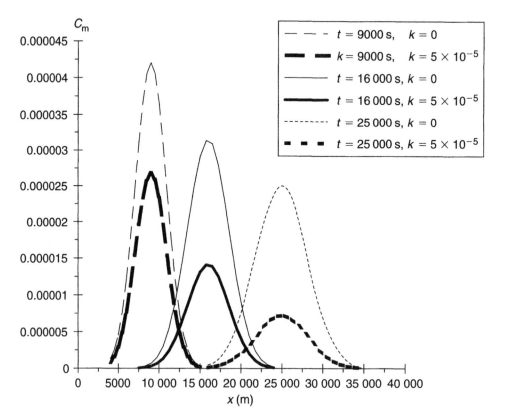

Fig. 9.5 Spread of a sudden concentration increase at the origin, solutions of equation (9.11) for $k = 0$ and 5×10^{-5} (assuming $M/A = 0.2$, $K = 200$ and $V = 1$).

Remarks
1. At a distance x from the injection point, the maximum mass concentration equals:

$$C_{max} = \frac{M}{A\sqrt{4\pi K \dfrac{x}{V}}} \exp\left(-k\frac{x}{V}\right)$$

The result leads to equation (8.7) for $k = 0$ (no reaction).
2. The standard deviation of the contaminant cloud equals: $\sigma = \sqrt{2Kt}$.

Sudden increase in mass concentration at the origin

The concentration is initially zero everywhere. At the initial time $t = 0$, the concentration is suddenly raised to C_0 at the origin $x = 0$ and held constant: $C_m(0, t \geq 0) = C_0$. This is a steady situation and the dispersion equation becomes:

$$V\frac{\partial C_m}{\partial x} = K\frac{\partial^2 C_m}{\partial x^2} - kC_m \tag{9.13}$$

with the boundary condition: $C_m = 0$ at $x = +\infty$. A solution of the dispersion equation is:

$$C_m = C_0 \exp\left(-\frac{kx}{V}\left(\frac{2}{\alpha}\left(\sqrt{\alpha + 1} - 1\right)\right)\right) \tag{9.14a}$$

where:

$$\alpha = \frac{4Kk}{V^2}$$

and C_0 is an integration constant. Neglecting the existence of the initial zone, the initial concentration C_0 is equivalent to the steady release of $\dot{M}$ units of mass per unit time into a river (discharge Q):

$$C_0 = \frac{\dot{M}}{Q}\left(\frac{2}{\alpha}\left(\sqrt{\alpha + 1} - 1\right)\right) \tag{9.15}$$

Remarks
1. $\alpha = 4Kk/V^2$ is a dimensionless coefficient of reaction–dispersion. For BOD, α is typically of the order: $\alpha < 0.4$.
2. In natural rivers and using equations (7.6) and (8.8), it yields:

$$\alpha = \frac{4Kk}{V^2} = 0.0264\,\frac{W^2 k}{\varepsilon_t}$$

where W is the channel width and ε_t is the transverse mixing coefficient.

9.3.3 Discussion

The one-dimensional dispersion analysis (equation (9.10)) assumes that the contaminant is fully mixed across the channel: i.e. it does not apply to the initial zone (Fig. 7.1). For a steady flow in rivers, the distance required for cross-sectional mixing is about (Section 5.3):

$$L \sim 0.4VW^2/\varepsilon_t \qquad \text{Complete mixing of side discharge}$$

while the characteristic distance of contaminant decay is V/k.[1] The latter exceeds the cross-sectional mixing distance (i.e. $x' = 0.4$) only if the dimensionless coefficient of reaction–dispersion α is less than about 0.066.

   If the contaminant concentration decays before the end of the initial zone (i.e. $x' < 0.4$), equation (9.10) is not suitable, and the contaminant concentration must be computed numerically (e.g. Fischer *et al.* 1979). Practically, equation (9.10) may be applied *only for* $\alpha < 0.066$.

   If the longitudinal dispersion term is very small (i.e. $K \, \partial^2 C_m / \partial^2 x \ll k C_m$), the dimensionless coefficient of reaction–dispersion α is very small. Considering a steady river flow and contaminant injection, the solution of equation (9.13) becomes:

$$C_m = C_o \exp\left(-\frac{kx}{V}\right) \qquad \text{for } \alpha \text{ very small} \tag{9.14b}$$

Equation (9.14b) is a reasonable prediction of the downstream contaminant concentration for very small values of α.

Remarks

1. Fischer *et al.* (1979, p. 147) stated that $\alpha < 0.06$ for the decay distance to be greater than the length of the initial zone. The above result (i.e. $\alpha < 0.066$) is more precise, although it must be emphasized that equations (7.6) and (8.8) are approximate (Chapters 7 and 8).
2. For small values of α, the longitudinal dispersion term in equation (9.10) is very small (i.e. $K \, \partial^2 C_m / \partial x^2 \ll k C_m$).
3. In practice, effluent discharges are rarely steady. Typical daily fluctuations in sewage output lead to gradients of concentration of the discharged material, and these gradients are subsequently smoothed by the process of longitudinal dispersion. Hence equations (9.14a) and (9.14b) are seldom applicable.

9.4 Transport with reaction

9.4.1 Basic equation

A particular case is the transport of reactive contaminant from a continuous source at origin, under steady flow conditions, and in the absence of longitudinal dispersion. That is, the dispersion term in equation (9.10) is assumed negligible:

$$K \frac{\partial^2 C_m}{\partial x^2} \approx 0$$

---

[1]Remember that the time required for decay of the contaminant to a factor e^{-1} is $1/k$, during which the fluid is advected to a distance V/k.

For steady flow conditions, the dispersion equation becomes:

$$V \frac{\partial C_m}{\partial x} = -k_r C_m \tag{9.16}$$

where k_r is the decay rate constant (s^{-1}). The integration of the transport equation (9.16) yields:

$$C_m = C_o \exp\left(-\frac{k_r x}{V}\right) \tag{9.17}$$

where $C_o = \dot{M}/Q$, $\dot{M}$ is the mass flow rate of contaminant at $x = 0$ and Q is the steady river discharge.

9.4.2 Application to dissolved oxygen content (DOC) in natural streams

Equation (9.16) may be applied to estimate the self-cleaning capacity of a waterway receiving wastes. The water in the channel has the capacity to absorb organic matter and pollutants: i.e. through the micro-organisms that degrade certain non-conservative pollutants of organic origin (e.g. wastewater).

In a waterway, the sources of dissolved oxygen (DO) are the re-oxygenation at the free surface and the photosynthesis. The sinks of oxygen include the biochemical oxygenation of organic matter, the decomposition of sludge deposits and the respiration of aquatic plants and life forms. In a first approximation, the budget of the DO is primarily a function of the re-oxygenation rate and the biochemical oxygenation demand (BOD) (Fig. 9.6). At steady state, the DO balance may be expressed as:

$$V \frac{\partial C_m}{\partial x} = k_a (C_{sat} - C_m) - k_r \, \text{BOD} \tag{9.18}$$

where x is the distance in the flow direction, C_m is the DO concentration, BOD is the biochemical oxygen demand, C_{sat} is the concentration of dissolved gas in water at equilibrium (Appendix B, Section 9.6), k_a is the re-oxygenation rate constant and k_r is the BOD decay rate.

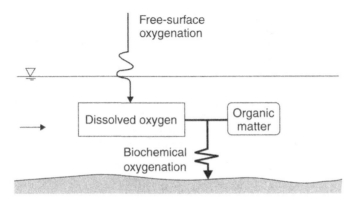

Fig. 9.6 Simplified analysis of DO balance in a waterway.

In equation (9.18), the right handside term includes the re-oxygenation term $(+k_a(C_{sat} - C_m))$ plus the biochemical oxygenation demand $(-k_r \text{BOD})$. Equation (9.18) may be integrated by a method of superposition. It yields:

$$C_m = C_{sat} - (C_{sat} - C_o)\exp\left(-\frac{k_a x}{V}\right) - \frac{k_r}{k_a - k_r}\text{BOD}_o\left(\exp\left(-\frac{k_r x}{V}\right) - \exp\left(-\frac{k_a x}{V}\right)\right) \quad (9.19)$$

where C_o is initial oxygen mass concentration (at $x = 0$), BOD_o is the initial biochemical demand (at $x = 0+$) and assuming that the effluent (wastewater) is released at the origin ($x = 0$).

Remarks

1. The BOD is the amount of oxygen used by micro-organisms in the process of breaking down organic matter in water.
2. The above analysis of waterway self-cleaning capacity is sometimes called sag analysis or DO sag analysis. It was first proposed by Streeter and Phelps in 1925.[2] Graf and Altinakar (1998, pp. 567–572) and Metcalf and Eddy (1991, pp. 1216–1220) discussed the analysis in more details.
3. The oxidation of BOD consumes oxygen and it is an oxygen sink for the ambient waters. The most widely used parameter in surface waters is the BOD of consumed oxygen during a period of 5 days at a temperature of 20°C in obscurity, denoted BOD_5. Metcalf and Eddy (1991, pp. 71–82) argued however that BOD_5 is only a representative index of oxygen consumption. *It is neither accurate nor precise.*
4. The BOD consumption decays as equation (9.17):

$$\text{BOD} = \text{BOD}_o \exp(-k_r t)$$

It yields the following relationship between the initial BOD_o and the initial BOD measured at 5 days:

$$\text{BOD}_o = \frac{(\text{BOD}_5)_o}{1 - \exp(-k_r \times 5 \times 86400)}$$

where k_r is expressed in s^{-1} and 5 days equal 86400 s.

5. Assuming that the effluent volume flow rate is negligible compared to the river discharge, and that the BOD of the stream is very small, the initial biochemical demand BOD_o equals:

$$\text{BOD}_o = \frac{\dot{M}}{VA}$$

where $\dot{M}$ is the effluent BOD mass discharge, V is the stream velocity and A is the stream cross-sectional area.

[2]Streeter, H.W., and Phelps, E.B. (1925), *US Publ. Heath Bulletin*, Vol. 146.

6. In the general case, the initial BOD equals:

$$\mathrm{BOD_o} = \frac{\dot{M} + \mathrm{BOD}(x < 0)\, Q}{Q_{\mathrm{effluent}} + Q}$$

where Q_{effluent} is the effluent volume flow rate, BOD $(x < 0)$ is the BOD of the river upstream of the effluent injection location and Q is the river flow rate.

Re-oxygenation rate constant and decay rate

Typical values of the decay rate constant of organic matter contained in wastewater are listed in Table 9.1. For the re-aeration rate constant, there are numerous correlations and Table 9.2 summarizes some. Accurate estimates of concentration of dissolved gas in water at equilibrium C_{sat} and molecular diffusivity D_{m} are given in Appendices B and C.

While Table 9.1 provides accepted estimates of the decay rate constant for BOD, the calculations of the re-aeration rate constant are most often empirical and inaccurate. In practice, the correlation of Gualtieri *et al.* (2002) may provide a robust estimate for clear-water flows.

Table 9.1 Typical BOD decay rate

Effluent (1)	k_{r} (day^{-1}) (2)	BOD$_{\mathrm{o}}$ (mg/L) (3)
Strong wastewater	0.39	250–400
Weak wastewater	0.35	110–150
Primary effluent	0.35	75–150
Secondary effluent	0.12–0.23	15–75
Tap water	<0.12	<1

References: Fair *et al.* (1968), Barnes *et al.* (1981), Liu *et al.* (1997).

Table 9.2 Empirical correlations of re-aeration rate constant in open channels (neglecting 'white waters')

Reference (1)	k_{a} (2)	Remarks (3)
Fair *et al.* (1968)	$11.0\dfrac{V}{d^{5/3}}$	k_{a} in day^{-1}, V in m/s and d in m
Gualtieri *et al.* (2002)	$1.1\left(\dfrac{D_{\mathrm{m}}^{2/3}\sigma}{d\rho\nu^{5/3}}\right)\sqrt[3]{\sin\theta}\ \sqrt[3]{\dfrac{g\rho^3\nu^4}{2\sigma^3}}$	Dimensionally correct relationship, k_{a} in s^{-1}; based upon a large number of USGS field experiments in natural streams ν: kinematic viscosity σ: surface tension
Simplified expression	$\dfrac{k_{\mathrm{L}}}{d}$	Rough estimate based upon geometrical considerations and liquid film coefficient estimate (see Appendix A) k_{a} in s^{-1}; k_{L} in m/s and d in m

Notes: D_{m}: molecular diffusivity (Appendix C, Section 9.7).

Discussion: aeration rate constant in 'white-water' flows

The re-oxygenation constant rate k_a is the product of the liquid film coefficient k_L by the specific surface area:

$$k_a = k_L a$$

where a is the air–water interface area per unit volume of flowing fluid.[3] The liquid film coefficient is primarily a function of the fluid and gas properties only (Appendix C, Section 9.7).

In air–water flows (i.e. 'white waters'), the re-oxygenation rate is drastically enhanced by the cumulative surface area of entrained bubbles (Appendix A and Fig. 9.8). For example, in a 2 m deep river with a depth-averaged specific interface area of $200\,\text{m}^2/\text{m}^3$, the re-aeration rate constant k_a is 400 times greater than predicted by clear-water flow correlations shown in Table 9.2.

Remarks

1. In natural channels, the re-aeration rate constant is enhanced by secondary currents and surface renewal induced by large eddies. For example, Gulliver and Halverson (1989), Tamburrino and Gulliver (1990).
2. In bubbly air–water flows, the specific interface area equals:

$$a = 6\frac{C}{d_{ab}}$$

where d_{ab} is the air bubble diameter and C is the void fraction (also called air concentration). For example, $a = 400\,\text{m}^{-1}$ for $C = 0.2$ and $d_{ab} = 3\,\text{mm}$.
3. The temperature influence on the aeration rate constant is often approximated by:

$$k_a(T) = k_a(T = 20°C)\, 1.024^{T-20}$$

where T is the temperature in Celsius and $k_a(T = 20°C)$ is the reference aeration rate constant at 20°C. However the influence of temperature must be properly accounted for through the fluid properties (e.g. ρ, μ, σ, D_m) which are temperature dependent.
4. Most aeration rate calculations neglect the effects of wind speed and cross-sectional shape. The wind speed is little relevant in flowing waters. The cross-sectional shape has little effect for wide channels.

9.4.3 DO sag analysis

Equation (9.19) is plotted in Fig. 9.7 for one particular example. The shape is typical. Downstream of the effluent injection (i.e. $x = 0$), the biochemical decomposition of the waste-water induces a reduction in DOC. However DO is replenished through surface aeration at a rate proportional to the DO deficit (i.e. $(C_{sat} - C_m)$). At a certain point, the re-aeration rate equals exactly the BOD consumption. Downstream, the aeration rate is larger than the BOD

---

[3]In clear-water flows, a is the air–water interface area per unit volume of water. In turbulent air–water flows, a is the air–water surface area per unit of (flowing) air and water.

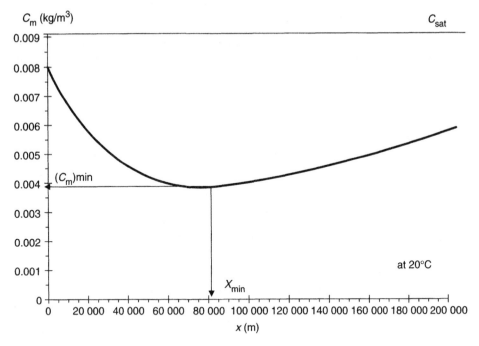

Fig. 9.7 Evolution of DO concentration in streams downstream of effluent injection: $BOD_0 = 0.0198\,kg/m^3$, $C_0 = 0.080\,kg/m^3$, $Q = 38\,m^3/s$, $V = 0.22\,m/s$, $d = 3.5\,m$, 20°C.

consumption and the DO content increases with distance. Far downstream, the waters becomes more and less saturated in DO. This may occur at a large downstream distance.

Important parameters of the DO sag curve are the minimum DO concentration $(C_m)_{min}$ and the distance X_{min} where it takes place. These may be estimated from equation (9.19) for $\partial C_m/\partial x = 0$. It yields:

$$X_{min} = \frac{V}{k_a - k_r}\ln\left(\frac{k_a}{k_r}\right)\left(1 - \frac{(C_{sat} - C_0)(k_a - k_r)}{k_r\,BOD_0}\right) \qquad (9.20)$$

$$(C_m)_{min} = \frac{k_r}{k_a}\,BOD_0\,\exp\left(-\frac{k_r X_{min}}{V}\right) \qquad (9.21)$$

Remarks
1. It is important to remember that the BOD consumption decays as equation (9.17), while the re-aeration rate is proportional to the DO deficit $(C_{sat} - C_m)$.
2. The time of arrival of the minimum DO content is $t_{min} = X_{min}/V$.
3. The 'sag analysis' has limited use because equations (9.18) and (9.19) neglect some important phenomena. In particular, they neglect several sources and sinks of DO: e.g. photosynthesis and 'white waters', the decomposition of sludge deposits and respiration of aquatic plants and life forms. Further equations (9.18)–(9.21) assume constant k_a, k_r, V and Q, and steady flow conditions.

9.5 Appendix A – Air–water mass transfer in air–water flows

Air–water flows in rivers and hydraulic structures have great potential for aeration enhancement of flow because of the large interfacial area generated by entrained bubbles (Part 4, Chapter 17; Chanson 1997, Fig. 7.8).

The mass transfer rate of a chemical across an interface varies directly as the coefficient of molecular diffusion and the negative gradient of gas concentration. If the chemical of interest is volatile (e.g. oxygen, nitrogen, chlorine, methane), the transfer is controlled by the liquid phase, and the gas transfer of the dissolved chemical across an air–water interface is usually rewritten as:

$$\frac{\partial C_m}{\partial t} = -k_L a(C_m - C_{sat}) \tag{9A.1}$$

where k_L is the liquid film coefficient, a is the specific surface area defined as the air–water interface area per unit volume of air and water, C_m is the local dissolved gas concentration and C_{sat} is the concentration of dissolved gas in water at equilibrium (Appendix B) (e.g. Gulliver 1990, Chanson and Toombes 2000, Chanson 2002). Mass transfer at bubble interface is a complex process. Kawase and Moo-Young (1992) reviewed several correlations for the liquid film coefficient calculations in turbulent gas–liquid flows. They showed that the mass transfer coefficient k_L is almost constant regardless of bubble sizes and flow situations. The transfer coefficient may be estimated by:

$$k_L = 0.28 D_{gas}^{2/3} \left(\frac{\mu_w}{\rho_w}\right)^{-1/3} \sqrt[3]{g} \quad d_{ab} < 0.25\,\text{mm} \tag{9A.2a}$$

$$k_L = 0.47 \sqrt{D_{gas}} \left(\frac{\mu_w}{\rho_w}\right)^{-1/6} \sqrt[3]{g} \quad d_{ab} > 0.25\,\text{mm} \tag{9A.2b}$$

for gas bubbles affected by surface active impurities, where μ_w and ρ_w are the dynamic viscosity and density of the liquid, D_{gas} is the coefficient of molecular diffusion (Appendix C, Section 9.7), d_{ab} is the gas bubble diameter, g is the gravity acceleration, all variables being expressed in SI units. Equation (9A.2) was compared successfully with more than a dozen of experimental studies.

Equation (9A.1) is very general. Importantly, it includes the effect of air bubble entrainment and the drastic increase in interfacial area. Experimental measurements in supercritical flows down a flat chute recorded local specific interface area of up to $110\,\text{m}^2/\text{m}^3$ (m^{-1}) with depth-averaged (bulk) interface area ranging from 10 to $21\,\text{m}^{-1}$. Larger specific interface areas were recorded in developing shear flows. Local interface areas of up to $400\,\text{m}^{-1}$ were observed in hydraulic jumps and maximum specific interface areas of up to $550\,\text{m}^{-1}$ [4] were measured in plunging jet flows (Chanson 1997, Chanson and Toombes 2000). These examples illustrate the potential for aeration enhancement in the presence of white water.

Chanson and Toombes (2000) combined equations (9A.1) and (9A.2) with measured air–water interfacial areas. They deduced aeration efficiency, in terms of DO, of about 30–40% for a 24 m long stepped cascade with a 1.4 m total drop. The result was verified with DO measurements. The very large aeration rate derives from the strong flow aeration and sustained air–water interface areas down the entire chute (Fig. 9.8).

---

[4]That is, nearly twice the surface area of two basketball courts in one cubic metre of air and water.

(a)

(b)

Fig. 9.8 Air–water flows in streams and rivers. (a) 'White waters' at sabo works on the Oyana River, Japan on 2 November 2001. (b) Hopetown Falls, Otway Range, Victoria (Australia) (courtesy of Dr Richard Manasseh).

9.6 Appendix B – Solubility of nitrogen, oxygen and argon in water

Dissolved gas concentrations in water are usually expressed in mass of dissolved chemical per unit volume or kg/m^3. Another unit is the ppm (parts per million). The conversion is: $1\,ppm = 1\,mg/L$.

Solubility of oxygen

The solubility of oxygen in water at equilibrium with water saturated air $C_{sat}(Pstd)$ at standard pressure (i.e. $Pstd = 1\,atm = 1.013\,25 \times 10^5\,Pa$) is calculated as:

$$\ln(1000C_{sat}(Pstd)) = -139.344\,11 + \frac{1.575\,701 \times 10^5}{TK}$$
$$- \frac{6.642\,308 \times 10^7}{TK^2} + \frac{1.243\,800 \times 10^{10}}{TK^3} - \frac{8.621\,949 \times 10^{11}}{TK^4}$$
$$- Chl\left(3.1929 \times 10^{-2} - \frac{19.428}{TK} + \frac{3.8673 \times 10^3}{TK^2}\right) \tag{9B.1}$$

where C_{sat} (Pstd) is the solubility of oxygen at standard pressure[5] in kg/m^3, TK is the temperature in Kelvin and Chl is the chlorinity in ppt.[6] The saturation concentration of DO at non-standard pressure is:

$$C_{sat}(P) = C_s(Pstd)\,P\left(\frac{\left(1 - \frac{P_v}{P}\right)(1 - Teta(P))}{(1 - P_v)(1 - Teta)}\right) \tag{9B.2}$$

where P is the absolute pressure in atm (within 0–2 atm), P_v is the partial pressure of water vapour in atm and TK is the temperature in Kelvin. In equation (9B.2), the expression of Teta is:

$$Teta = 0.000975 - 1.426 \times 10^{-5}TC + 6.436 \times 10^{-8}TC^2 \tag{9B.3}$$

where TC is the temperature in Celsius. The partial pressure of water vapour may be computed as:

$$\ln(P_v) = 11.8571 - \frac{3840.70}{TK} - \frac{216\,961}{TK^2} \tag{9B.4}$$

where P_v is in atm.
References: Apha (1985, 1989), Bowie *et al.* (1985).

---

[5]The standard pressure Pstd (also called standard atmosphere or normal pressure) at sea level equals:

$$Pstd = 1\,atm = 360\,mm\,of\,Hg = 101\,325\,Pa$$

where Hg is the chemical symbol of mercury.
[6]The chlorinity is defined in relation to salinity as: Salinity $= 1.806\,55 \times$ Chlorinity.

Temperature (°C)	C_{sat} (Pstd) $\times 10^{-3}$ [Chl = 0] (kg/m^3)	Dissolved oxygen ($\times 10^{-3}$)	
		[Chl = 10 ppt] (kg/m^3)	[Chl = 20 ppt] (kg/m^3)
(1)	(2)	(3)	(4)
0	14.621	12.388	11.355
5	12.770	11.320	10.031
10	11.238	10.058	8.959
15	10.084	9.027	8.079
20	9.0982	8.174	7.346
25	8.263	7.457	6.728
30	7.559	5.845	6.197
35	6.950	6.314	5.734
40	6.412	5.842	5.321

Reference: Bowie *et al.* (1985).

Volumetric solubility of nitrogen, oxygen and argon

Weiss (1970) proposed an expression of the volumetric solubility of nitrogen, oxygen and argon at one atmospheric total pressure (i.e. standard pressure):

$$\ln(C_{sat}(\text{Pstd})) = A1 + A2\,\frac{100}{TK} + A3\ln\left(\frac{TK}{100}\right) + A4\,\frac{TK}{100}$$
$$+ \text{Sal}\left(B1 + B2\,\frac{TK}{100} + B3\left(\frac{TK}{100}\right)^2\right) \tag{9B.5}$$

where C_{sat}(Pstd) is the solubility in mL/L, TK is the temperature in Kelvin and Sal is the salinity in ppt. The constants A1, A2, A3, A4, B1, B2 and B3 are summarized in the next table for nitrogen (N$_2$), oxygen (O$_2$) and argon (Ar).

Gas	A1	A2	A3	A4	B1	B2	B3
(1)	(2)	(3)	(4)	(5)	(6)	(7)	(8)
Nitrogen	−172.4965	248.4262	143.0738	−21.7120	−0.049781	0.025018	−0.0034861
Oxygen	−173.4292	249.6339	143.3483	−21.8492	−0.033096	0.014259	−0.0017000
Argon	−173.5146	245.4510	141.8222	−21.8020	−0.034474	0.014934	−0.0017729

Reference: Weiss (1970).

9.7 Appendix C – Molecular diffusion coefficients in water (after Chanson 1997a)

The gas–liquid diffusivity of oxygen and nitrogen in water is:

Temperature (°C)	$D(O_2) \times 10^{-9}$ Oxygen (m^2/s)	$D(N_2) \times 10^{-9}$ Nitrogen (m^2/s)
(1)	(2)	(3)
10	1.54	1.29
25	2.20	2.01
40	3.33	2.83
55	4.50	3.80

Reference: Ferrell and Himmelblau (1967).

The data of Ferrell and Himmelblau (1967) can be correlated as:

$$D(O_2) = 1.16793 \times 10^{-27} TK^{7.3892} \qquad \text{Oxygen} \qquad (9C.1)$$

$$D(N_2) = 5.567 \times 10^{-11} TK - 1.453 \times 10^{-8} \qquad \text{Nitrogen} \qquad (9C.2)$$

where D is the molecular diffusivity in m^2/s and TK is the temperature in Kelvin.

9.8 Exercises

1. Dye profiles were measured at the Manawatu River, below Palmerston North (New Zealand) (Rutherford 1994, pp. 271–273 and 226–229). The table below shows cross-sectional averaged dye concentrations at one measurement site located 5.0 km downstream of the injection point. (Preliminary calculations suggested that the site location was immediately downstream of the initial zone.) Estimate the time of passage of the centroid and the temporal variance.

Time (h)	C_m (mg/m^3)	Time (h)	C_m (mg/m^3)	Time (h)	C_m (mg/m^3)	Time (h)	C_m (mg/m^3)
2.106	0.833	3	35.8	3.93	10.8	5.1	2.17
2.19	9.58	3.12	31.9	4	10.17	5.2	2
2.27	11.8	3.18	29.6	4.1	8.92	5.3	1.75
2.37	14.9	3.27	29.2	4.2	8.08	5.4	1.67
2.445	21.9	3.345	25.3	4.31	7	5.5	1.38
2.52	26.9	3.43	23.46	4.39	6.04	5.75	1.17
2.62	29.2	3.51	21.7	4.51	5	6	0.83
2.675	35	3.59	17.9	4.71	3.92	6.25	0.625
2.78	36.4	3.68	16.4	4.79	3.25	6.5	0.375
2.85	36.3	3.755	14.17	4.89	2.83	6.75	0.33
2.93	36.2	3.84	12.8	5	2.3	7	0.25

2. Dye tests are conducted in a small stream. At a measurement station located downstream of the mass slug injection point, the concentration versus time may be approximated as:

t (h)	<1	1–1.5	1.5–2	2–2.5	>2.5
C_m (kg/m^3)	0	2×10^{-6}	2.2×10^{-6}	3×10^{-7}	0

(a) Calculate the 'real' concentration in the initial cell ($i = 0$) at $t = 5$ h.
(b) Predict the concentration versus time curve at a measurement station located 3 km downstream using an ADZ model with two cells (i.e. $i = 0$ and 1) (Fig. 9.2(c)).
(c) Calculate the mass concentration at $x = 3$ km and $t = 24$ h.
Assume a flow velocity of 0.15 m/s, a flow cross-section of 8.9 m^2 and a conservative contaminant. Neglect the cell time delay (i.e. $\Delta t = 0$).
3. A study of turbulent dispersion in a stream is conducted for the following design flow conditions:

$Q = 15$ m^3/s, average width: 8.5 m, average bed slope: 0.000 65, gravel bed: $k_s = 10$ mm.

A 6 km long reach is divided into five elements (sub-reaches, $i = 0$–4) of equal length. Uniform equilibrium flow conditions take place along the 6 km test section. A slug of dye ($M = 1.5$ kg) is suddenly injected in the first sub-reach (cell no. 0).

(a) Calculate the mass concentration versus time 6 km downstream of the injection point. (Neglect the initial zone. Discuss the approximation.)

(b) Calculate the mass concentration versus time in the cell no. 4 neglecting the cell time delay (i.e. $\Delta t = 0$).

(c) Calculate the mass concentration versus time in the cell no. 4 assuming $\Delta t = 20$ min. Present your results on the same graph and discuss the differences.

4. A mass slug of dye tracer (7.5 kg) is released during a small flood in Norman Creek, on the channel centreline from a foot bridge, to estimate the longitudinal dispersion coefficient. The stream, in Southern Brisbane, has the following channel characteristics during the flood event:

Flow depth: 0.73 m, width: 55 m, bed slope: 0.002, short grass: $k_s = 3$ mm.

(a) Calculate the maximum tracer concentration and its arrival time 13 500 m downstream of the injection point assuming a conservative tracer.

(b) The wrong tracer was selected and found to be reactive (decay rate constant: 0.8 day^{-1}). At 13.5 km downstream of the injection point, calculate the maximum tracer concentration and the mass concentration at $t = 1$ h after injection.

Assume uniform equilibrium flow conditions. Neglect vertical mixing and assume a slowly meandering stream.

5. During the same flood event, in Norman Creek (Exercise 4), a reactive contaminant is accidentally released from the channel centreline as a mass slug. The decay rate constant is 400 day^{-1}. May you use the longitudinal dispersion theory? In the affirmative, calculate the maximum tracer concentration at 3.5 km downstream of the injection point (slug mass: 1 kg).

6. A stream flows with an average velocity of 1 m/s and a flow rate of 8.5 m^3/s. The water is 95% saturated in DO and the water temperature is 10°C. The BOD$_5$ is 0.001 kg/m^3. A water treatment plant discharge some effluent into the stream at a rate of 1.5 m^3/s and the BOD$_5$ is 200 g/m^3. The decay rate constant and the re-aeration rate constant were previously estimated to be 0.2 and 0.5 day^{-1} respectively.

Calculate the minimum DOC downstream of the injection point and its location. Estimate the BOD$_5$ for a sample taken at that location. *Assume the effluent to be at the same temperature as the river. Assume further that the wastewater spreads almost instantaneously across the entire section.*

9.9 Exercise solutions

1. Manawatu River.
   Time of passage of the centroid = 3 h 17 min 30 s Temporal variance = 1.47×10^8 s^2.

2. The problem is solved by assuming that the concentration distribution at the upstream site corresponds to the superposition of three mass slugs:

M (kg)	$2 \times 10^{-6} \times 3000 \times 8.9$	$2.2 \times 10^{-6} \times 3000 \times 8.9$	$3 \times 10^{-7} \times 3000 \times 8.9$
T (s)	4500 (1.25 h)	6300	8100

The volume of each cell is 3000 m $\times$ 8.9 m^2. The cell residence time is $\Delta T = 26\,700/(0.15 \times 8.9) = 20\,000$ s.

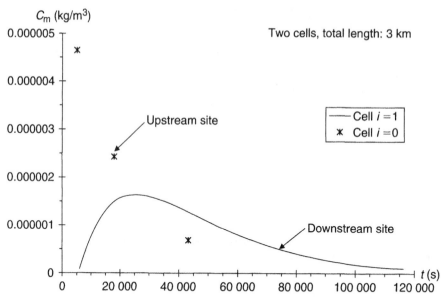

Fig. 9.9 Concentration versus time at $x = 0$ and 3 km (two-cell ADZ model).

In the initial cell ($i = 0$), $C_m(t = 5\,\text{h}) = 2.43 \times 10^{-6}\,\text{kg/m}^3$. (Note the difference with the 'approximated' concentration distribution.)

In the second cell ($i = 1$), $C_m(t = 24\,\text{h}) = 3.2 \times 10^{-7}\,\text{kg/m}^3$. The concentration versus time distribution at $x = 3$ km is plotted below, as well some data points at the upstream site (Fig. 9.9).

3. Uniform equilibrium flow conditions yield: $d = 1.26\,\text{m}$, $V = 1.4\,\text{m/s}$, $V_* = 0.08\,\text{m/s}$, $\varepsilon_t = 0.059\,\text{m}^2/\text{s}$, $K = 15.8\,\text{m}^2/\text{s}$, length of initial zone = 680 m (hence the initial may be neglected in first approximation for an analysis over the 6 km reach).

Sub-reach volume = 16 026 m³ (i.e. Vol = LWd), cell residence time = 1068 s (i.e. $\Delta T = \text{Vol}/Q$).

ADZ model calculations are performed using equation (9.9b) for $n = 4$ (i.e. five cells). The results are presented below.

Remark: Note the discrepancy between Taylor's dispersion model and the ADZ model. In the latter, the sudden mass slug injection is diluted instantaneously in the initial cell ($i = 0$) where: $C_m(i = 0,\ t = 0) = 1.5/16\,026 = 9.036 \times 10^{-5}\,\text{kg/m}^3$, before being advected downstream (Fig. 9.10).

4. Hydraulic calculations: $Q = 95\,\text{m}^3/\text{s}$, $V = 2.37\,\text{m/s}$, $V_* = 0.12\,\text{m/s}$, $\varepsilon_t = 0.052\,\text{m}^2/\text{s}$, $K = 2160\,\text{m}^2/\text{s}$, initial zone length: 13.8 km (almost the study reach length):

$$\alpha = \frac{4Kk}{V^2} = 0.014$$

	T (s)	$C_{\text{max}} \times 10^{-5}$ (kg/m³)	$C_m(t = 1\,\text{h}) \times 10^{-6}$ (kg/m³)	$C_m(t = 2\,\text{h}) \times 10^{-5}$ (kg/m³)
Conservative tracer	5700	1.5	9.1	1.2
Reactive tracer	5700	1.35	8.3	1

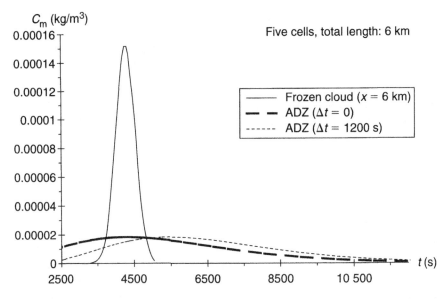

Fig. 9.10 Concentration versus time at $x = 6$ km (five-cell ADZ model). Comparison between Taylor's dispersion model (solid line) and the ADZ model (dotted/dashed lines).

where the conservative contaminant calculations are calculated using equation (8.9) (concentration versus time at a fixed location $x = 13.5$ km) and equation (9.11) is re-arranged (using the frozen cloud approximation) as:

$$C_{\mathrm{m}} = \frac{M}{A\sqrt{4\pi K \dfrac{x}{V}}} \exp\left(-\frac{(x - Vt)^2}{4K \dfrac{x}{V}} - kt\right) \tag{9.11}$$

Note: The flood flow conditions are near-critical: i.e. $Fr = 0.9$. Free-surface undulations are likely to develop, and possibly to enhance mixing and dispersion. The results of longitudinal dispersion tests under such conditions are probably *not* representative.

5. The characteristic distance of contaminant decay, $V/k = 512$ m, is less than the initial zone. The longitudinal dispersion theory cannot be applied.
6. $BOD_o = 48.8$ mg/L (weighted average, at $t = 0$)
   $C_o(x < 0) = 0.95 \times 11.2 = 10.7$ mg/L (river, upstream injection point)
   $C_o(x = 0+) = 9.07$ mg/L (river, immediately downstream of injection point, assuming effluent DO = 0)
   $X_{\min} = 236$ km
   $C_{\min} = 0.007$ kg/m^3. Note that this is a very low value (compare with Australian Standards)
   $BOD(x = X_{\min}) = BOD_o \exp(-k_r X_{\min}/V) = 0.028$ kg/m^3
   $BOD_5(x = X_{\min}) = BOD(x = X_{\min})(1 - \exp(-k_r 5 \times 86\,400)) = 0.018$ kg/m^3

10

Mixing in estuaries

Summary

In this chapter, the basics of mixing and dispersion in estuaries are presented. After some definitions and basic concepts, simple applications are developed.

10.1 Presentation

An estuary is a water body where the tide meets a river flow and where mixing of freshwater and seawater occurs. Estuaries may be classified as a function of the salinity distribution and density stratification (Fig. 10.1). A salt-wedge estuary develops when the river flows into a sea with low tidal range (i.e. $< 2\,\text{m}$). The fresher surface waters flow over a denser bottom layer, called the *salt wedge*. The strong density gradient at the wedge interface inhibits mixing between freshwater and saltwater. Shear stresses acting on the interface induced however some saltwater entrainment into the freshwater flow. The loss in saltwater from the wedge, without gain of freshwater (because of stratification), induces a residual saltwater flow, sketched in Fig. 10.1(a). With moderate tidal ranges ($2–4\,\text{m}$), the tidal flow enhances mixing and the estuary becomes partially mixed. Vertical mixing (sketched in Fig. 10.1(b)) produces larger bottom residual flows than in a salt-wedge estuary. For larger tidal ranges, the estuary becomes well mixed vertically. There is very little variation in salinity with depth and the residual velocity is seaward at all depths.

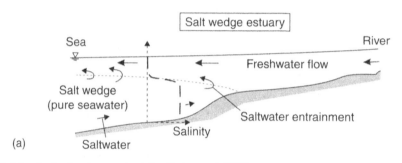

Fig. 10.1 Salinity distributions in estuaries (after Fischer *et al.* 1979).

The above classification does not express the unique features of each estuary (Fig. 10.2), and it does not account for seasonal changes, nor for differences between neap and spring tides. The study of mixing in estuary remains a complex process (Ippen 1966, Fischer *et al.* 1979, Lewis 1997).

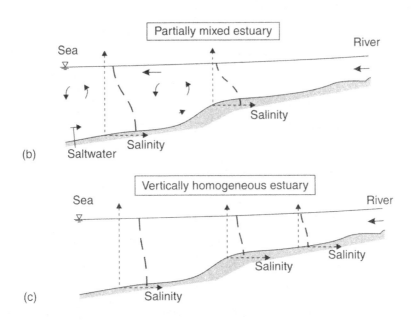

(b)

(c)

Fig. 10.1 (*Contd*)

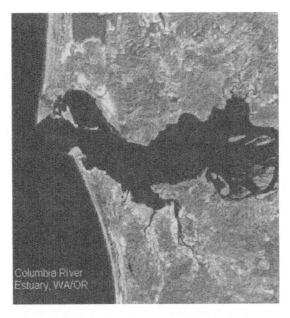

Fig. 10.2 Examples of estuaries. (a) Estuary of the Columbia River, USA (courtesy of the Coastal and Hydraulics Laboratory, U.S. Army Corps of Engineers).

(b-i)

(b-ii)

Fig. 10.2 (*Contd*) (b) Estuary of the Flora River, Dahouet (Côtes d'Armor, France) – looking upstream. The fjord-like estuary is an harbour for fishing boats and small ships. A new marina was installed in the old swamp (background). The river arrives on the top left of the photograph. The tidal range may exceed 11 m in spring tides and the harbour is empty at low tides. The estuary is well mixed. (b-i) At low tide on 8 September 2000. (b-ii) At high tide on 3 September 2000.

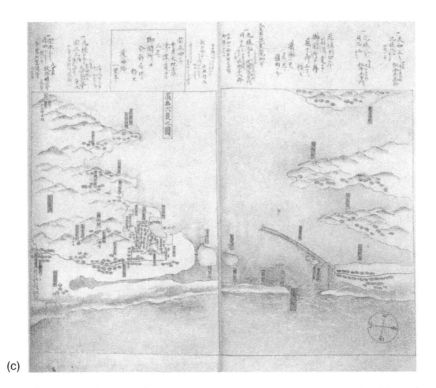

(c)

(d)

Fig. 10.2 (c) Hamanako Lake, Hama-matsu (Japan). Old sketch around AD 1707 (?) of Arai Check point (on left) and Hamanako Lake Estuary. The Pacific Ocean is in the bottom foreground. Tidal range <2 m. (d) Permanently open Bree Estuary, South Africa (courtesy of the CERM, South Africa). Western Cape Estuary.

(e)

(f)

Fig. 10.2 (*Contd*) (e) Permanently open Sundays River Estuary in the Eastern Cape, South Africa (courtesy of the CERM, South Africa). The photograph shows the extensive flood tidal delta inside the mouth. (f) Old harbour of Nagoya (Japan) on 18 November 2001. Site of Atsuta terminal of ferry between Atsuta posting station (next to bell house, foreground right) and Kuwana town on the old Tokai-do highway. Looking south at the harbour discharging into the shallow waters of Ise Bay.

Notes

1. The residual flow is the velocity field obtained by averaging the velocity over the full tidal cycle.
2. The salinity is often expressed in parts per million (ppm) or parts per thousand (ppt). The conversion is:

$$1\,\text{ppm} = 1 \times 10^{-3}\,\text{ppt} = 1\,\text{mg/L}$$

3. The salinity is defined as the amount of dissolved salts in water. In surface waters of open oceans, the salinity ranges from 33 to 37 ppt. An average value of 35 ppt is typical.

   Note however that regions of high evaporation may have higher surface salinities (e.g. parts of the Red Sea have surface salinity up to 41 ppt) while regions of high precipitation have lower surface salinities. In shoreline regions close to large freshwater sources, the salinity may be further reduced by dilution: e.g. areas of the Baltic Sea have salinity values below 10 ppt.

4. The chlorinity is defined in relation to salinity as:

$$\text{Salinity} = 1.80655 \times \text{chlorinity}$$

   Chlorinity is basically defined as the number of grams of chlorine, bromine and iodine contained in 1 kg of seawater, assuming that the bromine and iodine are replaced by chlorine.

5. In the field, the salinity is often measured in terms of the chloride ion content or the electrical conductivity.

Seawater properties

Seawater is a complex mixture of 96.5% water, 2.5% salts and smaller amounts of other substances. The most abundant salts are sodium chloride (NaCl, 29.536 ppt), sulphate (SO$_4$, 2.649 ppt), magnesium (Mg, 1.272 ppt), calcium (Ca, 0.400 ppt) and potassium (K, 0.380 ppt) (Riley and Skirrow 1965, Open University Course Team 1995).

Seawater density, viscosity and surface tension differ little with freshwater properties (Table 1.1, Chapter 1). Typical values of density, dynamic viscosity and surface tension are respectively: $\rho = 1024 \, \text{kg/m}^3$, $\mu = 1.22 \times 10^{-3} \, \text{Pa s}$ and $\sigma = 0.076 \, \text{N/m}$.

10.2 Basic mechanisms

The analysis of mixing in estuaries is more complicated than in rivers. Mixing in estuaries is affected by a combination of three agents: the wind, the tide and the river. In real estuaries, these three effects are superposed, although one or two may dominate. For example, the estuary of the Flora River (Fig. 10.2(b)) is dominated by tidal and river interactions. Many estuaries are affected by seasonal changes. For example, a flood may stratify a previously well-mixed estuary, while cyclonic winds may mix the estuary. An example is the Hamanako Lake (Fig. 10.2(c)). The saltwater lake system is strongly stratified in summer, but field measurements showed rapid destratification (within 24 h) during typhoons, as the result of mixing induced by stormy winds. An estuary may be strongly affected by floods. Figure 10.2(e) shows the extent of the flood channel.

10.2.1 Mixing caused by winds

Wind may or may not play a major role in estuary mixing. Its effect depends primarily upon the currents induced. Currents are produced by momentum transfer from the atmospheric boundary layer to the sea at the free surface. Considering an uniform wind blowing over a

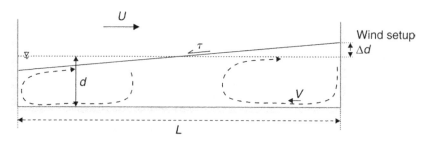

Fig. 10.3 Sketch of wind setup for a well-mixed system.

free surface, the wind exerts a drag onto the water surface. The wind will pull floating objects in the wind direction: e.g. the dispersion of an oil spill is strongly affected by the wind direction. In turn, the wind shear may induce a recirculation pattern. The actual stress on the surface (i.e. wind stress) is often expressed as:

$$\tau = \frac{1}{2} C_d \rho_{air} U_{10}^2 \tag{10.1}$$

where U_{10} is the wind speed measured $10\,m$ above the water surface, ρ_{air} is the air density and C_d is a dimensionless drag coefficient.

For the ideal water body sketched in Fig. 10.3, the dashed line depicts the free surface at rest in absence of wind. When the wind is blowing, a shear force acts on the water surface and the surface tilts with a wind setup in the downstream wind direction and a wind setdown on the upstream side, as shown in Fig. 10.3. The wind setup Δd may be derived from the balance of forces acting on the water:

$$\frac{1}{2} \rho g W (d - \Delta d)^2 - \frac{1}{2} \rho g W (d + \Delta d)^2 + \tau W L = 0 \tag{10.2}$$

where ρ is the water density, W is the water width normal to the wind direction and L is the fetch. It yields:

$$\Delta d = \frac{C_d}{4} \frac{\rho_{air}}{\rho} \frac{U_{10}^2 L}{gd} \tag{10.3}$$

For a given wind speed, the wind setup increases with the fetch. It is more important in shallow waters than in deep waters as $\Delta d \propto 1/d$. Equations (10.2) and (10.3) were developed for a steady or quasi-steady state. In practice, the wind must blow for a certain time to produce a wind setup.

Application: recirculation current

The wind shear stress may induce the development of recirculation cell(s), as sketched in Fig. 10.3 for a well-mixed system. The number of cells is a function of the fetch length, bathymetry and topography. For an ideal two-dimensional water body (e.g Fig. 10.3), a bottom recirculation current is generated by the pressure difference across the fetch ($2\rho g \Delta d$) resulting from the wind setup. The bottom current direction is opposed to the wind direction and the velocity magnitude V may be estimated from

Darcy's law:

$$\rho g\left(2\Delta d\right) = f\,\frac{L}{D_{\mathrm{H}}}\,\rho\,\frac{V^2}{2}$$

where f is the Darcy–Weisbach friction factor (Appendix A in Chapter 5) and the hydraulic diameter D_{H} of the bottom current is about: $D_{\mathrm{H}} \sim 2d$ assuming that the current flows in the lower half of an infinitely wide water body. It yields:

$$V \approx \sqrt{\frac{4g}{f}\,\frac{\Delta d}{L}\,D_{\mathrm{H}}}$$

Combining with equation (10.3), the recirculation current velocity becomes:

$$V \approx \sqrt{\frac{2C_{\mathrm{d}}}{f}\,\frac{\rho_{\mathrm{air}}}{\rho}\,U_{10}}$$

Notes

1. The drag coefficient is about $C_{\mathrm{d}} \sim 0.002$ for shallow waters independently of the wind speed (Fischer *et al.* 1979, p. 162). For deep water reservoirs, Novak *et al.* (2001, p. 183) recommended $C_{\mathrm{d}} \sim 0.006$.
2. The fetch is the distance over which the wind acts on the water body.
3. The wind must blow over the fetch for a certain time to produce the wind setup (and setdown). The time increases with increasing fetch length and decreasing wind speed. Novak *et al.* (2001, p. 183) gave typical values of 1 h for a 3 km long fetch and of 3 h for a 20 km long fetch, for a 11 m/s wind speed.
4. A related form of wind setup is the storm surge. A storm surge refers to any departure from normal water level resulting from the action of storms. In coastal zones, a wind blowing onshore (i.e. toward the coast) may induce water levels near the coast that are higher than normal and cause flooding of low-lying areas. Such an event is called a positive storm surge. By contrast, a wind blowing offshore may induce water levels lower than normal (i.e. negative storm surge).

   Most literature deals with positive surges because these are usually more dramatic (Ippen 1966, Gourlay and Apelt 1978). For example, in East Pakistan, 200 000 lives were lost in 1970 when a 6 m positive surge covered the Ganges delta. In North Queensland, cyclone 'Mahina' caused a large storm surge in Bathurst Bay on 5–6 March 1899; despite some confusion, the magnitude of the storm surge was estimated to be between 12 and 15 m. Fifty-four vessels were wrecked, including several pearl fishing boats (Fig. 10.4), and over 300 lives were lost.

   Negative surges are important also. For example, the British Meteorological Office has a special service warning big tankers of negative surges in the English Channel to avoid accidental grounding caused by the lower sea levels.

Discussion: The Sea of Reeds

In the Bible, a wind-setup effect allowed Moses and the Hebrews to cross shallow water lakes and marshes (Sea of Reeds) during their exodus. At the end of wind setdown the returning waters crushed the pursuing Egyptian army (Exodus 13–15). Although there is some controversy among scholars, the Sea of Reeds is believed to be just North of the

Gulf of Suez in the Eastern Nile Delta. This region of shallow marshes is sometimes called the Great Bitter Lakes region or Lake Timsah area. Note that the Sea of Reeds ('papyrus') is often mistaken for the Red Sea.

Scientistic studies suggested that the wind setup/setdown was caused by a strong tropical storm. The strong winds could have created a 1–4 m wind setdown on the shallow part of the marshes. The wind was maintained all night (Exodus 14:21). In the morning, the water level returned to normal (Exodus 14:27): i.e. the drop in water level lasted for 12 h at most.

Fig. 10.4 The pearling station at Goode Island, North Queensland (Australia) (courtesy of the Queensland State Emergency Service Far North). All the boats were lost in the 1899 cyclone Mahina. From 'The Pearling Disaster', 1899.

10.2.2 Mixing caused by tides

The tides generate an oscillation in water elevations at the downstream end of a river system. The change in downstream water level creates a backwater effect. For example, the rising tide (flood tide) may induce a reversal in flow direction in the estuary and in the lower reach of the river, while the flow direction is seaward during the ebb tide (declining tide).

Figure 10.5 shows field measurements in a small estuary, with free-surface elevations at the river mouth and 2 km upstream of the mouth, and velocities at 2 km upstream of the river mouth. At 2 km from the mouth, the maximum and minimum water levels are observed slightly after the high and low tides (at the river mouth). The information on tide reversal must travel upstream the river estuary with a celerity of about $\sqrt{gd}$ in first approximation where d is the water depth (Chapter 12). Note that the velocities are about zero at high and low tides, although the data exhibit some scatter (Fig. 10.5).

Remarks
1. The ebb is the flow of the declining tide.
2. The flood is the flow of the rising tide.
3. The tides are waves caused by the combined effects of the gravitational attraction between the Earth and the Sun and between the Earth and the Moon, and the rotation of the Earth around its axis. Predictions of the tide at most ports are published annually in tide tables. Tides may be *diurnal* ($T \approx 24$ h 50 min) or *semi-diurnal* ($T \approx 12$ h 25 min).
4. The Earth rotates with respect to the Moon with a period of $T \approx 24$ h 50 min. It is a constant for each coastal location.

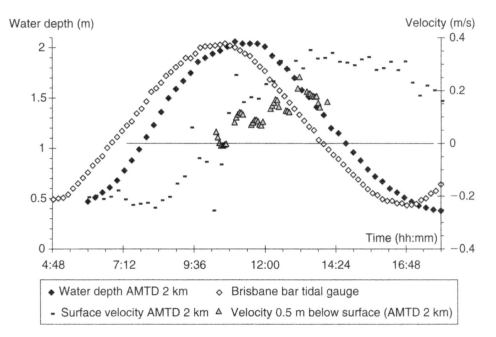

Fig. 10.5 Experimental observations of water elevation and velocities in an estuary (after Chanson et al. 2003). Water depth, surface velocity and velocity measured 0.5 m beneath the free-surface in a meander of Eprapah Creek at 2.0 km upstream of the river mouth on 4 April 2003 – comparison with the tidal heights at the river mouth (Brisbane bar).

Shear effect in estuaries

In an estuary, the flow appears like a river but it goes back and forth with the tide. Friction of the tidal flow on the estuary boundaries generates turbulence and leads to turbulent mixing. The effects of flow oscillation on the longitudinal dispersion coefficient may be expressed as:

$$K = K_\infty f\left(\frac{T}{T_c}\right) \tag{10.4}$$

where K_∞ is the longitudinal dispersion coefficient for a relatively long tidal period ($T \gg T_c$), T is the tide period, $T_c = W^2/\varepsilon_t$ is the cross-sectional mixing time (or time scale for transverse mixing), W is the waterway width and ε_t is the rate of transverse mixing.

Shear flow dispersion has a maximum effect where the tide period T is similar to the time scale for transverse mixing T_c. For such conditions, the maximum dispersion coefficient equals:

$$K(T \approx T_c) = 0.016\, V T \tag{10.5}$$

where V is the mean tidal flow velocity. Equation (10.5) was developed assuming an uniform, long estuary with quasi-homogeneous density distribution. It provides an useful estimate of the dispersion in constant-density sections of an estuary, but it accounts for only the effect of (unsteady) shear flow.

The same analysis shows also much smaller dispersion coefficients in tidal flows than in similar steady river flows (paragraph 10.4). For short tidal periods ($T \ll T_c$), the concentration distribution does not have time to respond to the changes in velocity profiles, and the longitudinal dispersion coefficient tends towards 0 (Fischer et al. 1979).

Remarks

1. The effects of flow oscillation on the longitudinal dispersion coefficient were developed analytically by Fischer *et al.* (1979, pp. 94–99). The derivation was applied to tidal estuaries (Fischer *et al.* 1979, pp. 234–237).
2. Most estuaries are wider than deep, and hence the characteristic mixing time T_c is defined in terms of the channel width and rate of transverse mixing.

Tidal pumping

The term 'tidal pumping' refers to residual circulation effects. Although not obvious to a casual observer, the residual circulation is the velocity field obtained by averaging the velocity over the tidal cycle. In an estuary, the residual circulation is a combination of (1) Coriolis effect (i.e. effect of the Earth's rotation) and (2) some dissymmetry in tidal fluctuations of depth-averaged velocity and depth-averaged salinity between rising and declining tidal flows (sometimes called phase shift or Stokes drift, Lewis 1997).

In Northern Hemisphere, Earth's rotation deflects flood tide currents toward the left bank (when looking downstream, toward the sea), and ebb currents towards the right bank, resulting in a net clockwise circulation. In the Southern Hemisphere, the flood current is deflected toward the right bank and the ebb flow toward the left bank.

An example of tidal dissymmetry is the flow through a narrow inlet mouth (e.g. Figs 10.2(c) and 10.6). The flood tide (rising tide) flows into the estuary like an orifice flow, forming a confined jet. During retreating tide, the ebb flow comes around all the inlet mouth, in the form of a two-dimensional potential flow around a sink (e.g. Vallentine 1969). Another example is the net flow around a series of islands and in a braided estuary: e.g. the Arai canal system (Fig. 10.6). Remember that friction is proportional to the square of the velocity. As the seaward flow (ebb flow) is faster than the landward flow, the friction will be in average greater during the ebb flow. As a result, the net flow is often landward in the narrower channel. Fischer *et al.* (1979, pp. 237–241) illustrated further examples of tidal dissymmetry between ebb and flood flows.

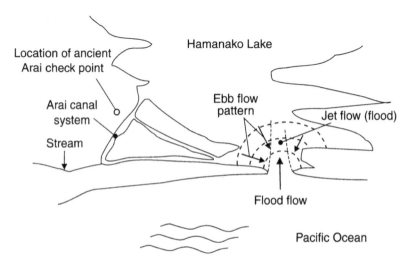

Fig. 10.6 Sketch of tidal flow dissymmetry at Hamanako Lake inlet (see also Fig. 10.2(c)).

Remarks

1. The Coriolis force per unit mass (i.e. Coriolis acceleration) equals:

$$\Gamma_{\text{Coriolis}} = 2V\omega \sin \phi$$

where V is the velocity of the fluid/object subjected to the Coriolis force, ω is the angular velocity of the Earth ($\omega = 2\pi/T$), T is the earth rotation period ($T = 24\,\text{h}$) and ϕ is the latitude. If x is the coordinate in the flow direction, z is the vertical coordinate positive upwards and ϕ is positive in Northern Hemisphere, the Coriolis acceleration applies in the y-direction: i.e. to the left (when looking downstream) in the Northern Hemisphere and to the right in Southern Hemisphere.

2. Gustave Gaspard Coriolis (1792–1843) was a French mathematician and engineer of the 'Corps des Ponts-et-Chaussées' who first described the Coriolis force.

3. Located on the Enshu Coast along the Pacific Ocean, the Arai township is located about half-distance between Tokyo and Osaka (Japan). It is on the right bank of the Hamanako Lake estuary (Figs 10.2(c) and 10.6). During the Edo period, a main check point was located at Arai, controlling the traffic of weapons and movement of people on the road to Edo (Tokyo).

Tidal trapping

Estuaries, like rivers, are affected by 'dead zones'. The role of such zones are enhanced by tidal action. The propagation of the tide in an estuary is a balance between the inertia of the water mass, the pressure force due to the slope of the water surface and the retarding force of bottom friction. As the tide changes, small dead zones have little momentum and the flow direction will change as soon as the water level begins to drop. In contrast, the flow in the main channel has an initial momentum and the current will continue to flow against the opposing pressure gradient. This process will enhance longitudinal dispersion induced by the dead zones.

10.2.3 Mixing caused by the river

One, or more, rivers enter an estuary and deliver a freshwater discharge Q. The river is a source of buoyancy flux $\Delta\rho g Q$, where $\Delta\rho$ is the density difference between the sea and the river. In an estuary, a dimensionless measure of density stratification is the estuary Richardson number defined as:

$$Ri_t = \frac{\Delta\rho g}{\rho} \frac{Q}{WV_t^3}$$

where W is the channel width and V_t is the rms tidal velocity (Fischer *et al.* 1979). If Ri_t is very small, the estuary is well mixed (Fig. 10.1(c)), and density effects may be neglected. If Ri_t is very large, the estuary is strongly stratified, and the flow motion is dominated by density currents. For example, when a river discharges into an estuary connected to a tideless sea (e.g. Mediterranean Sea), the freshwater flows over the saltwater as an undiluted freshwater layer. Saltwater intrudes underneath the freshwater flow in the form of a wedge (Fig. 10.1(a)).

Remarks

1. Field observations suggest that the transition between well mixed to strongly stratified estuary occurs for $0.08 < Ri_t < 0.8$ (Fischer et al. 1979).
2. Fischer et al. (1979) suggested a modified estuary Richardson number defined as: $Ri_t = (\Delta\rho \times g/\rho)Q/(W V_*^3)$, where V_* is the rms shear velocity.
3. rms stands for root mean square.

Application: Stratification of the Pimpama Creek Estuary

Pimpama Creek, Queensland is a small water system in Albert Shire, North of the Gold Coast. The catchment is very flat and made of acid soils. The creek is blocked by a weir at adopted middle thread distance (AMTD) measured upstream from the mouth 3.8 km (Fig. 7.2(c)). The structure is designed to prevent salt intrusion into the waterway. Gates are only open during floods periods to relieve upstream flooding.

Detailed field measurements were conducted by the writer on 19 December 2002 around 1:30 p.m. at AMTD 3.25 km, located downstream of the weir system. The time corresponded to the mid ebb tide. The observations are given below.

The data showed a strong stratification of the estuary, although the system was mainly saltwater (i.e. no brackish water) because of the weir gates were closed, preventing mixing between saltwater and freshwater. The top layer was warmer and saturated in oxygen. From 1.2 m below the free surface, the water was cooler and had low dissolved oxygen contents. On 19 December 2002, the tidal range was particularly important. At high tide, the river banks and mangroves were submerged. These waters were warmed up by the sun. During the ebb, the warmer, lighter waters flowed above the denser deep-channel waters and the phenomenon induced the marked stratification.

Depth (m) (1)	Temperature (°C) (2)	Dissolved oxygen content (%) (3)	Turbudity NTU (4)	Conductivity mS/cm (5)	pH (6)	Remarks (7)
0.2	29.95	1.061	7	50.5	7.9	
0.4	29.9	1.061	8	50.5	7.9	
0.6	29.9	1.042	9	50.6	7.8	
0.8	29.8	1.051	9	50.7	7.8	
1	29.8	1.055	9	50.8	7.9	
1.2	29.6	1.018	9	50.9	7.8	
1.4	29.1	0.846	10	51.5	7.8	
1.6	28.8	0.621	12	51.7	7.7	
1.8	28.6	0.553	13	51.8	7.6	
2	28.5	0.545	16	51.8	7.6	
2.2	28.2	0.596	20	51.8	7.7	Just above the bottom

Notes: Depth measured below the free surface; data collected during early ebb flow.

10.2.4 Discussion: mixing induced by tidal bores

When a river mouth has a flat, converging shape and when the tidal range exceeds 6–9 m, the river may experience a tidal bore (Fig. 10.7). A tidal bore is basically a series of waves

Fig. 10.7 Examples of tidal bores. (a) Tidal bore on the Daly River, Northern Territory, Australia (courtesy of Gary and Rhonda Higgins). (b) Tidal bore at Batang Lupar, Malaysia (courtesy of Mr Lim Hiok HWA, Department of Irrigation and Drainage, Sarawak). (c) Undular tidal bore of the Dordogne River on 27 September 2000 at 5:00 p.m. Looking upstream at the murky waters after the bore passage (foreground), the surfers riding the bore and the glossy free surface in background.

propagating upstream as the tidal flow turns to rising. It is a positive surge (Chapter 12). As the surge progresses inland, the river flow is reversed behind it (e.g. Lynch 1982, Chanson 2001).

The best historically documented tidal bores are probably those of the Seine River (France) and Qiantang River (China). The *mascaret* of the Seine River was documented first during the 7th and 9th Centuries AD, and in writings from the 11th to 16th Centuries (Malandain 1988). It was locally known as 'la Barre'. The Qiantang River bore, also called Hangzhou bore, was early mentioned during the 7th and 2nd Centuries BC. It was described in 8th Century writings and later in a 16th Century Chinese novel.[1] The bore was then known as 'The Old Faithful' because it kept time better than clocks. A tidal bore on the Indus River might have wiped out the fleet of Alexander the Great (Malandain 1988, Jones 2003). Another famous tidal bore is the 'pororoca' of the Amazon River observed by V.Y. Pinzon and C.M. de La Condamine during the 16th and 18th Centuries respectively. The Hoogly (or Hooghly) bore on the Gange was documented in 19th Century shipping reports. Smaller tidal bores occur on the Severn River near Gloucester, England, on the Garonne and Dordogne Rivers, France, at Turnagain Arm and Knik Arm, Cook Inlet (Alaska), in the Bay of Fundy (at Petitcodiac and Truro), on the Styx and Daly Rivers (Australia) and at Batang Lupar (Malaysia) (Fig. 10.7).

A tidal bore may affect shipping industries. For example, the *mascaret* of the Seine River had had a sinister reputation. More than 220 ships were lost between 1789 and 1840 in the Quilleboeuf-Villequier section (Malandain 1988). The height of the *mascaret* bore could reach up to 7.3 m and the bore front travelled at a celerity of about 2–10 m/s. Even in modern times, the Hoogly and Hangzhou bores are hazards for small ships and boats.

Tidal bores induce strong turbulent mixing in the estuary and river mouth. The effect may be felt along considerable distances. With appropriate boundary conditions, a tidal bore may travel far upstream: e.g. the tidal bore on the Pungue River (Mozambique) is still about 0.7 m high about 50 km upstream of the mouth and it may reach 80 km inland. Mixing and dispersion in a tidal bore affected estuary are not comparable to well-mixed estuary processes. Instead the effects of the tidal bore must be accounted for and the bore may become the predominant mixing process. The effect on sediment transport was studied at Petitcodiac and Shubenacadie Rivers (Canada), in the Sée and Sélune Rivers (France), Ord River (Australia), Turnagain Arm inlet (Alaska) and on the Hangzhou bay (China) (e.g. Tessier and Terwindt 1994, Bartsch-Winkler *et al.* 1985, Wolanski *et al.* 2001, Chen *et al.* 1990). The arrival of the bore front is associated with intense bed shear and scour. Behind sediment material is advected upwards by large scale turbulent structures evidenced in Fig. 10.7(c). Sediment suspension behind the bore is sustained by strong long-lasting wave motion. At the Dee River (UK), Dr E. Jones observed more than 230 waves, also called whelps or *éteules*. Murphy's (1983) photograph showed more than 30 well-formed undulations behind the Amazon *pororoca*. At the Dordogne River (France), the writer observed an intense wave motion lasting more than 20 min after the bore passage (Chanson 2001). Mixing and dispersion in a tidal bore affected estuary is not comparable to well-mixed estuary processes (e.g. Appendix C in Chapter 7 and Appendix A, Section 10.6).

The impact of tidal bores on the ecology is acknowledged. In the Amazon River, piranhas eat matter in suspension after the passage of the bore (Cousteau and Richards 1984). At Turnagain Arm inlet, bald eagles and eagles were seen fishing behind the bore, while beluga whales were observed playing in the bore as it formed near the mouth of the arm (Bartsch-Winkler and Lynch 1988, Molchan-Douthit 1998). In the same estuary, a moose tried unsuccessfully

---

[1] 'Outlaws of the Marsh', by Shi Nai'an and Luo Guanzhong. *Foreign Languages Press*, Beijing 1980. Written in the 16th Century, the novel described historical events from the period AD 1100–1130.

to outrun the bore; he was caught and disappeared (Molchan and Douthit 1998). In the Severn River, the bore impacted on sturgeons in the past and on elvers (young eels) today (Witts 1999, Jones 2003). In the Bay of Fundy, Rulifson and Tull (1999) studied the impact of bores on striped bass spawning.

A tidal bore is a very fragile process. The bore development is closely linked with the tidal range and river mouth shape (Chapter 12). Once formed, the bore existence relies upon the exact momentum balance between the initial and new flow conditions. A small change in boundary conditions and river flow may affect adversely the bore existence. Dredging and river training yielded the disappearance of several tidal bores: the *mascaret* of the Seine River (France) no longer exists, the Colorado River bore (Mexico) is drastically smaller. Although the fluvial traffic gained in safety in each case, the ecology of the estuarine zones were adversely affected. The tidal bores of the Couesnon (France) and Petitcodiac (Canada) rivers almost disappeared after construction of an upstream barrage. Natural events may also affect a tidal bore. During the 1964 Alaska earthquake (magnitude 8.5), the inlet bed at Turnagain and Knik Arms subsided by 2.4 m. Since smaller bores have been observed. Also at Turnagain and Knik Arm inlets, strong and winds (opposing the flood tide) were seen to strengthen the bore. On the other side, the construction of the Ord River dam (Australia) induced siltation of the river mouth and appearance of a bore (Wolanski *et al.* 2001). The bore disappeared since following large flood flows in 2000 and 2001 which scoured the river bed.

10.3 Applications

10.3.1 Salt wedges

In a tideless sea, the estuary is characterized by strong stratification and saltwater intrusion into the river system (Fig. 10.8). The saltwater layer is clearly definable, limited in space and it underlies the freshwater. This is called a *salt wedge*.

The steady form of the salt wedge in a tideless sea is the *arrested salt wedge*. Its form was analysed theoretically and the result was demonstrated experimentally. The shape of the

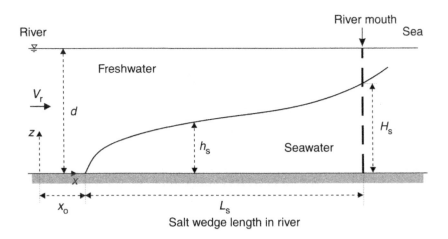

Fig. 10.8 Definition sketch of an arrested salt wedge.

saline wedge may be estimated as:

$$\frac{h_s}{H_s} = \frac{0.001926 + 3.560\,\dfrac{x-x_0}{L_s}}{1 + 8.741\,\dfrac{x-x_0}{L_s} - 6.118\left(\dfrac{x-x_0}{L_s}\right)^2} \qquad \text{for } x_0 < x < x_0 + L_s \qquad (10.6)$$

where h_s is the wedge height at a distance x, H_s is the wedge height at the river mouth, x is the distance in the river flow direction, x_0 is the location of the most upstream wedge intrusion and L_s is the wedge length into the river (Fig. 10.8). Importantly equation (10.6) was found to be independent of the seawater salinity, river velocity, water depth, channel width and fluid viscosity.

The length of the salt wedge and the wedge height at the river mouth may be correlated by:

$$\frac{L_s}{d} = 6.0\left(\frac{1}{\nu}\sqrt{\frac{\Delta\rho}{\bar{\rho}}gd^3}\right)^{1/4}\left(2Fr\sqrt{\frac{\bar{\rho}}{\Delta\rho}}\right)^{-5/2} \qquad (10.7)$$

$$\frac{H_s}{d} = 0.8927 - 5.932\exp\left(\frac{-2.43}{\left(2Fr\sqrt{\bar{\rho}/\Delta\rho}\right)^{0.2424}}\right) \qquad (10.8)$$

where Fr is the river flow Froude number (i.e. $Fr = V_r/\sqrt{gd}$), ν is the freshwater kinematic viscosity, ρ is the freshwater density, $\Delta\rho$ is the density difference between the saltwater and freshwater,[2] and $\bar{\rho}$ is the average density of the two liquids[3] (Fig. 10.8).

Practically the above development is limited to steady and ideal flow conditions. In natural river systems, the salt wedge may respond to changes in river flow and to tidal oscillations.

Application

The theory of arrested salt wedge may be applied to design a vertical barrier preventing salt intrusion in a river system at a location x such as ($x_0 < x < L_s$), the minimum height h of the salt barrier above the river bottom must satisfy:

$$\frac{h}{H_s} > \frac{h_s}{H_s} = \frac{0.001926 + 3.560\,\dfrac{x-x_0}{L_s}}{1 + 8.741\,\dfrac{x-x_0}{L_s} - 6.118\left(\dfrac{x-x_0}{L_s}\right)^2}$$

For $x < x_0$, a vertical barrier is not needed.

Other means to arrest a salt wedge may include a water curtain or an air curtain (e.g. Nakai and Arita 2002).

Remarks

1. The theory of salt wedge was developed by Keulegan (in Ippen 1966, pp. 546–574) who verified it experimentally.

---

[2] That is, the saltwater density equals $\rho + \Delta\rho$.
[3] That is, $\bar{\rho} = \rho + \Delta\rho/2$.

2. In equation (10.7), the term $(1/v)\sqrt{\Delta\rho g d^3/\bar{\rho}}$ is analogous to a Reynolds number, called densimetric Reynolds number. The term $2Fr\sqrt{\bar{\rho}/\Delta\rho}$ is some densimetric Froude number, also called river flow parameter.

3. For the height of the wedge at the river mouth, Keulegan obtained an analytical expression:

$$\frac{H_s}{d} = 1 - \frac{1}{2^{2/3}}\left(2Fr\sqrt{\frac{\bar{\rho}}{\Delta\rho}}\right)^{2/3} \tag{10.9}$$

Agreement between theory (equation (10.9)) and data was satisfactory when:

$$2Fr\sqrt{\frac{\bar{\rho}}{\Delta\rho}} \sim 1$$

However, equation (10.8) provides a more accurate estimate for the range of flow conditions investigated in laboratory by Keulegan:

$$0.10 \leqslant 2Fr\sqrt{\frac{\bar{\rho}}{\Delta\rho}} \leqslant 1.5$$

4. Equations (10.6) and (10.8) are best fits of a series of detailed experimental results.

5. Garbis Hovannes Keulegan (1890–1989) was an Armenian mathematician who worked as hydraulician for the US Bureau of Standards since its creation in 1932.

10.3.2 Steady vertical circulation

Considering a well-mixed estuary, in a steady state case, the depth-averaged density increases with increasing distance x seaward. The slope of the mean water surface must counterbalance the mean density gradient.

In the horizontal direction, the Navier–Stokes equation becomes:

$$\frac{\partial V}{\partial x} + V\frac{\partial V}{\partial x} = -\frac{1}{\rho}\left(\frac{\partial P}{\partial x} + v_T\frac{\partial^2 V}{\partial z^2}\right) \tag{10.10}$$

where v_T is the eddy viscosity or momentum exchange coefficient in turbulent flow, P is the pressure, x is the horizontal coordinate and z is the vertical coordinate positive upward (Fig. 10.9(a)). Note that equation (10.10) was developed assuming a negligible vertical velocity component. For a steady flow and assuming that the advective term is small (i.e. $V \times \partial V/\partial x \sim 0$), the motion equation becomes:

$$0 = -\frac{1}{\rho}\left(\frac{\partial P}{\partial x} + v_T\frac{\partial^2 V}{\partial z^2}\right) \tag{10.11}$$

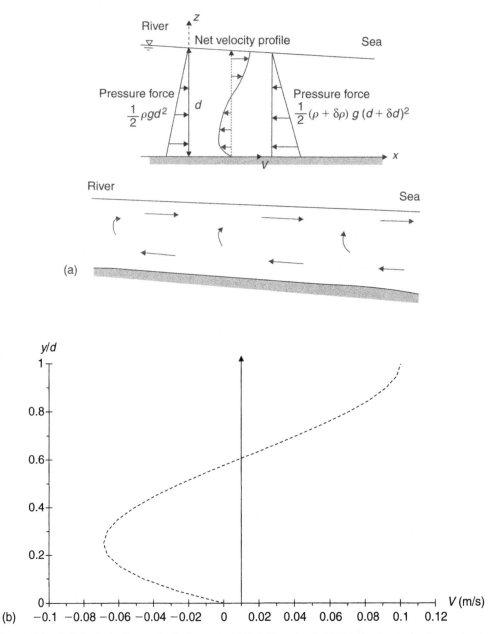

Fig. 10.9 Vertical circulation in a well-mixed estuary. (a) Definition sketch. (b) Velocity distribution in an estuarine system (equation (10.15)) assuming a 0.1 m/s surface velocity.

At an elevation z, the pressure equals the weight of the water column above:

$$P(z) = \int_{y=z}^{y=d} \rho g \, dy \qquad (10.12)$$

where ρ is the local fluid density and using the atmospheric pressure as a reference (i.e. $P(z = d) = 0$). Combining equations (10.11) and (10.12), it yields:

$$-g \frac{\partial d}{\partial x} - \frac{g}{\rho} \frac{\partial \rho}{\partial x} + \nu_T \frac{\partial^2 V}{\partial z^2} = 0 \qquad (10.13)$$

assuming a well-mixed estuary and where ρ is the depth-averaged density:

$$\rho = \frac{1}{d} \int_{z=0}^{z=d} \rho \, dz$$

In first approximation, the integration of equation (10.13) yields:

$$\frac{\partial d}{\partial x} = -\frac{3d}{8\rho} \frac{\partial \rho}{\partial x} \qquad (10.14)$$

where d is the water depth. Equation (10.14) expresses the relationship between the longitudinal mean density gradient and the free-surface slope. It derives from momentum considerations and it yields:

$$\frac{\partial d}{\partial x} \propto -\frac{\partial \rho}{\partial x}$$

The forces acting on a small vertical slice of the estuary are the pressure forces, the boundary friction and possibly the wind stress. At steady state, the solution of the motion equation gives the velocity distribution:

$$V(z) = \frac{gd^3}{48\rho\nu_T} \frac{\partial \rho}{\partial x} \frac{z}{d} \left(-8 \left(\frac{z}{d} \right)^2 + 15 \frac{z}{d} - 6 \right) \qquad (10.15)$$

where the momentum exchange coefficient ν_T is assumed constant with depth. Equation (10.15) predicts a steady circulation pattern, sketched in Fig. 10.9(a) and plotted in Fig. 10.9(b) assuming a 0.1 m/s surface velocity. Note that a surface velocity $V(z = d)$ of about 0.1 m/s is enough to generate a recirculation flux much greater than the river flow.

Basically equation (10.15) characterizes a circulation pattern in the estuary, induced by the longitudinal gradient in salinity and density. In turn, a weak salinity stratification produces a net mass transport across a vertical section which increases the stratification. Note that the circulation flow is more important in deep waters.

Discussion: density-driven currents at the Strait of Gibraltar

The Mediterranean Sea looses more water by evaporation that it receives freshwater from rain and rivers. It is saltier than the Atlantic Ocean. The result is a strong Atlantic surface water current (flowing westward) at the Strait of Gibraltar, while a deep, dense

Mediterranean current, flows westward, near the bottom. The deep, westward current takes place about 120–150 m below the free surface.

The Strait of Gibraltar was called the Pillars of Hercules (Fretum Herculeum) by the Ancients. Gibraltar is considered to have been one of the two Pillars of Hercules, the other being Mount Hacho, on the African Coast. The straight is 58 km long and it narrows to 13 km in width between Point Marroquí (Spain) and Point Cires (Morocco). The average water depth is 365 m.

In the Antiquity, Phoenician sailors used heavy, ballasted sea-anchors (sank in the bottom current) to pull their boats into the Atlantic Ocean. Jacques Cousteau repeated the experiment with the Calypso in the 1980s.[4] During World War II, German submarines used the surface water current and rising tide to pass silently the Strait of Gibraltar into the Mediterranean Sea.

Notes

1. The Navier–Stokes equation was first derived by L. Navier in 1822 and S.D. Poisson in 1829 by an entirely different method. They were later derived by Barré de Saint-Venant in 1843 and G.G. Stokes in 1845.
2. Louis Navier (1785–1835) was a French engineer who primarily designed bridge but also extended Euler's equations of motion. Siméon Denis Poisson (1781–1840) was a French mathematician and scientist. He developed the theory of elasticity, a theory of electricity and a theory of magnetism. Adhémar Jean Claude Barré de Saint-Venant (1797–1886), French engineer, developed the equations of motion of a fluid particle in terms of the shear and normal forces exerted on it. George Gabriel Stokes (1819–1903), British mathematician and physicist, is known for his research in hydrodynamics and a study of elasticity.
3. Equations (10.14) and (10.15) were developed for an idealized, two-dimensional, horizontal estuary, assuming a steady state, no wind shear stress and negligible freshwater inflow compared to the recirculation flow (Lewis 1997, p. 66).
4. The integration was performed assuming a constant momentum exchange coefficient with depth (i.e. $\nu_T(z)$ = constant).

10.4 Turbulent mixing and dispersion coefficients in estuaries

Estuaries are very complex systems and there are fundamental differences between salt-wedge estuaries, partially mixed estuaries and well-mixed estuaries (Fig. 10.1). Although estimate of mixing and dispersion coefficients should rely upon field tests, experimental observations are difficult and there is little systematic data on mixing and dispersion coefficients in estuaries. Appendix A regroups a number of observations of mixing and dispersion coefficients in estuarine zones. A small number of empirical correlations are shown in Table 10.1. It is however extremely difficult to apply these.

Overall the measured values of mixing and dispersion coefficients are relatively small compared to mixing coefficients in rivers. By comparison, field observations in a

[4] Cousteau, J.Y., and Paccalet, Y. (1987). Méditerranée: la mer blessée. *Flammarion*, 192 pp.

Table 10.1 Empirical estimates of mixing coefficients in estuarine zones

Coefficient (1)	Correlation (2)	Remarks (3)
Vertical mixing coefficient	$\dfrac{\varepsilon_v}{dV_*} = 0.067$	Well-mixed river flow
	$\dfrac{\varepsilon_v}{d\Delta V} = 0.0025$	Well-mixed river flow. At mid-depth, where ΔV = amplitude of depth-averaged current. After Bowden
Longitudinal dispersion coefficient	$K = \dfrac{V_s^2 d^2}{240\varepsilon_v}$	Well-mixed system. V_s = amplitude of surface tidal current. Assuming bed velocity is zero. After Bowden

Fig. 10.10 Ino-Hana Lake on 1 April 1999, shortly after a very strong wind storm during which a boat sunk and a nearby road was overtopped by wind waves – looking North.

meandering reach of the Missouri River yielded: $\varepsilon_t = 0.12\,\text{m}^2/\text{s}$ and $K = 1500\,\text{m}^2/\text{s}$ for $V = 1.55\,\text{m/s}$, $d = 2.7\,\text{m}$, $W = 200\,\text{m}$, $V_* = 0.074\,\text{m}$ (Fischer *et al.* 1979, pp. 110 and 126).

10.5 Applications

10.5.1 Application no. 1: Ino-hana Lake, Hama-matsu (Japan)

Ino-hana Lake is a shallow water lake near Hama-matsu (Japan) (Fig. 10.10). This saltwater body is connected to Hamanako Lake (Fig. 10.2(c)) and it is alimented in freshwater by several small streams. The lake is about 3 km long (North–South direction) and 1.5 km wide. The water depth is about 6 m. In summer months, the lake may be strongly stratified and the bottom waters become depleted in oxygen (table below).

During typhoons (i.e. cyclones), the wind storm and freshwater inflow induce a rapid mixing of the lake. This is illustrated in the table (below) showing field observations prior to a typhoon (9 September 2001), at the peak of the rainfall (11 September 2001 at 1:00 a.m.), and at the end of the event (12 September 2001 at 6:00 p.m.). Salinity and dissolved oxygen data are summarized in Fig. 10.11.

Date	Depth (m)	Salinity (ppt)	Density (kg/m³)	Dissolved oxygen (% saturated)	Mean wind speed (m/s) daily average	Maximum wind speed (m/s) daily average	Rainfall intensity (mm/h)	Remarks
(1)	(2)	(3)	(4)	(5)	(6)	(7)	(8)	(9)
9/09/2001	−1	25.2	1015.1	145	2.1	7.44	0	Strongly
at 0:00 a.m.	−2	26.5	1016.3	135				stratified
	−3	27.7	1016.9	90				lake.
	−4	29.05	1018.5	30				Before
	−5	30.1	1019.2	9				storm
	−6	30.8	1019.7	2				
11/09/2001	−1	22.5	1014	97	4.27	11.7	9.9	Peak of
at 1:00 a.m.	−2	25.2	1015.5	95				rainfall
	−3	25.7	1016	92				
	−4	26.1	1016.2	85				
	−5	27	1016.8	60				
	−6	30	1019	15				
12/09/2001	−1	18.2	1001	100	2.96	9.9	0	After
at 6:00 p.m.	−2	18.3	1005	100				maximum
	−3	18.8	1009	94				wind and
	−4	19.9	1012	70				rainfall
	−5	28.1	1017.8	49				
	−6	29.5	1018.5	18				

Note: Data courtesy of Pr S. Aoki, Toyohashi University of Technology (Japan).

On 11 September 2001 at 1:00 a.m., calculate the wind setup and bottom recirculation current for a North–South wind. Perform the calculations for both mean and maximum wind speed. Compare the results. (Assume a 10 mm bottom roughness.)

Solution
The wind setup is calculated using equation (10.3):

$$\Delta d = \frac{C_d}{4} \frac{\rho_{air}}{\rho} \frac{U_{10}^2 L}{gd} \qquad (10.3)$$

assuming $C_d = 0.002$ for shallow waters. The depth-averaged water density equals 1016.25 kg/m³. In the North–South direction, the fetch is 3 km long. The results are

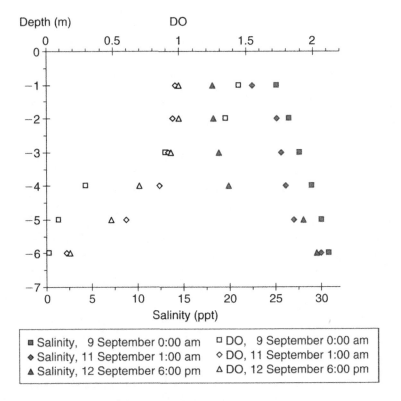

Fig. 10.11 Field observations (salinity and dissolved oxygen) at Ino-Hana Lake during the typhoon on 9–12 September 2001 (data courtesy of Pr S. Aoki). DO: dissolved oxygen.

summarized below:

	Mean wind speed	Maximum wind speed
Wind speed (m/s)	4.27	11.7
Wind setup Δd (m)	0.5 mm	4 mm
Recirculation velocity V (m/s)	0.07	0.18

Significant differences are noted between the mean wind speed conditions and the maximum wind speed conditions. With mean wind speed conditions, the lake waters recirculated completely in about 24 h (i.e. $T = 2 \times 3000/0.07$ s).

Discussion
In April 1999, the writer saw a strong wind storm on Ino-Hana Lake during which a major arterial road was overtopped by wind waves and the traffic was interrupted (Fig. 10.10). A boat sunk during the same event.

10.5.2 Application no. 2: Eprapah Creek, Queensland (Australia)

Eprapah Creek, Queensland is a small water system in Redlands Shire, South of Brisbane (Australia). The creek is characterized by low dissolved oxygen contents, high turbidity and

Fig. 10.12 Eprapah Creek, Redlands, Queensland on 24 June 1999 (courtesy of the Waterways Scientific Services, Queensland Environment Protection Agency). Note exotic waterweeds. The creek is characterized by low dissolved oxygen contents, high turbidity and nutrient concentrations.

nutrient concentrations. Field measurements were conducted in Eprapah Creek on 7 February 1996 in the estuarine zone (Fig. 10.12). The data are summarized below.

Date (1)	AMTD (km) (2)	depth (m) (3)	Salinity (ppt) (g/L) (4)
07-02-96	0	0.2	34.43
	0	1.5	34.4
	2	0.2	28.73
	2	2	33.46
	2	3	33.63
	2.4	0.2	27.31
	2.4	2	29.09
	2.7	0.2	21.86
	2.7	2	26.96
	2.7	3	26.99

Note: Depth: measured below the free surface.
Source: Data courtesy of Queensland Environment Protection Agency.

Notes
1. Turbidity refers to cloudiness caused by very small particles of silt, clay and other substances suspended in water.
2. AMTD is the distance upstream measured from the river mouth, following the riverbed. AMTD = 0 is the river mouth.

(a) Calculate the slope of the mean water surface to counterbalance the mean density gradient between AMTD = 2000 and 2700 m. (b) Predict the velocity profile at the river mouth

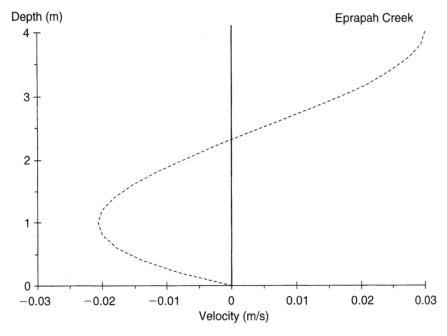

Fig. 10.13 Estimated velocity distribution at AMTD 2.4 km (Eprapah Creek).

and at AMTD = 2400 m. *Neglect the bed slope. Assume an average water depth of 4 m and $\nu_T = 0.003\ m^2/s$.*

Solution

First the data must be analysed to obtain the depth-averaged density. It yields:

Date (1)	AMTD (km) (2)	Depth-averaged density (kg/m^3) (3)
07-02-96	0	1023.6
	2	1022
	2.4	1019.3
	2.7	1017

The water density is almost constant between AMTD 2 km and the river mouth (AMTD = 0). Between AMTD 2 and 2.7 km, the average density gradient is: $\partial\rho/\partial x = 7.0 \times 10^{-3}\,\text{kg/m}^4$. In turn this implies a water level slope $\partial d/\partial x = -1.0 \times 10^{-5}$ to compensate for the mean density gradient. Assuming a reasonably well-mixed system, the steady vertical circulation yields a surface velocity of 0.03 m/s. The velocity profile is plotted in Fig. 10.13.

Discussion

Chanson *et al.* (2003) conducted a similar analysis at Epapah Creek based upon some field work in April 2003. Their results implied also some residual circulation leading to a renewal of the estuarine waters in about 1 week.

10.5.3 Application no. 3: Strait of Gibraltar

The Strait of Gibraltar is 13 km wide at the narrowest point and the average water depth is 365 m. The Mediterranean saline waters have a salinity of about 38 ppt, corresponding to a

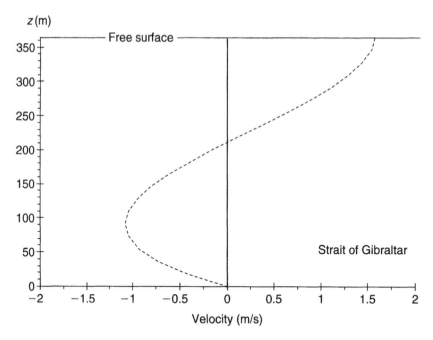

Fig. 10.14 Estimated velocity distribution in the Strait of Gibraltar assuming a well-mixed system.

water density of about $1030.5 \, kg/m^3$. The North Atlantic waters have a salinity of about 36 ppt, corresponding to a density of $1027.5 \, kg/m^3$. Assuming that the water density difference occurs over a 23 km long stretch, calculate the velocity profile at mid-distance of the straight. *Assume a reasonably well-mixed system. Use $v_T = 0.8 \, m^2/s$.*

Solution
Assuming a well-mixed system, the density gradient equals: $\partial\rho/\partial x = 0.00013 \, kg/m^4$, implying a water depth gradient $\partial d/\partial x = -1.7 \times 10^{-5}$, where x is the West–East direction (towards the Mediterranean Sea).

At mid-distance, the average water density is $1029 \, kg/m^3$. The vertical circulation calculations yield a free-surface velocity of $+1.6 \, m/s$, and a maximum negative velocity of $-1.1 \, m/s$ at 275 m below the free surface. The velocity profile is plotted in Fig. 10.14.

Discussion
The Strait of Gibraltar is a strongly stratified system. Hence the analysis of vertical circulation for well-mixed system does not apply. Further there is net flux of fresh seawater from the Atlantic Ocean into the Mediterranean Sea to compensate for the water losses by evaporation.

Nonetheless, the above results are close to field measurement. Scientific observations indicated that, near the surface the inflow may have speeds as high as 2 m/s, and the outflow reaches speeds of more than 1 m/s at about 275 m below the surface.[5]

---

[5] *Reference*: Encyclopedia Britannica (1997).

10.6 Appendix A – Observations of mixing and dispersion coefficients in estuarine zones

Site (1)	Vertical mixing coefficient ε_v (m²/s) (2)	Transverse mixing coefficient ε_t (m²/s) (3)	Longitudinal dispersion coefficient K (m²/s) (4)	Reference (5)	Remarks (6)
Cordova Bay, British Columbia, Canada	–	$\dfrac{\varepsilon_t}{dV_*} \sim 0.42$	–	Ward' re-analysis quoted by Fischer et al. (1979, p. 252)	Well-mixed estuary
Duwamish water way, Washington, USA	0.005	–	–	Partch and Smith's, experiment quoted by Fischer et al. (179, p. 250)	Salt-wedge estuary with strongly stratified surface layer
Fraser Estuary, British Columbia, Canada	–	$\dfrac{\varepsilon_t}{dV_*} \sim 0.44$ to 1.61	–	Ward' experiment quoted by Fischer et al. (1979, p. 252)	Well-mixed estuary
Gironde Estuary, France	–	$\dfrac{\varepsilon_t}{dV_*} \sim 1.03$	–	Ward' re-analysis quoted by Fischer et al. (1979, p. 252)	Well-mixed estuary In absence of tidal bore
Kinsale, Ireland	–	0.22^1 0.40^2	2.85^1 7.17^2	Elliott et al.'s work quoted by Lewis (1997, p. 216): ^{1}tidal current: 0.35 m/s, $d = 5.5$ m; ^{2}tidal current: 0.25 m/s, $d = 5.5$ m	Well-mixed estuary system
Mersey River, UK	0.05 to 0.071	–	160 to 360	Bowden's and Bowden and Gilligan's experiments quoted by Fischer et al. (1979, pp. 242 and 263)	Dispersion my tidal trapping (dead zones) primarily
Plym Estuary, UK	0.0016 to 0.0014	–	0.0292 to 0.0107	Lewis (1997, p. 222)	Well-mixed estuary
Potomac River, USA	–	–	6 to 20	Mass slug injection tests. Heitling and O'Connell's experiment quoted by Fischer et al. 912979, p. 236)	Tidal period $\gg T_c$

(Contd)

Site (1)	Vertical mixing coefficient ε_v (m²/s) (2)	Transverse mixing coefficient ε_t (m²/s) (3)	Longitudinal dispersion coefficient K (m²/s) (4)	Reference (5)	Remarks (6)
San Francisco Bay, USA	–	$\dfrac{\varepsilon_t}{dV_*} \sim 1.0$[1]	200[2]	[1]'Ward' re-analysis quoted by Fischer et al. (1979, p. 252) [2]Glenne and Selleck's experiment quoted by Fischer et al. (1979, p. 263)	Well-mixed estuary
Severn River, UK	0.0024 to 0.011[1]	0.021 to 0.014[1]	54 to 540[2]	[1]Elliott et al.'s work quoted by Lewis (1997, p. 296) [2]Bowden's experiment quoted by Fischer et al. (1969, p. 263)	Well-mixed estuary. River affected by a tidal bore
Strangford Lough, Ireland	–	0.16[1] 0.19[2]	1.19[1] 2.13[2]	[1]Elliott et al.'s work quoted by Lewis (1997, p. 216): [1]tidal current: 0.5 m/s, $d = 10$ m; [2]tidal current: 0.12 m/s, $d = 6.5$ m	Well-mixed system
Tees Estuary, UK	0.001 to 0.0016	–	0.0547 to 0.0580	Lewis (1997, p. 222)	Well-mixed estuary
Thames River, UK	–	–	53 to 84[1] 338[2]	[1]Bowden's experiments quoted by Fischer et al. (1969, p. 263) [2]For high river flow	

10.6.1 Field observations of mixing in tidal bores

Field measurements in tidal bores are scarce. Kjerfve and Ferreira (1993) and Wolanski *et al.* (2001) reported observations of sediment mixing immediately behind bores in the Rio Mearim (Brazil) and Ord River (Australia) respectively. Bartsch-Winkler and Lynch (1988) dropped bags of dye in the Turnagain Arm bore. Rulifson and Tull (1999) discussed the longitudinal dispersion of fish eggs in tidal bore affected rivers in the Bay of Fundy (Canada). Kjerfve and Ferreira (1993) presented quantitative measurements of salinity and temperature changes behind a bore. Their data highlighted a sharp jump in water properties about 18 min after the bore passage at two locations, while a rapid change in salinity was observed 42 min after the bore passage at a more upstream location.

Two fascinating experiments were conducted by M. Partiot in the Seine River mouth (in Bazin 1865b, pp. 640–641). The experiments highlighted different flow patterns next to the surface and at deeper depths. On 13 September 1855, in front of the Chapel Barre-y-Va (downstream of Caudebec-en-Caux), two floats were introduced in the river flow (a) at the surface and (b) next to the bottom (3.3 m beneath the surface). When the undular bore arrived, the surface float (a) continued to flow downstream for 130 s after the bore passage and flowed upstream afterwards, while the bottom float (b) flowed downstream only 90 s after the bore passage. On 25 September 1855, in front of Vallon de Caudebecquet, three floats were introduced (a) at the surface, (b) 1.5 m beneath the surface and (c) next to the bottom, all in the middle of the river. At the undular surge arrival, the float (a) started to run upstream 145 s after the bore passage, while the floats (b) and (c) flowed upstream 60 s after the bore passing.

10.7 Exercises

1. The estuary of the Flora River is 240 m long and 17 m wide. At high tide, the average water depth is 3.5 m. Calculate the wind setup for a 25 m/s wind blowing along the main axis of the estuary. Estimate the cross-sectional mixing time and compare the result with the tidal period. Estimate the bottom current assuming a bottom roughness $k_s = 5$ mm. *The tidal regime is semi-diurnal. The transverse mixing coefficient is assumed to be $0.007 \, m^2/s$.*

2. A lagoon on the North Coast of Papua New Guinea extends along 15 km of the shoreline and is about 250 m wide in average. The mean water depth is 0.9 m. Calculate the wind setup for a 22 m/s wind blowing parallel to the shoreline and perpendicular to the coast. Estimate the resonance frequencies of the lagoon. *The resonance frequency of a water body is the 'sloshing' frequency of the seiche.*

3. The estuary of the Loup River, Nice (France) is affected by a salt-wedge system. The river width is about 25 m. For a 4.8 m³/s flow rate, the water depth is about 1.5 m. Calculate the length of the salt wedge, the height of the wedge at the river mouth, and the saline wedge height at 90 m upstream of the river mouth. *Use Keulegan's theory. Assume Var River density of $1012 \, kg/m^3$ (because of suspended sediments).*

4. A river flow into a tideless sea. The river flow velocity is 0.15 m/s and the mean water depth is 1.35 m. A salt barrier is to be built 50 m upstream of the river mouth to prevent salt intrusion into the river system and on the water table. Calculate the minimum salt barrier height.

5. A reasonably well-mixed estuary is about 1200 m long. The depth-averaged density varies between 1005 and 1022 kg/m³ over that distance, and the average water depth 2.4 m. Calculate the required longitudinal free-surface gradient. Plot the velocity profile in the

Fig. 10.15 Cabbage Tree Creek, Brisbane, Queensland on 12 February 2003 near AMTD 2.0 km – looking upstream.

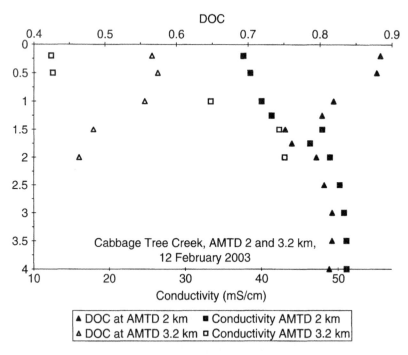

Fig. 10.16 Vertical distribution of conductivity and dissolved oxygen content at Cabbage Tree Creek on 12 February 2003. DOC: dissolved oxygen content.

estuary (assuming $\nu_T = 0.001$ m^2/s). Estimate the free-surface velocity and the maximum negative velocity.

6. The Knysna Bay Estuary in the Southern Cape (South Africa) is 230 m wide with an average channel depth of approximately 10 m. The longitudinal gradient in average density

$\partial \rho / \partial x$ is about 0.0027. Near the river mouth, the average density is $1016\,\text{kg/m}^3$ and the free-surface velocity is $0.226\,\text{m/s}$. Plot the velocity profile. Calculate the mass flux in the upstream direction and the upper-surface mass flux in the downstream direction.

7. Cabbage Tree Creek, Queensland (Australia) is a small water system in the northern suburbs of Brisbane (Australia). The creek is affected by local traffic (trawlers and boats) in its estuarine zone. Field measurements were conducted in Cabbage Tree Creek on 12 February 2003 in the estuarine zone (Figs 10.15 and 10.16). The data are summarized below.

AMTD (km) (1)	Depth (m) below free surface (2)	Temperature (°C) (3)	Dissolved oxygen (%) (4)	Turbidity NTU (5)	Conductivity (mS/cm) (6)	pH (7)	Remarks (8)
2	0.2	26.8	0.884	12	37.6	7.9	From a drifting boat
2	0.5	26.8	0.879	12	38.5	7.9	
2	1	26.7	0.819	14	40	7.9	
2	1.25	26.8	0.803	14	41.3	7.95	
2	1.5	26.8	0.752	20	47.9	8.1	
2	1.75	26.85	0.761	17	46.3	8.1	
2	2	26.9	0.795	14	48.9	8.1	
2	2.5	26.9	0.806	15	50.2	8.2	
2	3	26.9	0.817	18	50.7	8.2	
2	3.5	26.8	0.817	19	51.1	8.2	
2	4	26.8	0.813	26	51.1	8.2	Just above the bottom
3.2	0.2	25.7	0.566	25	12.3	7	From a bridge
3.2	0.5	25.8	0.574	25	12.5	7.1	
3.2	1	26.5	0.556	20	33.3	7.5	
3.2	1.5	26.9	0.484	22	42.3	7.7	
3.2	2	26.8	0.464	35	43	7.7	Just above the bottom

Source: Field data collected by the writer and the Queensland Environment Protection Agency.

(a) Calculate the slope of the mean water surface to counterbalance the mean density gradient between AMTD $= 2000$ and $3200\,\text{m}$. (b) Predict the velocity profile at AMTD $2.8\,\text{km}$. *Neglect the bed slope. Assume an average water depth of $4.2\,\text{m}$ and $2.2\,\text{m}$ at AMTD 2.0 and $3.2\,\text{km}$ respectively, and $v_T = 0.003\,\text{m}^2/\text{s}$.*

10.8 Exercise solutions

1. Wind setup

$$\Delta d = \frac{C_d}{4} \frac{\rho_{\text{air}}}{\rho} \frac{U_{10}^2 L}{gd} = \frac{0.002}{4} \times \frac{1.2}{1020} \times \frac{25^2 \times 240}{9.81 \times 3.5} = 2.6\,\text{mm} \quad \text{equation (10.3)}$$

Cross-sectional mixing time: $T_c = W^2/\varepsilon_t = 17^2/0.007 = 11\,\text{h}\,28\,\text{min}$
Tidal period: $12\,\text{h}$ $\quad \Rightarrow \quad T \sim T_c$ maximum dispersion
Bottom recirculation current: $0.4\,\text{m/s}$

2. Wind setup (wind blowing parallel to the shoreline)

$$\Delta d = \frac{C_d}{4} \frac{\rho_{\text{air}}}{\rho} \frac{U_{10}^2 L}{gd} = \frac{0.002}{4} \times \frac{1.2}{1020} \times \frac{22^2 \times 15\,000}{9.80 \times 0.9} = 0.5\,\text{m}$$

Wind setup (wind blowing normal to the shoreline)

$$\Delta d = \frac{C_d}{4} \frac{\rho_{air}}{\rho} \frac{U_{10}^2 L}{gd} = \frac{0.002}{4} \times \frac{1.2}{1020} \times \frac{22^2 \times 250}{9.80 \times 0.9} = 8\,mm$$

The resonance period of water body is approximately $2L/\sqrt{gd}$ Hence:
    resonance frequency along major axis = 0.0001 Hz ($T = 2\,$h 48 min) and
    resonance frequency along minor axis = 0.006 Hz ($T = 168\,$s)

3. Salt wedge (Loup River)
   $H_s = 0.76\,$m, $L_s = 853\,$m, $h_s = 0.62\,$m (90 m upstream of the river mouth).
4. Salt barrier
   Salt wedge: $H_s = 0.74\,$m, $L_s = 1260\,$m.
   Salt barrier: minimum height, $h = 0.68\,$m (50 m upstream of the river mouth).
5. $\partial d/\partial x = -1.3 \times 10^5$
   Free-surface velocity = +0.04m/s, maximum negative velocity = $-0.027\,$m/s. (at $z/d = 0.3$)
6. The momentum exchange coefficient must be first calculated: $\nu_T = 0.024\,m^2/s$.
   At equilibrium, the mass flux in the downstream direction equals the mass flux in the upstream direction (i.e. $0.58\,m^2/s$).

Part 2 Revision exercises

1. The Waikato River, near Hamilton, New Zealand has the following channel characteristics: $Q = 150\,\text{m}^3/\text{s}$, $d = 2.5\,\text{m}$, width: $85\,\text{m}$, $V = 0.7\,\text{m/s}$, $V_* = 0.057\,\text{m/s}$.

   A first dye test will be conducted by injecting continuously a tracer (Lissamine Red 4B) at the side (125 g/s).

   (a) Calculate the distance from the injection point where the tracer concentration reach $10\,\text{mg/m}^3$ on the opposite bank. *Assume an infinitely wide rectangular channel for simplicity and neglect vertical mixing (for question (a) only).*

   In a second dye test, a mass slug of dye (2.5 kg) is suddenly released *on the centreline.*

   (b) Predict the dispersion coefficient and the length of the initial zone.

   A measurement station is located 15 km downstream of the injection point (second test).

   (c) Calculate the maximum tracer concentration at that station and the arrival time.

   (d) The detection limit of Lissamine Red 4B dye is $0.1\,\text{mg/m}^3$. Calculate the length of time during Lissamine dye will be detectable at the measurement station.

   (e) At the time when the tracer concentration is maximum at the station, calculate the cloud length with concentrations exceeding $0.5\,\text{mg/m}^3$.

   *   *   *

2. A chemical is accidentally released in a natural stream between 1:30 a.m. and 2:30 a.m. (at night) at a rate of 7 kg/h (top hat inflow). Calculate the period during which concentration exceed $1.5\,\text{mg/m}^3$, as well as the maximum concentration, at a site located 12 km downstream of the source. The river characteristics are: $Q = 5\,\text{m}^3/\text{s}$, width: $25\,\text{m}$, bed slope: 0.00025, gravel bed: $k_s = 25\,\text{mm}$. *Approximate the top hat discharge by four mass slugs of 1.75 kg released at 1:37 a.m., 1:52 a.m., 2:07 a.m. and 2:22 a.m.*

   *   *   *

3. Dye profiles were measured at the Manawatu River, below Palmerston North (New Zealand). The table below shows cross-sectional averaged dye concentrations at two measurement sites located respectively 5.0 and 6.4 km downstream of the injection point. (Preliminary calculations suggested that the site locations were downstream of the initial zone.)

   (a) Using the frozen cloud approximation, estimate the flow velocity and longitudinal dispersion coefficient between the two measurement sites.

   (b) Using a ADZ 'dead zone' model, predict the dye concentration versus time at the second measurement location, and compare it with the 'frozen cloud' dispersion model.

   Assume a conservative tracer.

Location (1)	Time (h) (2)	C_m (mg/m^3) (3)	Location (4)	Time (h) (5)	C_m (mg/m^3) (6)
Site 1	2.106	0.833	Site 2	2.85	1.96
Site 1	2.19	9.58	Site 2	2.93	5.96
Site 1	2.27	11.8	Site 2	3	10.6
Site 1	2.37	14.9	Site 2	3.12	14
Site 1	2.445	21.9	Site 2	3.18	21.15
Site 1	2.52	26.9	Site 2	3.22	23.92
Site 1	2.62	29.2	Site 2	3.24	24.33
Site 1	2.675	35	Site 2	3.295	27.25
Site 1	2.78	36.4	Site 2	3.345	25.8
Site 1	2.85	36.3	Site 2	3.379	32.7
Site 1	2.93	36.2	Site 2	3.404	33.3
Site 1	3	35.8	Site 2	3.48	33.5
Site 1	3.12	31.9	Site 2	3.53	34.29
Site 1	3.18	29.6	Site 2	3.596	35.1
Site 1	3.27	29.2	Site 2	3.638	34.7
Site 1	3.345	25.3	Site 2	3.722	34.5
Site 1	3.43	23.46	Site 2	3.76	33.7
Site 1	3.51	21.7	Site 2	3.83	32.8
Site 1	3.59	17.9	Site 2	3.915	29.4
Site 1	3.68	16.4	Site 2	3.965	29.8
Site 1	3.755	14.17	Site 2	4.14	24.3
Site 1	3.84	12.8	Site 2	4.2	23.5
Site 1	3.93	10.8	Site 2	4.26	21.9
Site 1	4	10.17	Site 2	4.33	21.6
Site 1	4.1	8.92	Site 2	4.39	19.9
Site 1	4.2	8.08	Site 2	4.5	16
Site 1	4.31	7	Site 2	4.97	9.5
Site 1	4.39	6.04	Site 2	5.133	6.5
Site 1	4.51	5	Site 2	5.44	4.5
Site 1	4.71	3.92	Site 2	5.52	3.29
Site 1	4.79	3.25	Site 2	5.84	3.25
Site 1	4.89	2.83	Site 2	6	2.17
Site 1	5	2.3	Site 2	6.25	1.58
Site 1	5.1	2.17	Site 2	6.5	1.17
Site 1	5.2	2	Site 2	6.75	0.96
Site 1	5.3	1.75	Site 2	7	0.83
Site 1	5.4	1.67			
Site 1	5.5	1.38			
Site 1	5.75	1.17			
Site 1	6	0.83			
Site 1	6.25	0.625			
Site 1	6.5	0.375			
Site 1	6.75	0.33			
Site 1	7	0.25			

Note: Data from Rutherford (1994, pp. 271–273).

*   *   *

4. Investigations are undertaken for a scheme to increase waste water effluent release in Hilliard Creek. To assist the environmental impact study, calculations are conducted for the following flow conditions:

Conditions	Q (m^3/s)	Average width (m)	Average bed slope	Bed roughness height (mm)	Water temperature (°C)	Dissolved oxygen content (%)	BOD$_5$ (kg/m^3)
1	1	2	0.00012	10	20	75	
2	1.5	2	0.00012	10	20	75	
3	12	3	0.00012	25	20	85	

The water treatment plant is to discharge some effluent into the creek at a rate of 0.45 m³/s with a BOD₅ is 270 g/m³. Calculate the minimum DOC downstream of the injection point and its location. Estimate the BOD₅ for a sample taken at that location. *Assume the effluent to be at the same temperature as the river. Assume further that the wastewater spreads almost instantaneously across the entire section.*

*   *   *

5. A chemical is accidentally released in a natural stream between 1:00 a.m. and 3:00 a.m. at a rate of 25 kg/h for the first 45 min and later 10 kg/h for the next 75 min. Calculate the period during which concentration exceeds 0.05 mg/m³ as well as the maximum concentration, at a site located 20 km downstream of the source. Plot the chemical concentration versus time on the same graph. The river characteristics are: $Q = 3.3 \, \text{m}^3/\text{s}$, width: 15 m, bed slope: 0.000 28, small gravel bed: $k_s = 25$ mm. *Approximate the chemical release by a succession of mass slugs and apply the method of superposition.*

Assignment solutions

1. (a) $\varepsilon_t = 0.085 \, \text{m}^2/\text{s}$ equation (7.6)

$$C_m = \frac{2\dot{M}}{Vd\sqrt{4\pi\varepsilon_t \dfrac{x}{V}}} \exp\left(-\frac{y^2}{4\varepsilon_t \dfrac{x}{V}}\right) \qquad \text{equation (7.7)}$$

by applying the principle of superposition and the method of images. (In the virtual stream, the injection rate is 2×125 g/s.) It yields: $x = 4065$ m.

(b) $K = 273 \, \text{m}^2/\text{s}$
Initial region: $L = 5900$ m

(c) $C_{max} = 1.37 \times 10^{-6} \, \text{kg/m}^3$
$t = 6$ h (frozen cloud method because $x' \sim 0.254 > 0.1$)

(d) 22360 s or 6.2 h (frozen cloud method) (equation (8.9))

(e) 9.6 km (frozen cloud method) (equation (8.10))

DISCUSSION
Field measurements were conducted in the Waikato River near Hamilton by injecting 36 kg of dye. At 22.8 km downstream of the injection point, the maximum concentration was about 3 mg/m³ and the peak was observed about 9.1 h after injection (Rutherford 1994). The dispersion coefficient K was estimated to be between 52 and 67 m²/s.

Although the calculations are of the same order of magnitude as the field data, the scatter between estimates and field observations must be acknowledged.

*   *   *

2. $d = 0.45$ m, $V = 0.44$ m/s, $V_* = 0.033$ m/s
$\varepsilon_t = 0.0088 \, \text{m}^2/\text{s}$, $K = 92.6 \, \text{m}^2/\text{s}$
$C_{max} = 1.09 \times 10^{-4} \, \text{kg/m}^3$ at 9:26 a.m.
$C_m > 1.5 \times 10^{-6} \, \text{kg/m}^3$ between 6:40 a.m. and 12:21 p.m.

DISCUSSION
The problem is solved by applying the method of superposition, for four mass slugs, and applying the frozen cloud approximation (Fig. R.1).

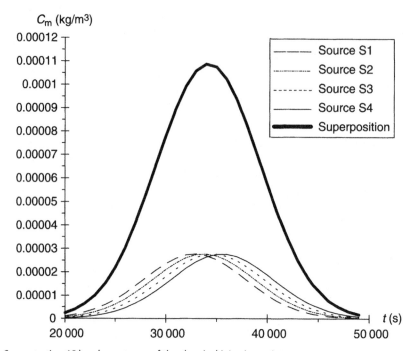

Fig. R.1 Concentration 12 km downstream of the chemical injection point.

* * * *

3. First the temporal moment must be computed at Sites 1 and 2:

	Time of passage of centroid (s)	Temporal variance (s²)	Velocity (m/s)
Site 1	11 793.8	1.47×10^8	–
Site 2	14 629.91	2.21×10^8	0.4936

Secondly the concentration versus time distribution at Site 1 is divided into a series of small mass slugs $\delta M(t)$ injected during a time interval δt such as:

$$C_{\mathrm{m}}(t) = \frac{\delta M(t)}{A V \delta t}$$

where A is the flow cross-sectional area.

Time (s) (1)	$\delta M/A$ (kg/m²) (2)
8000	0.011 652
10 000	0.033 257
12 000	0.025 254
14 000	0.012 055
16 000	0.005 932
18 000	0.002 478
20 000	0.001 473
22 000	0.000 718
24 000	0.000 348

Note the accuracy of the prediction is enhanced with increasing number of mass slugs.

Then the concentration versus time distribution at Site 2 is predicted by superposition of all small mass slugs injected at Site 1.

> **Note**
> Note that both Sites 1 and 2 are downstream of the initial zone. Hence the longitudinal dispersion of small mass slugs between Sites 1 and 2 is not affected by the initial zone. The principle of superposition is used for both 'frozen cloud' approximation and 'dead zone' models.

Using the 'frozen cloud' approximation, the solution of the longitudinal dispersion equation is:

$$C_m(t) = \frac{\delta M}{A\sqrt{4\pi K T_1}} \exp\left(\frac{(x_1 - Vt)^2}{4K T_1} \right) \quad \text{Concentration versus time at fixed } x_1 \text{ (equation (8.9)}$$

where $x_1 = 6400 - 5000 = 1400\,\text{m}$, $T_1 = 14\,630 - 11\,793 = 2836\,\text{s}$, $V = 0.4936\,\text{m/s}$.

> **Note**
> The concentration at Site 2 is the superposition of the nine mass slugs injected at Site 1. The dispersion coefficient is deduced from the best fit of the data with the prediction. Using nine mass slugs, best fit is achieved for $30 < K < 36\,\text{m}^2/\text{s}$.
>
> **Remark**
> Rutherford (1994, pp. 271–273) applied the 'frozen cloud' approximation and he integrated numerically the concentration versus time distribution at Site 1. He obtained a good data fit for $0.5 < V < 0.6\,\text{m/s}$ and $17.5 < K < 20\,\text{m}^2/\text{s}$. Rutherford added that 'the match in the tail was fairly good in this (case), but in many rivers the observed dye profile has a longer tail than the predicted one'.

The concentration versus time distribution, predicted by the 'frozen cloud' approximation is plotted in Fig. R.2 and compared with the data at Sites 1 and 2.

An ADZ 'dead zone' model with two cells is used. Each cell is 1400 m long. The residence time of each cell ΔT is about 2800 s. The concentration in the second cell (Site 2) is given by:

$$C_m(t) = \gamma(t - \Delta t)\frac{\delta M}{AL} \exp(-\gamma(t - \Delta t)) \quad \text{Two-cell system (9.9a)}$$

where $\gamma = 1/\Delta T$ ($k = 0$), Δt is the dead zone time delay, L is the cell length and t is the time from mass slug injection.

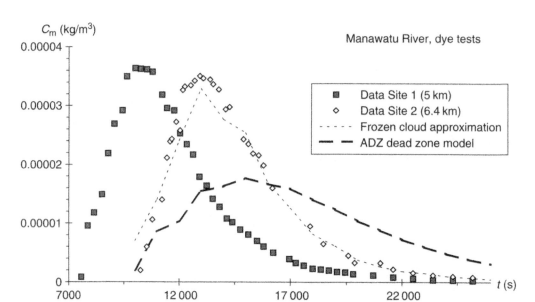

Fig. R.2 Concentration at Sites 1 and 2 km – comparison between field measurements, 'frozen cloud' approximation ($K = 34\,\text{m}^2/\text{s}$) and ADZ 'dead zone' model ($\Delta T = 2840\,\text{s}$, $\Delta t = 100\,\text{s}$).

Note

The concentration at Site 2 is the superposition of the nine mass slugs injected at Site 1. The dead zone time delay Δt is deduced from the best fit of the data with the prediction. Using nine small mass slugs and a two-cells model, a 'reasonable' best fit is difficult to achieve because the limited number of mass slugs and, to a lesser extent, of the selection of only two cells. (Remember that each mass slug is instantaneously diluted in the initial cell.)

Practically, the existence of dead zones implies that the turbulence is not homogeneous across the river, and that the time taken for contaminant particles to sample the entire flow is significantly enhanced (i.e. the length of the initial zone is increased). Dead zones are thought to explain long tails of tracer observed in natural rivers. That is, a 'dead zone' model will approximate reasonably well the tail of contaminant mass concentration distribution. (That is, the right handside distribution region in a concentration versus time distribution curve.)

Remark

Rutherford (1994, pp. 227–229) analysed experimental data recorded during the same test at $x = 2.7$ and $6.4\,\text{km}$. (The latter corresponds to Site 2 in the homework, but the former is located in the initial zone.) Rutherford integrated numerically equation (9.7) using a two-cell model with $\Delta T = 1440\,\text{s}$ (0.4 h) and $\Delta t = 6750$ (1.875 h).

PART 3
Introduction to Unsteady Open Channel Flows

Tidal bore of the Dordogne River on 27 September 2000, view from the left bank, looking upstream at the bore propagating into the quiescent river system.

This series of lectures is designed for undergraduate students in civil, environmental and hydraulic engineering, and professionals who want to expand their understanding of open channel hydraulics. It will first develop the basic equations of unsteady open channel flows. That is, the Saint-Venant equations and the method of characteristics. Later, simple applications are developed. For example, the propagation of waves, positive and negative surges, the dam break wave problem. At the end, simple numerical models are presented and explained.

Relevant Internet links

http://www.uq.edu.au/~e2hchans/photo.html#Tidal bores, mascaret, pororoca	Photographs of tidal bores
http://www.uq.edu.au/~e2hchans/mascaret.html	Tidal bore of the Seine River
http://boreriders.com/	Bore Riders Club
http://www.scvhistory.com/scvhistory/stfrancis.htm	St Francis dam catastroph, Santa Clarita Valley Historical Society
http://membres.lycos.fr/vitosweb/	Malpasset dam catastroph

11

Unsteady open channel flows: 1. Basic equations

Summary

The continuity and momentum equations are developed for one-dimensional unsteady open channel flows. They yield the Saint-Venant equations. The basic assumptions are detailed and the method of characteristics is later introduced.

11.1 Introduction

Common examples of unsteady open channel flows includes flood flows in rivers and tidal flows in estuaries, irrigation channels, headrace and tailrace channel of hydropower plants, navigation canals, stormwater systems and spillway operation. Figure 11.1 illustrates extreme unsteady flow conditions. Figure 11.1(a) shows the upstream propagation of a tidal bore. Surfers give the scale of the phenomenon. Figure 11.1(b) presents a dam break wave propagating down a flat stepped waterway.

In unsteady open channel flows (e.g. Fig. 11.1), the velocities and water depths change with time and longitudinal position. For one-dimensional applications, the relevant flow parameters (e.g. V and d) are functions of time and longitudinal distance. Analytical solutions of the basic equations are nearly impossible because of their non-linearity, but numerical techniques may provide approximate solutions for some specific cases.

The first major mathematical model of a river system was developed in J.J. Stoker for the Ohio and Mississippi systems (Stoker 1953). It was followed by important developments in the late 1950s in particular by A. Preissmann and J.A. Cunge in France. Mahmood and Yevdjevich (1975) regrouped major contributions to the topic. Yevdjevich (1975) and Montes (1998, pp. 470–471) summarized the historical developments in numerical modelling of unsteady open channel flows.

Notes

1. The name of Vujica Yevdjevich is sometimes spelled Jevdjevich or Yevdyevich.
2. J.J. Stoker, E. Isaacson and A. Troesch, from the Courant Institute, New York University, developed and implemented a numerical model for flood wave profiles in the Ohio and Mississippi rivers (Stoker 1953, Isaacson *et al.* 1954, 1956). James Johnston Stoker was a Professor at the Courant Institute, New York University. His book on water waves is a classical publication (Stoker 1957).
3. Alexandre Preissmann (1916–1990) was born and educated in Switzerland. From 1958, he worked on the development of mathematical models at Sogreah in Grenoble. Born and educated in Poland, Jean A. Cunge worked in France at Sogreah and he lectured at the Hydraulics and Mechanical Engineering School of Grenoble (ENSHMG).

(a-i)

(a-ii)

Fig. 11.1 Example of unsteady open channel flows. (a) Advancing tidal bore propagation into the Dordogne River (France) on 27 September 2000.

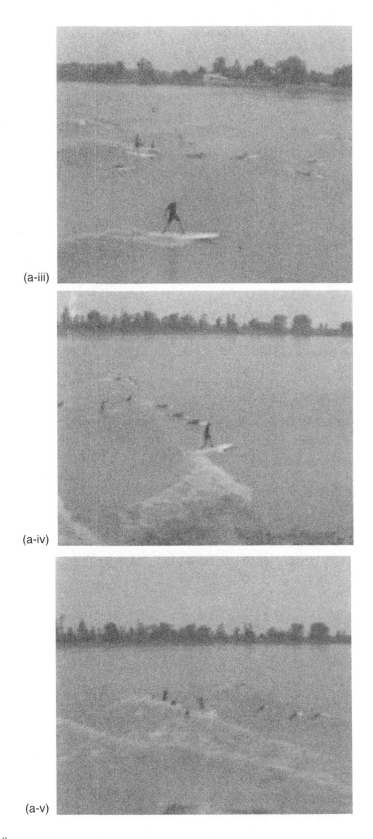

(a-iii)

(a-iv)

(a-v)

Fig. 11.1 (*Contd*)

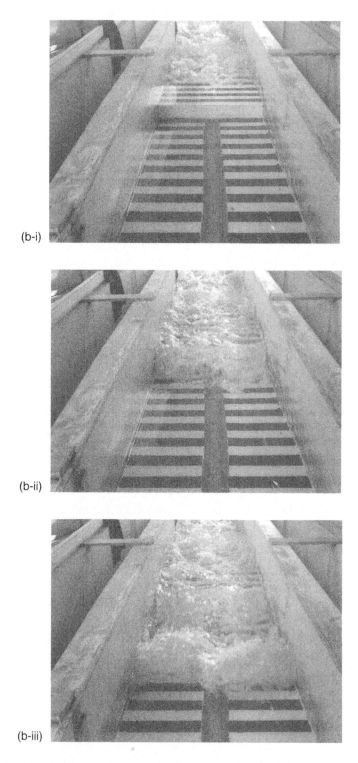

(b-i)

(b-ii)

(b-iii)

Fig. 11.1 (*Contd*) (b) Dam break wave down a stepped waterway (courtesy of Chye-Guan Sim and Frankie Tan): $\theta = 3.4°$, $W = 0.5\,\text{m}$, step height: $0.0715\,\text{m}$, $Q = 0.065\,\text{m}^3/\text{s}$. Looking upstream at the advancing wave.

11.2 Basic equations

11.2.1 Presentation

The basic one-dimensional unsteady open channel flow equations are called the Saint-Venant equations. They are named after the French engineer Adhémar Jean Claude Barré de Saint-Venant (1871a, b). These equations are based upon a number of *key, basic assumptions*: (1) *the flow is one dimensional*, the velocity is uniform in a cross-section and the transverse free-surface profile is horizontal; (2) the streamline curvature is very small and the vertical fluid accelerations are negligible; as a result, *the pressure distributions are hydrostatic*; (3) the *flow resistance* and turbulent losses are the *same as for a steady uniform equilibrium flow* for the same depth and velocity, regardless of trends of the depth; (4) *the bed slope is small* enough to satisfy the following approximations: $\cos\theta \approx 1$ and $\sin\theta \approx \tan\theta \approx \theta$; (5) *the water density is constant* and (6) the Saint-Venant equations were developed for fixed boundary channels: that is, *sediment motion is neglected*.

These fundamental assumptions are valid for any channel cross-sectional shape. Practically, however, the cross-sectional shape is indirectly limited by the assumptions of one-dimensional flow, horizontal transverse free surface and hydrostatic pressure.

With such basic hypotheses, the flow can be described at any point and any time by two variables: e.g. V and d, or Q and d, where V is the flow velocity, d is the water depth and Q is the water discharge. Basically the unsteady flow properties must be described by two equations: the conservation of mass (continuity) and conservation of momentum.

Notes

1. Adhémar Jean Claude Barré de Saint-Venant (1797–1886), French engineer of the 'Corps des Ponts-et-Chaussées', developed the equation of motion of a fluid particle in terms of the shear and normal forces exerted on it (Barré de Saint-Venant 1871a, b). His original development was introduced for both fluvial and estuarine systems.
2. The assumption (4) is valid within 0.1% for $\theta < 2.6°$, and within 1% for $\theta < 8.1°$.
3. The assumption (5) implies that sediment suspension and free-surface aeration are neglected.
4. The equations of conservation of momentum and conservation of energy are equivalent if the two relevant variables (e.g. V and d) are continuous functions. At a discontinuity (e.g. a hydraulic jump), the equivalence becomes untrue. Unsteady open channel flow equations must be based upon the continuity and momentum principles which are applicable to both continuous and discontinuous flow situations (see Chapter 2).

11.2.2 Integral form of the Saint-Venant equations

Considering the control volume defined by the cross-sections 1 and 2 in Fig. 11.2, located between $x = x_1$ and $x = x_2$, between the times $t = t_1$ and $t = t_2$, the continuity principle states

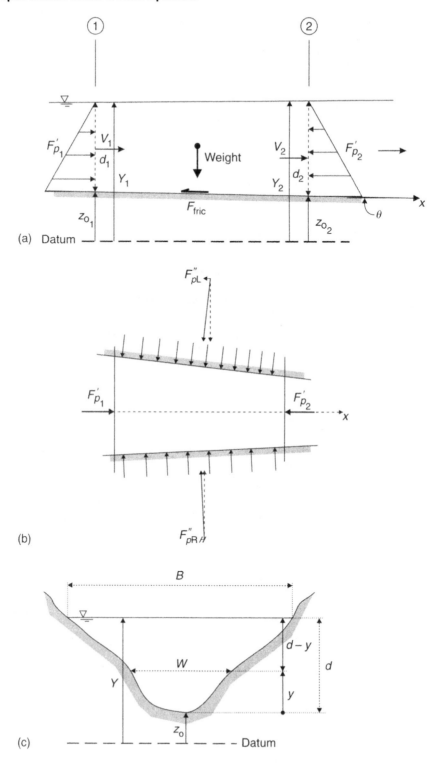

Fig. 11.2 Definition sketch: (a) side view, (b) top view and (c) cross-section.

that the net mass flux into the control volume equals the net mass increase of the control volume between the times t_1 and t_2. Assuming no lateral inflow, it yields:

$$\int_{t_1}^{t_2} (\rho V_1 A_1 - \rho V_2 A_2) dt + \int_{x_1}^{x_2} ((\rho A)_{t_2} - (\rho A)_{t_1}) \, dx = 0 \qquad (11.1)$$

where ρ is the fluid density, V is the flow velocity, A is the flow cross-sectional area, the subscripts 1 and 2 refer to the upstream and downstream cross-sections respectively (Fig. 11.2), and the subscripts t_1 and t_2 refer to the instants $t = t_1$ and $t = t_2$ respectively. Defining the total discharge $Q = VA$, and dividing equation (11.1) by the density ρ, the continuity equation becomes:

$$\int_{t_1}^{t_2} (Q_1 - Q_2) \, dt + \int_{x_1}^{x_2} (A_{t_2} - A_{t_1}) \, dx = 0 \qquad (11.2)$$

The application of the momentum equation in the x-direction states that the net change of momentum in the control volume between the instants t_1 and t_2 plus the rate of change of momentum flux across the volume equals the sum of the forces applied to the control volume in the x-direction. The net change of momentum in the control volume equals:

$$\int_{x_1}^{x_2} ((\rho VA)_{t_1} - (\rho VA)_{t_2}) \, dx \qquad (11.3)$$

while the rate of change of momentum flux across the volume is equal to:

$$\int_{t_1}^{t_2} ((\rho V^2 A)_1 - (\rho V^2 A)_2) \, dt \qquad (11.4)$$

The forces acting on the control volume contained between sections 1 and 2 are the pressure forces at sections 1 and 2, the pressure force components on the channel sidewalls if the channel width vary in the x-direction, the weight of the control volume, the reaction force of the bed (equal, in magnitude, to the weight in absence of vertical fluid motion) and the flow resistance opposing fluid motion. The pressure forces acting on sections 1 and 2 equal:

$$\int_{t_1}^{t_2} (F'_{P_1} - F'_{P_2}) \, dt = \int_{t_1}^{t_2} ((\rho I_1)_1 - (\rho I_1)_2) \, dt \qquad (11.5)$$

where

$$I_1 = \int_0^d (d - y) W \, dy$$

in which y is the distance measured from the bottom, d is the water depth and W is the channel width at the distance y above the bed (Fig. 11.2(c)).

When the channel width changes with distance, the pressure force components on the right and left sidewalls equal:

$$\int_{t_1}^{t_2} (F''_{pL} + F''_{pR}) \, dt = \int_{t_1}^{t_2} \int_{x_1}^{x_2} \rho g I_2 \, dx \, dt \qquad (11.6)$$

where

$$I_2 = \int_0^d (d - y) \left(\frac{\partial W}{\partial x} \right)_{y=\text{constant}} dy$$

Equation (11.6) states that, the water depth d being constant, an increase in wetted surface in the x-direction induces a positive sidewall pressure force component. The above expression of F_p'' is valid only for gradual variations in cross-section. For sudden changes in cross-sectional shapes, forces other than the hydrostatic pressure force take place.

The gravity force component in the flow direction (i.e. x-direction) is:

$$\int_{t_1}^{t_2} \int_{x_1}^{x_2} \rho g A S_o \, dx \, dt \qquad (11.7)$$

where the bed slope $S_o = \sin\theta$ may be approximated to: $S_o \approx \tan\theta = -\partial z_o/\partial x$ (assumption (4), Section 11.2.1).

Fluid motion is opposed by flow resistance and shear forces exerted on the wetted surfaces. The integration of the friction force F_{fric} between $t = t_1$ and $t = t_2$ is:

$$\int_{t_1}^{t_2} F_{\text{fric}} \, dt = \int_{t_1}^{t_2} \int_{x_1}^{x_2} \rho g A S_f \, dx \, dt \qquad (11.8)$$

where S_f is the friction slope defined as:

$$S_f = \frac{4\tau_o}{\rho g D_H} \qquad (11.9)$$

τ_o is the average boundary shear stress, and D_H is the hydraulic diameter.

Notes

1. The momentum was called *impetus* by the Ancients. For example, Leonardo da Vinci defined the *impetus* as: 'Impetus is a power created by movement and transmitted from the mover to the movable thing; and this movable thing has as much movement as the impetus has life' (McCurdy 1956, Vol. 1, p. 417). The term *impetus* as used by Leonardo da Vinci was a quantity proportional to the weight of the body and to its velocity (Levi 1995, p. 570). That is, the *impetus* was basically proportional to the momentum. Leonardo da Vinci added: 'Impetus is (also) termed derived movement' (McCurdy 1956, Vol. 1, p. 523).
2. Leonardo da Vinci (AD 1452–1519) was an Italian artist who extended his interest to medicine, science, engineering and architecture. In his notes, he described numerous flow situations and he commented the entrainment of air at waterfalls, plunging jet flows, drop structures, running waters and breaking waves (e.g. Chanson 1997a, pp. 327–329).
3. The momentum per unit volume equals ρV. The momentum flux equals $\rho V V A$.
4. The basic assumptions of the Saint-Venant equations (Section 11.2.1) imply that the vertical pressure distribution is hydrostatic: i.e. $P(y) = \rho g(d - y)$ for $0 \leqslant y \leqslant d$.
5. Simple geometrical considerations give:

$$W(x, y = d, t) = B(x, t)$$
$$\int_0^d W \, dy = A$$
$$\frac{\partial A}{\partial y} = B$$

where the free-surface width B and the flow cross-sectional area A are functions of both distance x and time t (Fig. 11.2).

6. The left bank is located on the left-hand side of an observer when looking downstream. Conversely, the right bank is on the right-hand side of an observer looking downstream.
7. Yen (2002) discussed specifically the definitions of the friction slope. One series of definitions is based upon the momentum principle (e.g. equation (11.9)) while the other is based upon the energy principle. All the definitions are equivalent for uniform equilibrium flow conditions.

Combining equations (11.3) to (11.8), and dividing by the density ρ which is assumed to be constant, the momentum equation becomes:

$$\int_{x_1}^{x_2} ((VA)_{t_1} - (VA)_{t_2}) \, dx + \int_{t_1}^{t_2} ((V^2A)_1 - (V^2A)_2) \, dt$$
$$= \int_{t_1}^{t_2} ((I_1)_1 - (I_1)_2) \, dt + \int_{t_1}^{t_2} \int_{x_1}^{x_2} gI_2 \, dx \, dt + \int_{t_1}^{t_2} \int_{x_1}^{x_2} gA(S_o - S_f) \, dx \, dt \qquad (11.10)$$

Equation (11.10) is the cross-sectional integration of the principle of momentum conservation for one-dimensional unsteady open channel flows. It states that the net change of momentum in control volume (i.e. unsteady term) plus the rate of change of momentum flux across the volume equal the pressure forces acting on sections 1 and 2, plus the pressure force components on the right and left sidewalls, plus the gravity force component in flow direction and minus the friction force F_{fric} acting on the wetted surface between $t = t_1$ and $t = t_2$.

Equations (11.2) and (11.10) form a system of two equations based upon the Saint-Venant equation assumptions (Section 11.2.1). They were developed without any additional assumption and some characteristic parameter (e.g. V, A, d) might not be continuous. If one or more parameter is discontinuous, equations (11.2) and (11.10) remain valid if the Saint-Venant hypotheses are respected.

11.2.3 Differential form of the Saint-Venant equations

The differential form of the Saint-Venant equations may be derived from the integral form (Section 11.2.2) if the relevant parameters are continuous and differentiable functions with respect to x and y. The Taylor-series expansion of each parameter follows:

$$A_{t_2} = A_{t_1} + \frac{\partial A}{\partial t} \Delta t + \frac{\partial^2 A}{\partial t^2} \frac{\Delta t^2}{2} + \cdots \qquad (11.11a)$$

$$Q_2 = Q_1 + \frac{\partial Q}{\partial x} \Delta x + \frac{\partial^2 Q}{\partial x^2} \frac{\Delta x^2}{2} + \cdots \qquad (11.11b)$$

where $\Delta t = t_2 - t_1$ and $\Delta x = x_2 - x_1$. Neglecting the second order term, the continuity equation (11.2) yields:

$$\int_{x_1}^{x_2} \int_{t_1}^{t_2} \left(\frac{\partial A}{\partial t} + \frac{\partial Q}{\partial x} \right) dt \, dx = 0 \qquad (11.12)$$

Similarly the momentum principle (equation (11.10)) becomes:

$$\int_{x_1}^{x_2} \int_{t_1}^{t_2} \left(\frac{\partial Q}{\partial t} + \frac{\partial (V^2A)}{\partial x} \right) dt \, dx = \int_{x_1}^{x_2} \int_{t_1}^{t_2} -g \left(\frac{\partial I_1}{\partial x} - I_2 - (S_o - S_f)A \right) dt \, dx \qquad (11.13)$$

If equations (11.12) and (11.13) are valid everywhere in the (x, t) plane, they are also valid over an infinitely small space $dxdt$, and the continuity and momentum principles yield:

$$\frac{\partial A}{\partial t} + \frac{\partial Q}{\partial x} = 0 \qquad \text{Continuity equation} \qquad (11.14)$$

$$\frac{\partial Q}{\partial t} + \frac{\partial}{\partial x}(V^2A + gI_1) = g(S_o - S_f)A + gI_2 \qquad \text{Momentum equation} \qquad (11.15a)$$

Based upon geometrical considerations and using Leibnitz rule, it can be proved that the terms I_1 and I_2 satisfy:

$$\frac{\partial}{\partial x}(gI_1) = gA\frac{\partial d}{\partial x} + gI_2 \qquad (11.16)$$

Replacing into the momentum equation (11.15a), it yields:

$$\frac{\partial Q}{\partial t} + \frac{\partial}{\partial x}(V^2A) + gA\frac{\partial d}{\partial x} = gA(S_o - S_f) \qquad \text{Momentum equation} \qquad (11.15b)$$

Note that the result (equation (11.15b)) is valid for prismatic and non-prismatic channels. That is, the pressure force contribution caused by a change in cross-sectional area at a channel expansion (or contraction) is exactly balanced by the pressure force component on the channel banks in the flow direction.

The continuity and momentum equations may be rewritten in terms of the free-surface elevation ($Y = d + z_o$) and flow velocity V only. It yields:

$$\frac{\partial Y}{\partial t} + \frac{A}{B}\frac{\partial V}{\partial x} + V\left(\frac{\partial Y}{\partial x} + S_o\right) + \frac{V}{B}\left(\frac{\partial A}{\partial x}\right)_{d=\text{constant}} = 0 \qquad (11.17)$$

$$\frac{\partial V}{\partial t} + V\frac{\partial V}{\partial x} + g\frac{\partial Y}{\partial x} + g\,S_f = 0 \qquad (11.18a)$$

The system of equations (11.17) and (11.18a) forms the basic *Saint-Venant equations*. They are also called *dynamic wave equations*. Equation (11.17) is the *continuity equation* while equation (11.18a) is called the *dynamic equation*.

Notes
1. A *prismatic* channel has a cross-sectional shape independent of the longitudinal distance along the flow direction. That is, the width $W(x, y, t)$ is only a function of y and it is independent of x and t.
2. In his original development, Barré de Saint-Venant (1871a) obtained the system of two equations:

$$\frac{\partial A}{\partial t} + \frac{\partial Q}{\partial x} = 0 \qquad (11.14)$$

$$\frac{1}{g}\frac{\partial V}{\partial t} + \frac{\partial}{\partial x}\left(\frac{V^2}{2g}\right) + \frac{\partial Y}{\partial x} + S_f = 0 \qquad (11.18b)$$

where Y is the free-surface elevation (i.e. $Y = d + z_o$).

3. Equation (11.18b) is dimensionless. The first term is related to the slope of flow acceleration energy line, the second term is the longitudinal slope of the kinetic energy line, the third one is the longitudinal slope of the free surface and the last term is the friction slope. In equation (11.18b), the two first terms are inertial terms. In steady flows, the acceleration term $\partial V/\partial t$ is zero.
4. The Saint-Venant equations may be expressed in several ways. For example the continuity equation may be written:

$$\frac{\partial A}{\partial t} + \frac{\partial Q}{\partial x} = 0 \tag{11.14}$$

$$B\frac{\partial d}{\partial t} + \frac{\partial Q}{\partial x} = 0$$

$$\frac{\partial Y}{\partial t} + \frac{A}{B}\frac{\partial V}{\partial x} + V\left(\frac{\partial Y}{\partial x} + S_o\right) + \frac{V}{B}\left(\frac{\partial A}{\partial x}\right)_{d=\text{constant}} = 0 \tag{11.17}$$

$$\frac{\partial d}{\partial t} + \frac{A}{B}\frac{\partial V}{\partial x} + V\frac{\partial d}{\partial x} + \frac{V}{B}\left(\frac{\partial A}{\partial x}\right)_{d=\text{constant}} = 0$$

$$\frac{\partial Y}{\partial t} + \frac{A}{B}\frac{\partial V}{\partial x} + V\left(\frac{\partial Y}{\partial x} + S_o\right) = 0 \tag{11.19}$$

Note that equation (11.19) equals equation (11.17) if the cross-section varies slowly with longitudinal distance which is one of the basic Saint-Venant assumptions.

The dynamic equation may be written:

$$\frac{\partial V}{\partial t} + V\frac{\partial V}{\partial x} + g\frac{\partial Y}{\partial x} + gS_f = 0 \tag{11.18a}$$

$$\frac{\partial V}{\partial t} + V\frac{\partial V}{\partial x} + g\frac{\partial d}{\partial x} + g(S_f - S_o) = 0$$

$$\frac{\partial Q}{\partial t} + \frac{\partial}{\partial x}(V^2A) + gA\frac{\partial d}{\partial x} + gA(S_f - S_o) = 0 \tag{11.15b}$$

$$\frac{1}{g}\frac{\partial V}{\partial t} + \frac{\partial}{\partial x}\left(\frac{V^2}{2g}\right) + \frac{\partial d}{\partial x} + (S_f - S_o) = 0$$

All these equations are basically equivalent.

Discussion

The system of equations (11.2) and (11.10) is the integral form of the Saint-Venant equations, while the system of equations (11.17) and (11.18a) is the differential form of the Saint-Venant equations.

The systems of two equations formed by equations (11.2) and (11.10), and equations (11.17) and (11.18a), are equivalent *if and only if all* the variables and functions are continuous and differentiable. In particular, for discontinuous solutions (e.g. hydraulic jump), equations (11.2) and (11.10) are applicable but equations (11.17) and (11.18a) might not be valid. The Saint-Venant equations were developed within very specific assumptions (Section 11.2.1). If these basic hypotheses are not satisfied, the Saint-Venant equations are not valid. For example, in an undular bore, the streamline curvature and vertical acceleration is important, and the pressure is not hydrostatic; in a mountain stream, the bed slope is not small; in sharp bends, the centrifugal acceleration may be important and the free surface is not horizontal in the radial direction.

11.2.4 Flow resistance estimate

The laws of flow resistance in open channels are essentially the same as those in closed pipes, although, in open channel, the calculations of the boundary shear stress are complicated by the existence of the free surface and the wide variety of possible cross-sectional shapes (Chapter 2). The head loss ΔH over a distance L along the flow direction is given by the Darcy equation:

$$\Delta H = f \frac{1}{D_\text{H}} \frac{V^2}{2g}$$

(11.20)

where f is the Darcy–Weisbach friction factor, V is the mean flow velocity and D_H is the hydraulic diameter or equivalent pipe diameter. In gradually varied flows, it yields:

$$V = \sqrt{\frac{8g}{f}} \sqrt{\frac{D_\text{H}}{4} S_\text{f}}$$

(11.21a)

where S_f is the friction slope. Equation (11.21a) may be rewritten as:

$$S_\text{f} = \frac{f}{8g \dfrac{D_\text{H}}{4}} V^2$$

(11.21b)

Discussion

In open channels, the Darcy equation (11.20) is the only sound method to estimate the friction loss. For various reasons, empirical resistance coefficients (e.g. Chézy coefficient, Gauckler–Manning coefficient) were and are still used. *Their use is highly inaccurate and most improper in man-made channels*. Most friction coefficients are completely empirical and they are limited to fully rough turbulent water flows. Liggett (1975) summarized nicely our poor understanding of the friction loss process: "*The (Chézy and Gauckler–Manning) equations express our continuing ignorance of turbulent processes*" (p. 45).

In man-made channels, flow resistance calculations must be performed with the Darcy friction factor only. In natural streams, the flow resistance may be expressed in terms of the Chézy equation:

$$V = C_\text{Chézy} \sqrt{\frac{D_\text{H}}{4} S_\text{f}}$$

where $C_\text{Chézy}$ is the Chézy coefficient (unit : $\text{m}^{1/2}/\text{s}$). The Chézy equation was first introduced as an *empirical correlation*. The Chézy coefficient ranges typically from $30\,\text{m}^{1/2}/\text{s}$ (small rough channel) up to $90 \, \text{m}^{1/2}/\text{s}$ (large smooth channel). Another *empirical formulation*, called the Gauckler–Manning formula, was developed for turbulent flows in rough channels:

$$V = \frac{1}{n_{\text{Manning}}} \left(\frac{D_{\text{H}}}{4} \right)^{2/3} \sqrt{S_{\text{f}}}$$

where n_{Manning} is the Gauckler–Manning coefficient (unit: $\text{s/m}^{1/3}$). The Gauckler–Manning coefficient is an empirical coefficient, found to be a function of the surface roughness primarily.

Remarks

1. Henri Philibert Gaspard Darcy (1805–1858) was a French civil engineer. He performed numerous experiments of flow resistance in pipes (Dàrcy 1858) and in open channels (Darcy and Bazin 1865), and of seepage flow in porous media (Darcy 1856). He gave his name to the Darcy–Weisbach friction factor and to the Darcy law in porous media.
2. James A. Liggett is an Emeritus Professor in Fluid Mechanics at Cornell University and a former Editor of the *Journal of Hydraulic Engineering*.
3. The friction slope may be expressed as a function of the velocity as:

$$S_{\text{f}} = \frac{f}{8g \dfrac{D_{\text{H}}}{4}} V |V|$$

$$S_{\text{f}} = \frac{1}{C_{\text{Chézy}}^2 \dfrac{D_{\text{H}}}{4}} V |V| \tag{11.21b}$$

$$S_{\text{f}} = \frac{n_{\text{Manning}}^2}{\left(\dfrac{D_{\text{H}}}{4} \right)^{4/3}} V |V|$$

The hydraulic diameter equals four times the cross-sectional area divided by the wetted perimeter. For a wide rectangular channel, $D_{\text{H}} = 4A/P_{\text{w}} \approx 4d$. The expressions of the friction slope yield then:

$$S_{\text{f}} = \frac{f}{8gd} V |V|$$

$$S_{\text{f}} = \frac{1}{C_{\text{Chézy}}^2 d} V |V| \tag{11.21c}$$

$$S_{\text{f}} = \frac{n_{\text{Manning}}^2}{d^{4/3}} V |V|$$

where $|V|$ is the magnitude of the flow velocity. The friction has the same sign as the velocity.

4. The *conveyance* (*débitance* in French) is defined as:

$$\text{Conveyance} = \frac{Q}{\sqrt{S_{\text{o}}}}$$

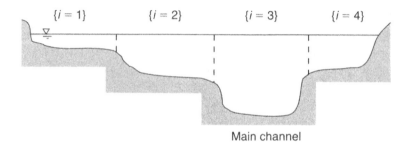

{i = 1}   {i = 2}   {i = 3}   {i = 4}

Main channel

Fig. 11.3 Sketch of a flood plain cross-section (with four sections).

Flood plain calculations

Considering a main channel with adjacent flood plains (Fig. 11.3), the flood plains and shallow-water zones are usually much rougher than the river channel. Assuming that the friction slope of each portion is the same, the flow resistance may be estimated from:

$$Q = \sum_{i=1}^{n} Q_i = \sum_{i=1}^{n} \sqrt{\frac{8g}{f_i}} A_i \sqrt{\frac{(D_{\mathrm{H}})_i}{4}} \sqrt{S_{\mathrm{f}}} \tag{11.22}$$

where Q is the total discharge in the cross-section, Q_i is the flow rate in the section $\{i\}$, and the subscript i refers to the section $\{i\}$. Conversely, if the discharge, cross-sectional shape and free-surface elevation are known, the friction slope may be estimated as:

$$S_{\mathrm{f}} = \frac{Q|Q|}{\left(\sum_{i=1}^{n} \sqrt{\frac{8g}{f_i}} A_i \sqrt{\frac{(D_{\mathrm{H}})_i}{4}} \right)^2} \tag{11.23}$$

where $|Q|$ is the magnitude of the flow rate. Equation (11.23) is general and that it takes into account the flow direction.

11.3 Method of characteristics

11.3.1 Introduction

The method of characteristics is a mathematical technique to solve a system of partial differential equations such as the Saint-Venant equations. The differential form of the Saint-Venant equations may be expressed in terms of the water depth d:

$$\frac{\partial d}{\partial t} + \frac{A}{B}\frac{\partial V}{\partial x} + V\frac{\partial d}{\partial x} + \frac{V}{B}\left(\frac{\partial A}{\partial x} \right)_{d=\text{constant}} = 0 \tag{11.24}$$

$$\frac{\partial V}{\partial t} + V\frac{\partial V}{\partial x} + g\frac{\partial d}{\partial x} + g(S_{\mathrm{f}} - S_{\mathrm{o}}) = 0 \tag{11.25}$$

where equations (11.24) and (11.25) are equivalent to equations (11.17) and (11.18a). Equation (11.24) is the continuity equation and equation (11.25) is the dynamic equation.

In an open channel, a small disturbance can propagate upstream and downstream. For a rectangular channel, the celerity of a small disturbance equals: $C = \sqrt{gd}$ (Henderson 1966, Chanson 1999a, 2004b). In a channel of irregular cross-section, the celerity of a small wave is: $C = \sqrt{g(A/B)}$, where A is the flow cross-section and B is the free-surface width (Fig. 11.2(c)). The celerity C characterizes the propagation of a small disturbance (i.e. small wave) relative to the fluid motion. For an observer standing on the bank, the absolute speed of the small wave is $(V + C)$ and $(V - C)$ where V is the flow velocity.

The differentiation of the celerity C with respect of time and of space satisfies respectively:

$$\frac{\partial}{\partial t}(C^2) = 2C\frac{\partial C}{\partial t} \approx g\frac{\partial d}{\partial t} \tag{11.26}$$

$$\frac{\partial}{\partial x}(C^2) = 2C\frac{\partial C}{\partial x} \approx g\frac{\partial d}{\partial x} \tag{11.27}$$

Replacing the terms $\partial d/\partial t$ and $\partial d/\partial x$ by equations (11.26) and (11.27) respectively in the Saint-Venant equations, it yields:

$$2\frac{\partial C}{\partial t} + 2V\frac{\partial C}{\partial x} + C\frac{\partial V}{\partial x} = 0 \tag{11.28}$$

$$\frac{\partial V}{\partial t} + 2C\frac{\partial C}{\partial x} + V\frac{\partial V}{\partial x} + g(S_f - S_o) = 0 \tag{11.29}$$

The addition of the two equations gives:

$$\left(\frac{\partial}{\partial t} + (V + C)\frac{\partial}{\partial x}\right)(V + 2C) + g(S_f - S_o) = 0 \tag{11.30}$$

while the subtraction of equation (11.29) from equation (11.28) yields:

$$\left(\frac{\partial}{\partial t} + (V - C)\frac{\partial}{\partial x}\right)(V - 2C) + g(S_f - S_o) = 0 \tag{11.31}$$

Equations (11.30) and (11.31) maybe rewritten as:

$$\frac{D}{Dt}(V + 2C) = -g(S_f - S_o) = 0 \qquad \text{Forward characteristic} \tag{11.32a}$$

$$\frac{D}{Dt}(V - 2C) = -g(S_f - S_o) \qquad \text{Backward characteristic} \tag{11.32b}$$

along their *respective* characteristic trajectories:

$$\frac{dx}{dt} = V + C \qquad \text{Forward characteristic C1} \tag{11.33a}$$

$$\frac{dx}{dt} = V - C \qquad \text{Backward characteristic C2} \tag{11.33b}$$

These trajectories are called characteristic directions or characteristics of the system. For an observer travelling along the forward characteristics (Fig. 11.4, equation (11.33a)), equation (11.32a) is valid at any point. For an observer travelling on the backward characteristics (equation 11.33b), equation (11.32b) is satisfied everywhere.

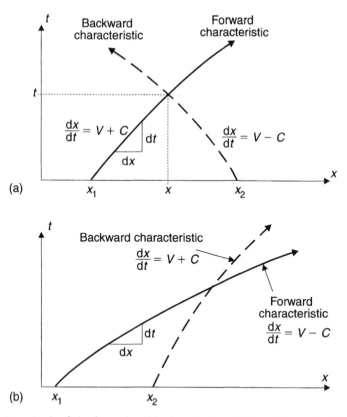

Fig. 11.4 Definition sketch of the forward and backward characteristics: (a) subcritical flow conditions and (b) supercritical flow conditions.

The system of four equations formed by equations (11.32) and (11.33) is equivalent to the system of equations (11.17) and (11.18a). *These four equations represent the characteristic system of equations that replaces the differential form of the Saint-Venant equations.* The families of forward (C1) and backward (C2) characteristics are shown in the (x, t) plane in Fig. 11.4 for sub- and supercritical flows.

Discussion

The absolute differential D/Dt of a scalar function $\Phi(x, t)$ equals:

$$\frac{D\Phi}{Dt} = \frac{\partial \Phi}{\partial t} + \frac{\partial x}{\partial t}\frac{\partial \Phi}{\partial x}$$

where $\partial x/\partial t$ is the velocity. This yields:

$$\frac{D\Phi}{Dt} = \frac{\partial \Phi}{\partial t} + V\frac{\partial \Phi}{\partial x}$$

Along a characteristic trajectory:

$$\frac{dx}{dt} = V \pm C$$

the absolute differential of the scalar function $\Phi(x, t)$ equals:

$$\frac{D\Phi}{Dt} = \left(\frac{\partial}{\partial t} + (V \pm C)\frac{\partial}{\partial x}\right)\Phi$$

The absolute differential D/Dt is also called absolute derivative.

Notes
1. The method of characteristics derived from the work of the French mathematician Gaspard Monge (1746–1818) and it was first applied by the Belgian engineer Junius Massau (1889, 1900) to solve graphically a system of partial differential equations. Today it is acknowledged to be the most accurate, reliable of all numerical integration techniques.
2. For an observer travelling along the forward characteristics C1 (Fig. 11.4), equation (11.32a) is valid at any point. For an observer travelling along the backward characteristics C2, equation (11.32b) is satisfied everywhere.
3. The forward characteristic trajectory is often called the *C1 characteristic* and positive characteristic, sometimes denoted C+. The backward characteristic is often called the *C2 characteristic* and negative characteristic, sometimes denoted C−.
4. For a horizontal, frictionless channel, the characteristic system of equations becomes:

$$V + 2C = constant \qquad \text{Along C1 characteristic}$$
$$V - 2C = constant \qquad \text{Along C2 characteristic}$$

The constants $(V + 2C)$ and $(V - 2C)$ are called the *Riemann invariants*.
5. If the term $g(S_f - S_o)$ is a constant along the channel and independent of the time, equation (11.28) becomes:

$$\frac{D}{Dt}\left(V + 2C + g(S_f - S_o)t\right) \qquad \text{Along C1 characteristic}$$

$$\frac{D}{Dt}\left(V - 2C + g(S_f - S_o)t\right) \qquad \text{Along C2 characteristic}$$

That is, the term $(V + 2C + g(S_f - S_o)t)$ is a constant along the forward characteristic C1 and the term $(V - 2C - g(S_f - S_o)t)$ is constant along the backward characteristic trajectory C2.

Discussion: graphical solution of the characteristic system of equations

Considering the case of a wave characterized by $S_o = S_f$, and for which the initial flow conditions (i.e. V, d) are known at each location x. Figure 11.5 illustrates four initial points D1–D4. The slope of the characteristic trajectories is known at each initial point: i.e. $dx/dt = V \pm C$. The forward and backward trajectories can be approximated by straight lines with slope $dt/dx = 1/(V + C)$ and $dt/dx = 1/(V - C)$ respectively, which intersect at the points E1–E3 (Fig. 11.5). The characteristic equations give the velocity and celerity at the next time step. For example, at the point E1:

$$V_{E1} + 2C_{E1} = V_{D1} + 2C_{D1} \qquad \text{Along C1 characteristic}$$

$$V_{E1} + 2C_{E1} = V_{D2} + 2C_{D2} \qquad \text{Along C2 characteristic}$$

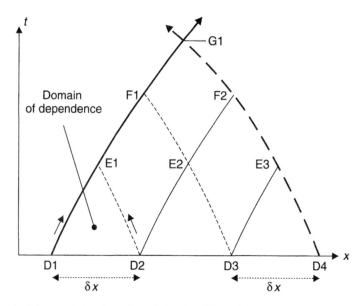

Fig. 11.5 Network of characteristic trajectories and domain of dependence.

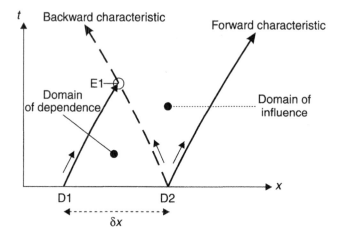

Fig. 11.6 Sketch of the region of influence.

Similarly at E2 and E3, and then at F1, F2 and ultimately at G1. When the distance δx tends to zero, the graphical solution approaches the exact solution of the Saint-Venant equations.

In the example illustrated in Fig. 11.5, the flow properties at the point G1 are functions of the initial flow properties everywhere in the reach D1–D4. The interval D1–D4 is called the *interval of influence*. In practice, an infinite number of characteristic curves crosses the interval D1–D4 and the region D1–G1–D4 is called the *domain of dependence* for the point G1. In the computational process (Fig. 11.6), the properties of the point D2, whose initial conditions are given, do influence a region delimited by the forward and backward characteristic trajectories through D2. This region is called the *domain of influence* (Fig. 11.6).

Note

When conducting a linear interpolation from a point where the flow properties are known (e.g. D1 and D2 in Fig. 11.5), the time step δt must be selected such as the point E1 is within the domain of influence of the interval D1–D2. That is, the time step must satisfy the conditions:

$$\delta t \leqslant \frac{\delta x}{|V + C|} \quad \text{and} \quad \delta t \leqslant \frac{\delta x}{|V - C|}$$

where $|V + C|$ is the absolute value of the term $(V + C)$. This condition is sometimes called the Courant condition or Courant–Friedrichs–Lewy (CFL) condition (Chapter 14, paragraph 14.1).

Application

Consider an infinitely long channel, flow measurements at two measurements stations given at $t = 0$:

	Station A	Station B
x (km)	1.5	4.2
d (m)	0.65	0.73
V (m/s)	+0.21	+0.22

In the (x, t) plane, plot the characteristics issuing from each gauging station. (*Assume straight lines.*) Calculate the location, time and flow properties at the intersection of the characteristics issuing from the two measurement stations. (*Assume* $S_f = S_o = 0$.)

Solution

The Saint-Venant equations become:

$$\frac{\mathrm{D}}{\mathrm{D}t}(V + 2C) = -g(S_f - S_o) = 0 \qquad \text{Forward characteristic} \qquad (11.32\text{a})$$

$$\frac{\mathrm{D}}{\mathrm{D}t}(V - 2C) = -g(S_f - S_o) = 0 \qquad \text{Backward characteristic} \qquad (11.32\text{b})$$

along respectively:

$$\frac{\mathrm{d}x}{\mathrm{d}t} = V + C \qquad \text{Forward characteristic C1}$$

$$\frac{\mathrm{d}x}{\mathrm{d}t} = V - C \qquad \text{Backward characteristic C2}$$

At $t = 0$, the basic flow properties are:

	Station A	Station B
x (km)	1.5	4.2
C (m/s)	2.52	2.67
$V + C$ (m/s)	+2.73	+2.89
$V - C$ (m/s)	-2.31	-2.54
$V + 2C$ (m/s)	+5.26	+5.57
$V - 2C$ (m/s)	-4.84	-5.13

Assuming straight characteristic lines, the equations of the forward characteristic issuing from station A and of the backward characteristic issuing from station B are respectively:

$$x = x_A + 2.73t \qquad \text{Forward characteristic issuing from station A}$$

$$x = x_B - 2.54t \qquad \text{Backward characteristic issuing from station B}$$

The characteristics intersect at $x = 2.90$ km for $t = 512$ s. The flow properties at intersection satisfy:

$$V + 2C = V_A + 2C_A \qquad \text{Along forward characteristic issuing from station A}$$

$$V - 2C = V_B - 2C_B \qquad \text{Along backward characteristic issuing from station B}$$

It yields: $V = +0.065$ m/s and $C = 2.60$ m/s ($d = 0.69$ m) at the intersection of the characteristics issuing from the two measurement stations assuming $S_f = S_o = 0$.

11.3.2 Boundary conditions

In summary, the method of characteristics is a mathematical technique to solve the system of partial differential equations formed by the continuity and momentum equations (i.e. Saint-Venant equations):

$$\frac{\partial Y}{\partial t} + \frac{A}{B}\frac{\partial V}{\partial x} + V\left(\frac{\partial Y}{\partial x} + S_o\right) + \frac{V}{B}\left(\frac{\partial A}{\partial x}\right)_{d=\text{constant}} = 0 \qquad (11.17)$$

$$\frac{\partial V}{\partial t} + V\frac{\partial V}{\partial x} + g\frac{\partial Y}{\partial x} + gS_f = 0 \qquad (11.18a)$$

by replacing the above two equations by a system of four ordinary differential equations:

$$\frac{D}{Dt}(V + 2C) = -g\,(S_f - S_o) \qquad \text{Forward characteristic} \qquad (11.32a)$$

$$\frac{D}{Dt}(V - 2C) = -g\,(S_f - S_o) \qquad \text{Backward characteristic} \qquad (11.32b)$$

along respectively:

$$\frac{dx}{dt} = V + C \qquad \text{Forward characteristic C1} \qquad (11.33a)$$

$$\frac{dx}{dt} = V - C \qquad \text{Backward characteristic C2} \qquad (11.33b)$$

The shape of the characteristic trajectories is a function of the flow conditions. Figure 11.7 illustrates four examples. Figure 11.7(a) presents a subcritical flow situation. The slope of the characteristics satisfies at each point:

$$\tan \Delta_1 = \frac{dt}{dx} = \frac{1}{V + C} \qquad \text{Along the forward (C1) characteristic}$$

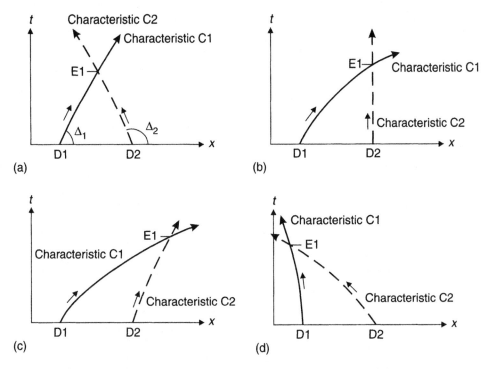

Fig. 11.7 Shape of characteristics as a function of the flow conditions: (a) subcritical flow ($Fr < 1$), (b) critical flow ($Fr = 1$), (c) supercritical flow ($Fr > 1$) and (d) supercritical flow in the negative x-direction ($Fr < -1$).

$$\tan \Delta_2 = \frac{\mathrm{d}t}{\mathrm{d}x} = \frac{1}{V - C} \qquad \text{Along the backward (C2) characteristic}$$

where Δ is the angle between the characteristics and the x-axis (Fig. 11.7(a)). Figure 11.7(b) shows a critical flow for which $V = C$ and $\Delta_2 = \pi/2$. Figure 11.7(c) and (d) illustrates both supercritical flow conditions.

Figure 11.7(c) corresponds to a torrential flow in the positive x-direction, while Fig. 11.7(d) shows a supercritical flow in the negative x-direction. In a supercritical flow (Fig. 11.7(c)), the velocity is greater than the celerity of small disturbances: i.e. $V > C$. As a result, $\tan \Delta_2$ is positive.

Remarks

1. The flow conditions (V, d) for which the specific energy is minimum for a given discharge and cross-sectional shape are called critical flow conditions (Chapter 2).
2. For a channel of irregular cross-sectional shape, the Froude number is usually defined as:

$$Fr = \frac{V}{\sqrt{g \dfrac{A}{B}}}$$

where V is the mean flow velocity, A is the cross-sectional area and B is the free-surface width. The Froude number may be written as the ratio of the flow velocity divided by the celerity of small disturbances:

$$Fr = \frac{V}{C}$$

At critical flow conditions, in a flat channel, the Froude number is unity.

Initial and boundary conditions

Assuming the simple case $g(S_f - S_o) = 0$, the characteristic system of equations is greatly simplified. That is, the Riemann invariants are constant along the characteristic trajectories:

$V + 2C = \text{constant}$ Forward characteristic issuing from station A

$V - 2C = \text{constant}$ Backward characteristic issuing from station B

Considering an *infinitely long reach*, the initial flow conditions ($t = 0$) are defined by two parameters (e.g. V and d). From each point D1 on the x-axis, two characteristics develop along which the Riemann invariants are known constants everywhere for $t > 0$ (Fig. 11.6). Similarly, at each point E1 in the ($x, t > 0$) plane, two characteristic trajectories intersect and the flow properties (V, d) are deduced from the initial conditions ($t = 0$). It can be mathematically proved that the Saint-Venant equations have a solution for $t > 0$ if and only if two flow conditions (e.g. V and d) are known at $t = 0$ everywhere.

Considering a *limited reach* ($x_1 < x < x_2$) with subcritical flow conditions (Fig. 11.8(a)), the flow properties at the point E2 are deduced from the initial flow conditions ($V(t = 0)$, $d(t = 0)$) at the points D1 and D2. At the point E1, however, only one characteristic curve intersects the boundary ($x = x_1$). This information is not sufficient to calculate the flow properties at the point E1: i.e. one additional information (e.g. V or d) is required. In other words, one flow condition must be prescribed at E1. The same reasoning applies at the point E3. In summary, in a bounded reach with subcritical flow, the solution of the Saint-Venant equations requires the knowledge of two flow parameters at $t = 0$ and one flow property at each boundary for $t > 0$.

Figure 11.8(b) shows a supercritical flow. At the point E2, two characteristics intersect and all the flow properties may be deduced from the initial flow conditions. But, at the upstream boundary (point E1), two flow properties are required to solve the Saint-Venant equations.

Types of boundary conditions

The initial flow conditions may include the velocity $V(x, 0)$ and water depth $d(x, 0)$, the velocity $V(x, 0)$ and the free-surface elevation $Y(x, 0)$, or the flow rate $Q(x, 0)$ and the cross-sectional area $A(x, 0)$.

For a limited river section, the prescribed boundary condition(s) (i.e. at $x = x_1$ and $x = x_2$) may be the water depth $d(t)$ or the free-surface elevation $Y(t)$, the flow rate $Q(t)$ or flow velocity $V(t)$ or a relationship between the water depth and flow rate. For the example shown in Fig. 11.8(a), several combinations of boundary conditions are possible: e.g. $\{d(x_1, t) \text{ and } d(x_2, t)\}$,

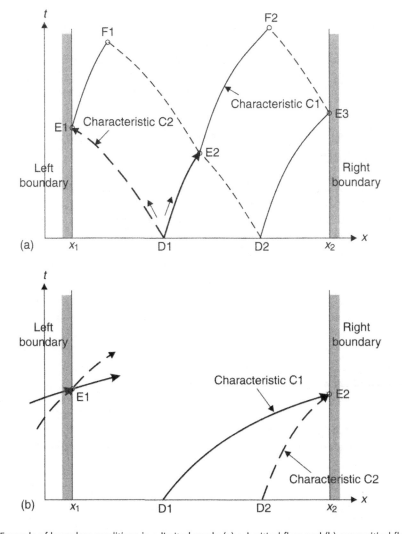

Fig. 11.8 Example of boundary conditions in a limited reach: (a) subcritical flow and (b) supercritical flow.

Table 11.1 Summary of initial and boundary conditions

Type of reach	Initial conditions	Boundary conditions
Infinitely long reach	Two parameters (e.g. V and d) everywhere	N/A (no lateral inflow)
Limited reach ($x_1 < x < x_2$) with subcritical flow conditions	Two parameters (e.g. V and d) everywhere	One flow property at each boundary for $t > 0$
Limited reach with supercritical flow conditions	Two parameters (e.g. V and d) everywhere	Two flow properties at upstream boundary for $t > 0$

$\{d(x_1,t)$ and $Q(x_2,t)\}$, $\{d(x_1,t)$ and $Q(x_2,t) = f(d(x_2,t))\}$, $\{Q(x_1,t)$ and $d(x_2,t)\}$ or $\{Q(x_1,t)$ and $Q(x_2,t)\}$.

   Table 11.1 summarizes various combinations of initial and boundary conditions. These are needed to solve both the Saint-Venant equations and the method of characteristics.

DISCUSSION

It is impossible to set $Q(x_1, t) = F(d(x = 0, t))$ unless critical flow conditions occur at the upstream boundary section ($x = x_1$).

For a set of boundary conditions $\{Q(x_1, t) \text{ and } Q(x_2, t)\}$, the particular case $Q(x_1, t) = Q(x_2, t)$ implies that the inflow equals the outflow. The mass of water does not change and the variations of the free surface is strongly affected by the initial conditions.

11.3.3 Application: numerical integration of the method of characteristics

The characteristic system of equations may be regarded as the conservation of basic flow properties (velocity, celerity) when viewed by observers travelling on the characteristic trajectories. At each intersection of forward and backward characteristics, there are four unknowns (x, t, V, C) which are the solution of the system of four differential equations. Such a solution is however impossible at a flow discontinuity (e.g. surge front).

The characteristic system of equations may be numerically integrated. Ideally a variable grid of point may be selected in the (x, t) plane where each point (x, t) is the crossing of two characteristics: i.e. a forward characteristic and a backward characteristic. The variable grid method requires however two types of interpolation because (1) the hydraulic and geometric properties of the channel are defined only at a limited number of sections and (2) the computed results are found at a number of (x, t) points unevenly distributed in the domain. In practice, a fixed grid method is preferred by practitioners: i.e. the Hartree method.

The Hartree scheme uses fixed time and spatial intervals (Fig. 11.9). The flow properties are known at the time $t = (n - 1)\,\delta t$. At the following time step $t + \delta t$, the characteristics intersecting at point M ($x = i\,\delta x, t = n\,\delta t$) are projected backward in time where they intersect the line $t = (n - 1)\,\delta t$ at points L and R whose locations are unknown (Fig. 11.9). Since x_L and x_R do not coincide with grid points, the velocity and celerity at points R and L must

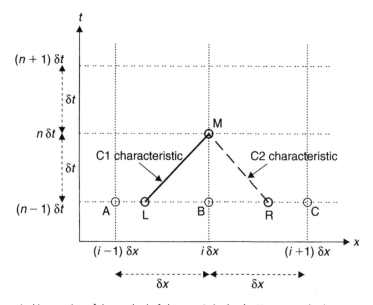

Fig. 11.9 Numerical integration of the method of characteristics by the Hartree method.

be interpolated between $x = (i - 1)\,\delta x$ and $x = i\,\delta x$, and between $x = i\,\delta x$ and $x = (i + 1)\delta x$ respectively. The discretization of the characteristic system of equations gives:

$$V_L + 2C_L = V_M + 2C_M + g\,(S_f - S_o)\,\delta t \quad \text{Forward characteristic} \qquad (11.34a)$$

$$V_R - 2C_R = V_M - 2C_M + g\,(S_f - S_o)\,\delta t \quad \text{Backward characteristic} \qquad (11.34b)$$

$$\frac{x_M - x_L}{\delta t} = V_L + C_L \qquad\qquad \text{Forward characteristic} \qquad (11.34c)$$

$$\frac{x_M - x_R}{\delta t} = V_R + C_R \qquad\qquad \text{Backward characteristic} \qquad (11.34d)$$

assuming $(S_f - S_o)$ constant during the time step δt. The system of four equations has four unknowns: V_M and C_M, and x_L and x_R, where the subscripts M, L and R refer respectively to points M, L and R defined in Fig. 11.9.

The stability of the solution of the method of characteristics is based upon the Courant condition. That is, the time step δt and the interval length δx must satisfy:

$$\frac{|V + C|}{\dfrac{\delta x}{\delta t}} \leqslant 1 \quad \text{and} \quad \frac{|V - C|}{\dfrac{\delta x}{\delta t}} \leqslant 1$$

where V is the flow velocity and C is the small wave celerity.

The solution of the Hartree scheme is artificially smoothened by interpolation errors to estimate the flow conditions (V, C) at points L and R. The errors are cumulative with time and the resulting inaccuracy limits the applicability of the method. In practice, finite difference methods offer many improvement in accuracy: e.g., Lax diffusive scheme, Preissmann-Cunge method (Chapter 14).

Notes

1. The *Hartree method* was named after the English physicist Douglas R. Hartree (1897–1958). The Hartree approximation to the Schrödinger equation is the basis for the modern physical understanding of the wave mechanics of atoms. The scheme is sometimes called the Hartree–Fock method after the Russian physicist V. Fock who generalized Hartree's scheme.
2. In open channel flows, the Hartree method is also called the *method of specified time intervals.*
3. When the flow properties at points L and R are linearly interpolated between points A and B, and between points B and C respectively (Fig. 11.9), it yields

$$V_L = \frac{V_B + \dfrac{\delta t}{\delta x}(C_B V_A - C_A V_B)}{1 + \dfrac{\delta t}{\delta x}(V_B + C_B - V_A - C_A)}$$

$$C_L = \frac{C_B + \dfrac{\delta t}{\delta x} V_L (C_A - C_B)}{1 + \dfrac{\delta t}{\delta x}(C_B - C_A)}$$

$$V_R = \frac{V_B + \frac{\delta t}{\delta x}(C_B V_C - C_C V_B)}{1 + \frac{\delta t}{\delta x}(V_C + C_C - V_B - C_B)}$$

$$C_R = \frac{C_B + \frac{\delta t}{\delta x} V_R (C_B - C_C)}{1 + \frac{\delta t}{\delta x}(C_B - C_C)}$$

If the term $(S_f - S_o)$ is estimated at points L and R, the discretization of the characteristic system of equations gives:

$$V_M = \frac{1}{2}\left(V_L + V_R + 2(C_L - C_R) - 2g\,\delta t \left(\frac{(S_f)_L + (S_f)_R}{2} - \frac{(S_o)_L + (S_o)_R}{2}\right)\right)$$

$$C_M = \frac{1}{4}\left(V_L - V_R + 2(C_L + C_R) - 2g\,\delta t \left(\frac{(S_f)_L - (S_f)_R}{2} - \frac{(S_o)_L - (S_o)_R}{2}\right)\right)$$

Note that the equations were derived assuming subcritical flow conditions as illustrated in Fig. 11.9. The interpolations have to be re-derived for supercritical flow conditions.

4. The above development based upon a linear interpolation was first proposed by Courant *et al.* (1952). A more accurate method is the parabolic interpolation presented by Montes (1998, pp. 492–495).

Application

A 2 km long irrigation channel is supplied by a large reservoir and controlled by a downstream radial gate (Fig. 11.10(a)). During a gate operation, the flow velocity, immediately upstream of the gate, increases linearly from 0 to 1 m/s in 2 min. The channel is concrete lined ($f = 0.015$), rectangular ($W = 12$ m), the invert slope is zero, and the water depth is maintained constant ($d = 2$ m) at the canal intake. Calculate the water depths at the downstream gate, at mid-distance and at the canal intake for the next 10 min.

Discussion

Initially the water in the canal is at rest and the free-surface elevation is that in the upstream reservoir.

Let select the x-direction positive in the downstream direction with $x = 0$ at the canal intake and $x = 2000$ m at the gate. The time origin ($t = 0$) is taken at the start of gate operation.

For $t > 0$, the boundary conditions are:

- constant depth at the upstream end and
- given flow velocity at the downstream end (i.e. gate).

At the upstream boundary, the flow velocity V is deduced from:

$$V_R - 2C_R = V - 2C + g(S_f - S_o)\,\delta t \qquad \text{Backward characteristic}$$

where C is deduced from the imposed water depth.

At the downstream boundary, the flow depth is calculated from:

$$V_L + 2C_L = V + 2C + g(S_F - S_o)\,\delta t \qquad \text{Forward characteristic}$$

where the flow velocity V is known.

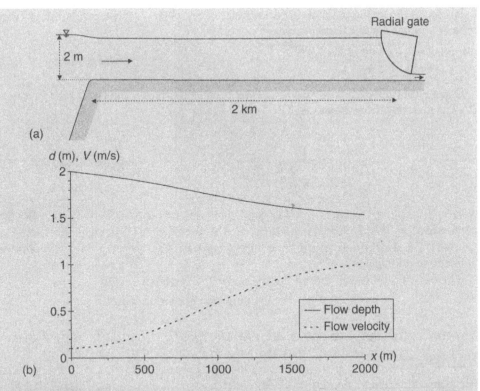

Fig. 11.10 Unsteady flow in an irrigation canal: (a) definition sketch and (b) flow depth and velocity at $t = 6$ min ($\delta x = 200$ m, $\delta t = 20$ s).

The canal length is divided into a number of spatial intervals: e.g. $\delta x = 2000/10$ m. The time step δt is selected so that the Courant condition is satisfied at all spatial locations at the current time step (e.g. $\delta t = 20$ s). The characteristic system of equations is solved by assuming that $(S_f - S_o)$ is a constant over a small time step δt.

Typical results at $t = 6$ min are presented in Fig. 11.10(b). They illustrate the progressive acceleration of the flow in the canal. At $t = 360$ s, the flow rate along the canal ranges from 2 m³/s at the upstream end to 15.3 m³/s at the gate.

11.4. Discussion

11.4.1 The dynamic equation

The differential form of the Saint-Venant equations may be written as:

$$\frac{\partial A}{\partial t} + \frac{\partial Q}{\partial x} = 0 \qquad \text{Continuity equation} \tag{11.14}$$

$$\frac{1}{g}\left(\frac{\partial V}{\partial t} + V\frac{\partial V}{\partial x}\right) + \frac{\partial d}{\partial x} + S_f - S_o = 0 \qquad \text{Dynamic equation} \tag{11.25b}$$

For *steady flows*, the continuity equation states $\partial Q/\partial x = 0$ (i.e. constant flow rate) in absence of lateral inflow and the dynamic equation yields:

$$\frac{\partial d}{\partial x} + \frac{1}{g}V\frac{\partial V}{\partial x} + S_f - S_o = 0 \qquad \text{Steady flows} \qquad (11.34)$$

Equation (11.34) is basically the backwater equation for flat channels (Chapter 2, paragraph 2.5.3) and it may be rewritten as:

$$\frac{\partial d}{\partial x}\left(1 - \beta\frac{V^2}{g\frac{A}{B}}\right) = S_o - S_f \qquad \text{Steady flows} \qquad (11.34b)$$

where β is the momentum correction coefficient which is assumed a constant (Henderson 1966, Chanson 1999a). Note a major difference between the backwater equation and equation (11.34b). The former derives from the energy equation but the latter was derived from the momentum principle.

For steady uniform equilibrium flows, the dynamic equation yields:

$$S_f - S_o = 0 \qquad \text{Steady uniform equilibrium flows} \qquad (11.35)$$

Simplification of the dynamic wave equation for unsteady flows

In the general case of unsteady flows, the dynamic equation (11.25b) may be simplified under some conditions, if the acceleration term $\partial V/\partial t$ and the inertial term $V(\partial V/\partial x)$ become small. For example, when the flood flow velocity increases from 1 to 2 m/s in 3 h (i.e. rapid variation), the dimensionless acceleration term $(1/g)(\partial V/\partial t)$ equals 9.4×10^{-6}; when the velocity increases from 1.0 to 1.4 m/s along a 10 km reach (e.g. reduction in channel width), the longitudinal slope of the kinetic energy line $(1/g)V(\partial V/\partial x)$ is equal to 4.9×10^{-6}. For comparison, the average bed slope S_o of the Rhône River between Lyon and Avignon (France) is about 0.7×10^{-3} and the friction slope is of the same order of magnitude; the average bed slope and friction slope of the Tennessee River between Watts Bar and Chickamaugo Dam is 0.22×10^{-3}; during a flood in the Missouri River, the discharge increased very rapidly from 680 to 2945 m³/s, but the acceleration term was <5% of the friction slope; on the Kitakami River (Japan), actual flood records showed that the dimensionless acceleration and inertial terms were <1.5% than the term $\partial d/\partial x$ (Miller and Cunge 1975, p. 189).

The dynamic equation may be simplified when some terms become negligible. Table 11.2 summarises various forms of the dynamic wave equation which may be solved in combination

Table 11.2 Simplification of the dynamic wave equation

Equation (1)	Dimensionless expression (2)	Remarks (3)
Dynamic wave equation	$\frac{1}{g}\left(\frac{\partial V}{\partial t} + V\frac{\partial V}{\partial x}\right) + \frac{\partial d}{\partial x} + S_f - S_o = 0$	Saint-Venant equation
Diffusive wave equation	$\frac{\partial d}{\partial x} + S_f - S_o = 0$	See Chapter 12, paragraph 12.6.
Kinematic wave equation	$S_f - S_o = 0$	See Chapter 12, paragraph 12.5 and Chapter 13.

with the unsteady flow continuity equation (Eq. (11.14)). Applications are presented in Chapters 12 and 13.

11.4.2 Limitations of the Saint-Venant equations

The Saint-Venant equations were developed for one-dimensional flows with hydrostatic pressure distributions, small bed slopes, constant water density, no sediment motion and assuming that the flow resistance is the same as for a steady uniform flow for the same depth and velocity.

Limitations of the Saint-Venant equation applications include two- and three-dimensional flows, shallow-water flood plains where the flow is nearly two-dimensional, undular and wavy flows, and the propagation of sharp discontinuities. Overall the assumptions behind the Saint-Venant equations are very restrictive and limit their applicability.

Flood plains
In two-dimensional problems involving flood propagation over inundated plains (Fig. 11.11(a)), the flood build-up is relatively slow, except when dykes break. The resistance terms are the dominant term in the dynamic equation. The assumption of one-dimensional flow becomes inaccurate.

Cunge (1975b) proposed an extension of the Saint-Venant equation in which flood plains are considered as storage volumes and with a system of equations somehow comparable to a network analysis.

Non-hydrostatic pressure distributions
The Saint-Venant equations are inaccurate when the flow is not one-dimensional and the pressure distributions are not hydrostatic. Examples include flows in sharp bends, undular hydraulic jumps, undular tidal bores (Fig. 11.11(b)). Other relevant situations include super-critical flows with shock waves, flows over a ski jump.

> ### Remarks
> 1. Undular bores (and surges) are positive surge characterized by a train of secondary waves (or undulations) following the surge front (Fig. 11.11(b)). They are called Boussinesq–Favre waves in homage to the contributions of J.B. Boussinesq and H. Favre. Classical studies of undular surges include Benet and Cunge (1971) and Treske (1994). Frazao and Zech (2002) and Cunge (2003) presented a pertinent discussion of the application of Saint-Venant equations to undular surges.
> 2. With supercritical flows, a flow disturbance (e.g. change of direction, contraction) induces the development of shock waves propagating at the free surface across the channel (e.g. Ippen and Harleman 1956). Shock waves are also called lateral shock waves, oblique hydraulic jumps, Mach waves, crosswaves, diagonal jumps. They induce flow concentrations. They create a local discontinuity in terms of water depth and pressure distributions.

Sharp discontinuities
The differential form of the Saint-Venant equations does not apply across sharp discontinuities. Figure 11.11(c) and (d) illustrates two examples: a hydraulic jump in a channelized stream

(a)

(b)

Fig. 11.11 Example of flow situations where the Saint-Venant equations are *not* applicable. (a) Flood of the Seine River at Duclair (15 km between Rouen and Caudebec, France) on 28 March 2001 (courtesy of Ms Nathalie Lemiere, Sequana-Normandie). The flooding was inflated by spring tides (coefficient 91) and strong winds. Looking downstream, note the barge and ferry attached to the quay. (b) Undular tidal bore in the Salmon River, Truro, Canada on 22 September 2001 (courtesy of Dr M.R. Gourlay).

(c)

(d)

Fig. 11.11 (*Contd*) (c) Hydraulic jump roller in a steep stream in Münich, English garden in August 2002 (courtesy of Ms Sasha Kurz), flow from bottom to top. (d) Small overflow over a stepped weir at Robina, Gold Coast on 3 February 2003.

and a weir. At a hydraulic jump, the momentum equation must be applied across the jump front (Fig. 11.11(c)). At the weir crest, the Bernoulli equation may be applied (Fig. 11.11(d)). If the weir becomes submerged, the structure acts as a large roughness and a singular energy loss.

Remarks

Another example of sharp transition is a sudden channel contraction. Assuming a smooth and short transition, the Bernoulli principle implies conservation of total head. Barnett (2002) discussed errors induced by the Saint-Venant equations for the triangular channel contraction.

(e)

(f)

Fig. 11.11 (*Contd*) (e) Extreme siltation of the Nishiyawa Dam Reservoir on the Hayakawa River (Japan, 1957) – looking at the dam wall in background with the fully silted reservoir in the foreground in November 1998. The reservoir became fully silted by gravel bed load in <20 years. It was dredged around 1988 (2 m depth) to resume hydro-electricity production. (f) Air entrainment at Chinchilla weir (Australia). Note self-aeration down chute and in hydraulic jump (foreground). The beige colour of water is caused by three-phase mixing (air, water and sediment).

Remarks

Figure 11.11(e) and (f) presents further flow situations when the Saint-Venant equations are not applicable. Figure 11.11(e) shows massive sediment motion in a Japanese river that lead to rapid reservoir siltation. Figure 11.11(f) illustrates free-surface aeration down a steep chute, while, at the chute toe, further air is entrapped at the discontinuity between the high-velocity flow and the receiving pool of water. For such a flow, the fluid density is not a constant and the Saint-Venant equations cannot be applied.

DISCUSSION

Air entrainment, or free-surface aeration, is defined as the entrainment/entrapment of un-dissolved air bubbles and air pockets that are carried away within the flowing fluid. The resulting air–water mixture consists of both air packets within water and water droplets sur-rounded by air. It includes also spray, foam and complex air–water structures. In turbulent flows, there are two basic types of air entrainment processes (Chapter 17). The entrainment of air packets can be localized or continuous along the air–water interface. Examples of *local aeration* include air entrainment by plunging jet and at hydraulic jump (Fig. 11.11(f) right). Air bubbles are entrained locally at the intersection of the impinging jet with the surround-ing waters. The intersecting perimeter is a singularity in terms of both air entrainment and momentum exchange, and air is entrapped at the discontinuity between the high-velocity jet flow and the receiving pool of water. *Interfacial aeration* (or continuous aeration) is defined as the air entrainment process along an air–water interface, usually parallel to the flow direc-tion: e.g. in chute flows (Fig. 11.11(f) left).

11.4.3 Summary

The Saint-Venant equations are the unsteady flow equivalent of the backwater equation. The latter was developed for steady gradually varied flows (Chapter 2) and it is derived from the energy equation, while the Saint-Venant equations are derived from the momentum equation. Both the backwater and Saint-Venant equations are developed for one-dimensional flows, assuming a hydrostatic pressure distribution, for gradually varied flows, neglecting sediment transport and assuming that the flow resistance is the same as for uniform equilibrium flow conditions for the same depth and velocity. The Saint-Venant equations assume further that the slope is small. Note that the backwater equation may account for the bed slope (Chapter 2) and a pressure correction coefficient may be introduced if the pressure is not hydrostatic (Henderson 1966, Chanson 1999a, 2004b).

The assumptions behind the Saint-Venant equations are very restrictive and limit the applic-ability of the results (see Section 11.4.2). Professor Liggett concluded: "*in the end it is neces-sary to 'calibrate' any mathematical model against field data before confidence can be placed in the computations*" (Liggett 1994, p. 305).

11.5 Exercises

1. Write the five basic assumptions used to develop the Saint-Venant equations.
2. Were the Saint-Venant equations developed for movable boundary hydraulic situations?
3. Are the Saint-Venant equations applicable to a steep slope?
4. Express the differential form of the Saint-Venant equations in terms of the water depth and flow velocity. Compare the differential form of the momentum equation with the back-water equation.
5. What is the dynamic wave equation? From which fundamental principle does it derive?
6. What are the two basic differences between the dynamic wave equation and the backwater equation?

7. Is the dynamic wave equation applicable to a hydraulic jump?
8. Is the dynamic wave equation applicable to an undular hydraulic jump or an undular surge?
9. Considering a channel bend, estimate the conditions for which the basic assumption of quasi-horizontal transverse free surface is no longer valid. *Assume a rectangular channel of width W much smaller than the bend radius r.*
10. Give the expression of the friction slope in terms of the flow rate, cross-sectional area, hydraulic diameter and Darcy friction factor only. Then, express the friction slope in terms of the flow rate and Chézy coefficient. Simplify both expressions for a wide rectangular channel.
11. Considering the flood plain with a main channel and an adjacent flood plain (e.g. Chanson 1999a, p. 90), develop the expression of the friction slope in terms of the total flow rate and respective Darcy friction factors.
12. Considering a long horizontal channel with the fluid initially at rest, a small wave is generated at the origin at $t = 0$ (e.g. by throwing a stone in the channel). What is the wave location as a function of time for an observer standing on the bank? (Consider only the wave travelling in the positive x-direction.) What is the value of $(V + 2C)$ along the forward characteristics for an initial water depth $d_o = 1.5\,\mathrm{m}$?
13. Considering a long channel, flow measurements at two gauging stations given at $t = 0$:

	Station 1	Station 2
Location x (km)	7.1	8.25
Water depth (m)	2.2	2.45
Flow velocity (m/s)	+0.5	+0.29

In the (x, t) plane, plot the characteristics issuing from each gauging station. (*Assume straight lines.*) Calculate the location, time and flow properties at the intersection of the characteristics issuing from the two gauging stations. (*Assume $S_f = S_o = 0$.*)
14. The analysis of flow measurements in a river reach gave:

	Station 1	Station 2
Location x (km)	11.8	13.1
Water depth (m)	0.65	0.55
Flow velocity (m/s)	+0.35	+0.55

at $t = 1\,\mathrm{h}$. Assuming a kinematic wave (i.e. $S_o = S_f$), plot the characteristics issuing from the measurement stations *assuming straight lines*. Calculate the flow properties at the intersection of the characteristics.
15. Considering a supercritical flow (flow direction in the positive x-direction), how many boundary conditions are needed for $t > 0$ and where?
16. What is the difference between the backwater equation, diffusive wave equation, dynamic wave equation and kinematic wave equation? Which one(s) does(do) apply to unsteady flows?
17. Are the basic, original Saint-Venant equations applicable to the following situations: (1) flood plains, (2) mountain streams, (3) the Brisbane River in Brisbane, (4) the Mississippi River near Saint Louis and (5) an undular tidal bore?

18. In an irrigation canal, flow measurements at several gauging stations give at $t = 0$:

	Gauge 1	Gauge 2	Gauge 3	Gauge 4	Gauge 5
Location x (km)	1.2	1.31	1.52	1.69	1.95
Water depth (m)	0.97	0.96	0.85	0.78	0.75
Flow velocity (m/s)	+0.51	+0.49	+0.46	+0.42	+0.405

The canal has a trapezoidal cross-section with a 1 m base width and 1V:2H sidewalls.

In the (x, t) plane, plot the characteristics issuing from each gauging station. (*Assume straight lines.*) Calculate the location, time and flow properties at the intersections of the characteristics issuing from the gauging stations. Repeat the process for all the domain of dependence. (*Assume* $S_f = S_o = 0$)

19. Downstream of a hydropower plant, flow measurements in the tailwater channel indicate at $t = 0$:

	Station 1	Station 2	Station 3	Station 4
Location x (m)	95	215	310	605
Water depth (m)	0.37	0.45	0.48	0.52
Flow velocity (m/s)	+1.55	+1.44	+1.36	+1.10

The canal has a rectangular cross-section with a 9 m width.

In the (x, t) plane, plot the characteristics issuing from each gauging station. (*Assume straight lines.*) Calculate the location, time and flow properties at the intersections of the characteristics issuing from the gauging stations. Repeat the process for all the domain of dependence. (*Assume* $S_f = S_o = 0$.)

11.6 Exercise solutions

3. In open channel flow hydraulics, a 'steep' slope is defined when the uniform equilibrium flow is supercritical (Chapter 5). The notion of steep and mild slope is not only a function of the bed slope but is also a function of the flow resistance.

A basic assumption of the Saint-Venant equations is a bed slope small enough such that $\cos\theta \approx 1$ and $\sin\theta \approx \tan\theta \approx \theta$. It is based solely upon the invert angle with the horizontal θ. The following table summarizes the error associated with the approximation with increasing angle θ.

θ (degree)	θ (radian)	$1 - \cos\theta$	$\sin\theta/\tan\theta$
0	0	0	1
0.5	0.008 727	3.81×10^{-5}	0.999 962
1	0.017 453	0.000 152	0.999 848
2	0.034 907	0.000 609	0.999 391
4	0.069 813	0.002 436	0.997 564
6	0.104 72	0.005 478	0.994 522
8	0.139 626	0.009 732	0.990 268
10	0.174 533	0.015 192	0.984 808
12	0.209 44	0.021 852	0.978 148
15	0.261 799	0.034 074	0.965 926
20	0.349 066	0.060 307	0.939 693
25	0.436 332	0.093 692	0.906 308

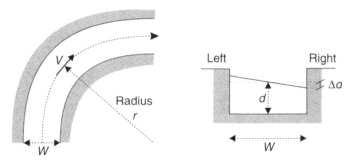

Fig. 11.12 Sketch of a circular channel bend.

6. The backwater equation derives from energy considerations for steady flow motion (Henderson 1966, Chanson 1999a, 2004b).
7. The dynamic wave equation is the differential form of the unsteady momentum equation. It might not be applicable to a discontinuity (e.g. a hydraulic jump), although the integral form of the Saint-Venant equations is (Section 11.2.2).
8. Both hydraulic jump and positive surge are a flow discontinuity, and the differential form of the unsteady momentum equation might not be applicable. However, the pressure distribution is not hydrostatic beneath waves, and this includes undular jumps and surges (e.g. Chanson 1995b, Montes and Chanson 1998). As the Saint-Venant equations were developed assuming a hydrostatic pressure distributions, they are not applicable to undular hydraulic jump flows not undular surges.
9. In a channel bend, the flow is subjected to a centrifugal acceleration acting normal to the flow direction and equal to V^2/r where r is the radius of curvature. The centrifugal pressure force induces a greater water depth at the outer bank than in a straight channel (Fig. 11.12).
In first approximation, the momentum equation applied in the transverse direction yields:

$$\frac{1}{2}\left(\rho g\,(d + \Delta d)^2 - \frac{1}{2}\rho g\,(d + \Delta d)^2 \right) = \rho \frac{V^2}{r}\,dW$$

assuming $W \ll r$ and a flat horizontal channel. The rise Δd in free-surface elevation is about:

$$\Delta d = \frac{V^2}{2rg}\,W$$

The change in water depth from the inner to outer bank is $<1\%$ if the channel width, curvature and water depth satisfy:

$$\frac{V^2}{rg}\frac{W}{d} < 0.01$$

Remark
When the bend radius is comparable to the channel width, the change in curvature radius across the width must be taken into account. A simple development shows that, for a horizontal transverse invert, the transverse variations in channel depth and depth-averaged

velocity satisfy respectively:

$$\frac{\partial d}{\partial r} = \frac{V^2}{gr}$$

$$\frac{\partial V}{\partial r} = -\frac{V}{r}$$

where r is the radial distance measured from the centre of curvature and V is the depth-averaged velocity (Henderson 1966, pp. 251–258). The results show that the water depth increases from the inner bank to the outer bank while the depth-averaged velocity is maximum at the inner bank. In practice, flow resistance affects the velocity field but maximum velocity is observed towards the inner bank.

10.
$$S_f = \frac{f}{8gd^3W^2}Q^2$$

$$S_f = \frac{Q^2}{C_{\text{Chézy}}^2 d^3 W^2}$$

where d is the water depth and W is the channel width.

12. A small disturbance will propagate with a celerity $C = \sqrt{gd_o}$. For a channel initially at rest, the wave location, for an observer standing on the bank is given by the forward characteristics trajectory:

$$\frac{dx}{dt} = V + C = 0 + \sqrt{gd_o} \qquad \text{Forward characteristics C1}$$

The integration gives:

$$x = \sqrt{gd_o}\, t$$

In a horizontal channel with the fluid initially at rest, $S_o = 0$ (horizontal channel) and $S_f = 0$ (no flow). Along the forward characteristics, $(V + 2C)$ is a constant and it is equal to 7.7 m/s.

Note that, at the wave crest, the water depth is $>d_o$, hence the wave celerity is $>\sqrt{gd_o}$. As a result, the velocity V becomes negative beneath the wave crest (Chapter 12, Section 12.2.1).

13. Answer: $x = 7.7$ km, $t = 120$ s, $V = +0.06$ m/s, $d = 2.34$ m.

The flow conditions correspond to a reduction in flow rate. At $x = 7.7$ km and $t = 120$ s, $q = 0.14$ m^2/s, compared to $q_1 = 0.77$ m^2/s and $q_2 = 0.71$ m^2/s at $t = 0$.

14. For a kinematic wave problem (i.e. $S_o = S_f$), the characteristic system of equations becomes:

$$\frac{D}{Dt}(V + 2C) = -g(S_f - S_o) = 0 \qquad \text{Forward characteristics}$$

$$\frac{D}{Dt}(V - 2C) = -g(S_f - S_o) = 0 \qquad \text{Backward characteristics}$$

along the characteristic trajectories:

$$\frac{dx}{dt} = V + C \qquad \text{Forward characteristics C1}$$

$$\frac{dx}{dt} = V - C \qquad \text{Backward characteristics C2}$$

At the intersection of the forward characteristics issuing from station 1 with the backward characteristics issuing from station 2, the trajectory equations satisfy:

$$x = x_1 + (V_1 + C_1)t$$

$$x = x_2 + (V_2 - C_2)t$$

where x and t are the location and time of the intersection. At the intersection, the flow properties satisfy:

$$V + 2C = V_1 + 2C_1$$

$$V + 2C = V_2 + 2C_2$$

where $C_1 = \sqrt{gd_1}$ and $C_2 = \sqrt{gd_2}$.

The solutions of the characteristic system of equations yields: $x = 12.6$ km, $t = 263$ s, $V = 0.80$ m/s, $C = 2.44$ m/s, $d = 0.61$ m and $q = 0.49$ m^2/s. As a comparison, $q_1 = 0.325$ m^2/s and $q_2 = 0.30$ m^2/s at $t = 0$. That is, the flow situation corresponds to an increase in flow rate.

18. At the latest (i.e. last) intersection of the characteristic curves, $x = 1.61$ km, $t = 39$ s, $V = +0.59$ m/s, $C = 2.35$ m/s. The flow conditions correspond to an increase in flow rate. At $x = 1.61$ km and $t = 39$ s, $Q = 1.58$ m^3/s and $d = 0.933$ m, compared to $Q_1 = 1.45$ m^3/s and $Q_5 = 0.929$ m^3/s at $t = 0$.

19. At the latest (i.e. last) intersection of the characteristic curves, $x = 421$ m, $t = 45$ s, $V = +0.59$ m/s, $C = 2.19$ m/s.

    The flow conditions correspond to a reduction in turbine discharge. At $x = 421$ m and $t = 45$ s, $Q = 4.28$ m^3/s, compared to $Q_1 = 5.16$ m^3/s and $Q_5 = 5.15$ m^3/s at $t = 0$.

12

Unsteady open channel flows: 2. Applications

Summary

In this chapter the Saint-Venant equations for unsteady open channel flows are applied. Basic applications include small waves and monoclinal waves, the simple wave problem, positive and negative surges. The dam break wave is treated in Chapter 13.

12.1 Introduction

In unsteady open channel flows, the velocities and water depths change with time and longitudinal position. For one-dimensional applications, the continuity and momentum equations yield the *Saint-Venant equations* (Chapter 11). The application of the Saint-Venant equations is limited by some *basic assumptions*:

1. the flow is one-dimensional,
2. the streamline curvature is very small and the pressure distributions are hydrostatic,
3. the flow resistance are the same as for a steady uniform flow for the same depth and velocity,
4. the bed slope is small enough to satisfy: $\cos\theta \approx 1$ and $\sin\theta \approx \tan\theta \approx \theta$,
5. the water density is a constant,
6. the channel has fixed boundaries and sediment motion is neglected.

With these hypotheses, the unsteady flow can be characterized at any point and any time by two variables: e.g. V and Y where V is the flow velocity and Y is the free-surface elevation. The unsteady flow properties are described by a system of two partial differential equations:

$$\frac{\partial Y}{\partial t} + \frac{A}{B}\frac{\partial V}{\partial x} + V\left(\frac{\partial Y}{\partial x} + S_{\mathrm{o}}\right) + \frac{V}{B}\left(\frac{\partial A}{\partial x}\right)_{d=\mathrm{constant}} = 0 \qquad (12.1)$$

$$\frac{\partial V}{\partial t} + V\frac{\partial V}{\partial x} + g\frac{\partial Y}{\partial x} + gS_{\mathrm{f}} = 0 \qquad (12.2)$$

where A is the cross-sectional area, B is the free-surface width, S_{o} is the bed slope and S_{f} is the friction slope (Fig. 11.2). Equation (12.1) is the continuity equation and equation (12.2) is the dynamic equation.

The Saint-Venant equations (12.1) and (12.2) cannot be solved analytically because of non-linear terms and of complicated functions. Examples of non-linear terms include the friction slope S_f while complicated functions include the flow cross-section $A(d)$ and free-surface width $B(d)$ of natural channels. A mathematical technique to solve the system of partial differential equations formed by the Saint-Venant equations is the *method of characteristics*. It yields a characteristic system of equations:

$$\frac{D}{Dt}(V + 2C) = -g(S_f - S_o) \qquad \text{Forward characteristic} \qquad (12.3\text{a})$$

$$\frac{D}{Dt}(V - 2C) = -g(S_f - S_o) \qquad \text{Backward characteristic} \qquad (12.3\text{b})$$

along:

$$\frac{dx}{dt} = V + C \qquad \text{Forward characteristic C1} \qquad (12.4\text{a})$$

$$\frac{dx}{dt} = V - C \qquad \text{Backward characteristic C2} \qquad (12.4\text{b})$$

where C is celerity of a small disturbance for an observer travelling with the flow: $C = \sqrt{gA/B}$. For an observer travelling along the forward characteristic, equation (12.3a) is valid at any point. For an observer travelling in the backward characteristic, equation (12.3b) is satisfied everywhere. The system of four equations formed by equations (12.3) and (12.4) represents the characteristic system of equations that replaces the differential form of the Saint-Venant equations.

Remark

Along the forward characteristic trajectory:

$$\frac{dx}{dt} = V + C$$

the absolute derivative of $(V + 2C)$ equals:

$$\frac{D}{Dt}(V + 2C) = \left(\frac{\partial}{\partial t} + (V + C)\frac{\partial}{\partial x} \right)(V + 2C) \qquad \text{Forward characteristic}$$

Conversely, the absolute derivative of $(V - 2C)$ along the backward characteristic is:

$$\frac{D}{Dt}(V - 2C) = \left(\frac{\partial}{\partial t} + (V - C)\frac{\partial}{\partial x} \right)(V - 2C) \qquad \text{Backward characteristic}$$

12.2 Propagation of waves

12.2.1 Propagation of a small wave

Considering a simple wave propagating in a horizontal channel initially at rest, the wave height is Δd and the wave propagation speed (or celerity) is U (Fig. 12.1). For an observer

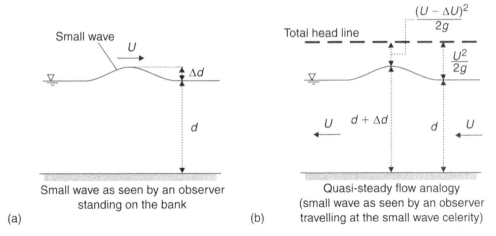

Fig. 12.1 Sketch of a small wave propagation in a fluid initially at rest. (a) View by an observer standing on the bank. (b) View by an observer travelling with the wave (quasi-steady flow analogy).

travelling with the wave, the continuity and energy equations may be written between an upstream location and the cross-section where the wave height is maximum. Neglecting energy loss, and assuming a prismatic rectangular channel, it yields:

$$U d = (U - \Delta U)(d + \Delta d) \qquad \text{Continuity equation} \qquad (12.5)$$

$$d + \frac{U^2}{2g} = d + \Delta d + \frac{(U - \Delta U)^2}{2g} \qquad \text{Bernoulli equation} \qquad (12.6)$$

where d is the water depth in the channel initially at rest.

After transformation, the wave celerity equals:

$$U = \sqrt{\frac{2g(d + \Delta d)^2}{2d + \Delta d}} \qquad (12.7a)$$

For a small wave, the celerity of the disturbance becomes:

$$U \approx \sqrt{gd}\left(1 + \frac{3}{4}\frac{\Delta d}{d}\right) \qquad \text{Small wave} \qquad (12.7b)$$

For an infinitely small disturbance ($\Delta d/d \ll 1$), equation (12.7) yields:

$$U = \sqrt{gd} = C \qquad \text{Infinitely small disturbance} \qquad (12.7c)$$

For a small wave propagating in an uniform flow (velocity V), the propagation speed of a small wave equals:

$$U \approx V + \sqrt{gd}\left(1 + \frac{3}{4}\frac{\Delta d}{d}\right) \qquad \text{Small wave} \qquad (12.8)$$

where U is the wave celerity for an observer standing on the bank and assuming that the wave propagates in the flow direction.

Notes
1. A celerity is a characteristic velocity or speed. For example, the celerity of sound is the sound speed.
2. In July 1870, Barré de Saint-Venant derived equation (12.7) and he compared it favourably with the experiments of H.E. Bazin (Barré de Saint-Venant 1871b, p. 239).
3. Henri Emile Bazin (1829–1917) was a French hydraulician and engineer, member of the French 'Corps des Ponts-et-Chaussées' and later of the Académie des Sciences de Paris. He worked as an assistant of Henri P.G. Darcy at the beginning of his career.
4. Henri Philibert Gaspard Darcy (1805–1858) was a French civil engineer. He studied at Ecole Polytechnique between 1821 and 1823, and later at the Ecole Nationale Supérieure des Ponts-et-Chaussées. He performed numerous experiments and he gave his name to the Darcy–Weisbach friction factor and to the Darcy law in porous media.

12.2.2 Propagation of a known discharge (monoclinal wave)

Considering the propagation of a known flow rate Q_2, the initial discharge is Q_1 before the passage of the monoclinal wave (Fig. 12.2). Upstream and downstream of the wave, the flow conditions are steady and the propagating wave is assumed to have a constant shape. The celerity of the wave must be greater than the downstream flow velocity. The continuity equation between sections 1 and 2 gives:

$$(U - V_1) A_1 = (U - V_2) A_2 \qquad (12.9)$$

where U is the wave celerity for an observer standing on the bank, V is the flow velocity, A is the cross-sectional area, and the subscripts 1 and 2 refer to the initial and new steady flow conditions respectively. After transformation, it yields:

$$U = \frac{Q_2 - Q_1}{A_2 - A_1} \qquad (12.10a)$$

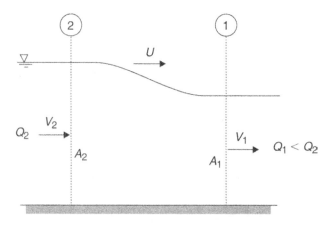

Fig. 12.2 Propagation of a monoclinal wave.

For a small variation in discharge, it becomes:

$$U = \frac{\partial Q}{\partial A} \tag{12.10b}$$

Using equation (12.10a), the flood discharge wave propagation is predicted as a function of the cross-sectional shape and flow velocity. Considering an uniform equilibrium flow in a wide rectangular channel, the discharge equals:

$$Q = \sqrt{\frac{8g}{f}} \, A \, \sqrt{d \, S_o}$$

For a small variation in flow rate, the celerity of a discharge wave is:

$$U = \frac{3}{2}V \tag{12.11}$$

where V is the flow velocity and assuming a constant Darcy friction factor f.

Notes
1. The propagation wave of a known discharge is known as a *monoclinal wave*.
2. Equation (12.10a) was first developed by A.J. Seddon in 1900 and used for the Mississippi River. It is sometimes called Seddon equation or Kleitz–Seddon equation.
3. The discharge wave celerity is a function of the steady flow rate Q.
4. Equation (12.10a) is highly dependent upon the flow resistance formula (e.g. Darcy–Weisbach, Gauckler–Manning). For example, for a wide rectangular channel, the flow resistance expression in terms of the Darcy friction factor gives:

$$U = \frac{\partial Q}{\partial A} = \frac{\dfrac{\partial Q}{\partial d}}{\dfrac{\partial A}{\partial d}} = \frac{\dfrac{3}{2}\sqrt{\dfrac{8g}{f}} \, \sqrt{d \, S_o} \, B}{B} = \frac{3}{2}V$$

Constant Darcy friction factor   (12.11)

while the Gauckler–Manning formula yields:

$$U = \frac{\partial Q}{\partial A} = \frac{5}{3}V \qquad \text{Constant Gauckler–Manning friction coefficient}$$

12.3 The simple wave problem

12.3.1 Basic equations

A simple wave is defined as a wave for which ($S_o = S_f = 0$) with initially constant water depth and flow velocity. In the system of the Saint-Venant equations, the dynamic equation becomes a kinematic wave equation (Section 11.4.1). The characteristic system of equations for a simple wave is:

$$\frac{D}{Dt}(V + 2C) = 0 \qquad \text{Forward characteristic}$$

$$\frac{D}{Dt}(V - 2C) = 0 \qquad \text{Backward characteristic}$$

along:

$$\frac{dx}{dt} = V + C \qquad \text{Forward characteristic C1}$$

$$\frac{dx}{dt} = V - C \qquad \text{Backward charcteristic C2}$$

Basically $(V + 2C)$ is a constant along the forward characteristic. That is, for an observer moving at the absolute velocity $(V + C)$, the term $(V + 2C)$ is constant. Similarly $(V - 2C)$ is constant along the backward characteristic. The characteristic trajectories can be plotted in the (x, t) plane (Fig. 12.3). They represent the path of the observers travelling on the forward and backward characteristics. On each forward characteristic, the slope of the trajectory is $1/(V + C)$ and $(V + 2C)$ is a constant along the characteristic trajectory. Altogether the characteristic trajectories form contour lines of $(V + 2C)$ and $(V - 2C)$. For the simple wave problem $(S_o = S_f = 0)$, a family of characteristic trajectories is a series of straight lines if any one curve of the family (i.e. C1 or C2) is a straight line (Henderson 1966).

In a simple wave problem, the initial flow conditions are uniform everywhere: i.e. $V(x, t = 0) = V_o$ and $d(x, t = 0) = d_o$. At $t = 0$, a disturbance (i.e. a simple wave) is introduced, typically at the origin ($x = 0$, Fig. 12.3). The wave propagates to the right along the forward characteristic C1 with a velocity $(V_o + C_o)$ where V_o is the initial flow velocity and C_o is the initial celerity: $C_o = \sqrt{gA_o/B_o}$ (Fig. 12.3). If $x = 0$ is not a boundary condition, a similar reasoning may be developed on the left side of the (x, t) plane following the negative characteristics issuing from the origin (e.g. Chapter 13). The disturbance must propagate into the undisturbed flow region with a celerity assumed to be C_o implying that the wave front is assumed small enough. In the (x, t) plane, the positive characteristic D1–E2 divides the flow region

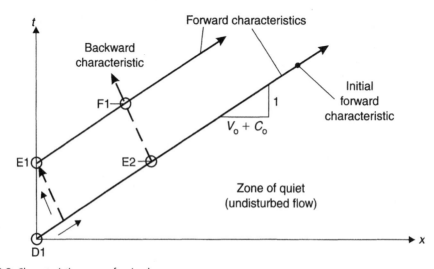

Fig. 12.3 Characteristic curves of a simple wave.

into a region below the forward characteristic which is unaffected by the disturbance (i.e. zone of quiet) and the flow region above where the effects of the wave are felt (Fig. 12.3).

In the *zone of quiet*, the flow properties at each point may be deduced from two characteristic curves intersecting the line ($t = 0$) for $x > 0$ where $V = V_o$ and $C = C_o$. In turn, the characteristic system of equations yields: $C = C_o$ and $V = V_o$ everywhere in the zone of quiet, also called undisturbed zone.

Considering a point E1 located on the left boundary (i.e. $x = 0$, $t > 0$), the backward characteristic issuing from this point intersects the first positive characteristic and it satisfies:

$$V_{E1} - 2C_{E1} = V_o - 2C_o \tag{12.12a}$$

Considering the forward characteristic issuing from the same point E1 (line E1–F1), the slope of the C1 characteristic satisfies:

$$\left(\frac{dt}{dx} \right)_{E1} = \frac{1}{V_{E1} + C_{E1}} = \frac{1}{3C_{E1} + V_o - 2C_o} \tag{12.12b}$$

At the boundary (i.e. point E1), one flow parameter (V or C) is prescribed. The second flow parameter and the slope of the forward characteristic at the boundary are then deduced from equation (12.12a).

Considering a point F1 in the (x, t) plane, both forward and backward characteristic trajectories intersect (Fig. 12.3). The flow conditions at F1 must satisfy:

$$V_{F1} + 2C_{F1} = V_{E1} + 2C_{E1} \qquad \text{Forward characteristic} \tag{12.13a}$$

$$V_{F1} - 2C_{F1} = V_o - 2C_o \qquad \text{Backward characteristic} \tag{12.13b}$$

This linear system of equations has two unknowns V_{F1} and C_{F1}. In turn, the flow properties can be calculated everywhere in the (x, t) plane, but in the zone of quiet where the flow is undisturbed.

Discussion

Henderson (1966, pp. 289–294) presented a comprehensive discussion of the simple wave problem.

Although the simple wave problem is based upon drastic assumptions, it may be relevant to a number of flow situations characterized by rapid changes such that the acceleration term $\partial V/\partial t$ is much larger than both bed and friction slopes. Examples include the rapid opening (or closure) of a gate and the dam break wave (Chapter 13).

Remarks

1. The forward characteristic issuing from the origin ($x = 0$, $t = 0$) is often called the *initial forward characteristic* or initial characteristic.
2. Stoker (1957) defined the wave motion under which the forward characteristics are straight as a 'simple wave'.
3. In the (x, t) plane (e.g. Fig. 12.3), the slope of a forward characteristic is:

$$\frac{dt}{dx} = \frac{1}{V(x,t) + C(x,t)} \qquad \text{Forward characteristic}$$

while the slope of a backward characteristic is:

$$\frac{dt}{dx} = \frac{1}{V(x,t) - C(x,t)} \qquad \text{Backward characteristic}$$

12.3.2 Application

Considering a simple wave problem where a disturbance originates from $x = 0$ at $t = 0$, and for which $x = 0$ is a known boundary condition, the calculations must proceed in a series of successive steps. These are best illustrated in the (x, t) plane (Fig. 12.3). For the subcritical flow sketched in Fig. 12.3, the basic calculations are:

Step 1 Calculate the initial flow conditions at $x > 0$ for $t = 0$. These include the initial flow velocity V_0 and the small wave celerity $C_0 = \sqrt{gA_0/B_0}$.

Step 2 Plot the initial forward characteristic.
The initial forward characteristic trajectory is:

$$t = \frac{x}{V_0 + C_0} \qquad \text{Initial forward characteristic}$$

The C1 characteristic divides the (x, t) space into two regions. In the undisturbed flow region (zone of quiet), the flow properties are known everywhere: i.e. $V = V_0$ and $C = C_0$.

Step 3 Calculate the flow conditions (i.e. $V(x = 0, t_0)$ and $C(x = 0, t_0)$) at the boundary $(x = 0)$ for $t_0 > 0$.
One flow condition is prescribed at the boundary: e.g. the flow rate is known or the water depth is given. The second flow property is calculated using the backward characteristics intersecting the boundary:

$$V(x = 0, t_0) = V_0 + 2(C(x = 0, t_0) - C_0) \qquad (12.12a)$$

Step 4 Draw a family of forward characteristics issuing from the boundary $(x = 0)$ at $t = t_0$.
The forward characteristic trajectory is:

$$t = t_0 + \frac{x}{V(x = 0, t_0) + C(x = 0, t_0)} \qquad \text{where } t_0 = t(x = 0)$$

Step 5 At the required time t, or required location x, calculate the flow properties along the family of forward characteristics.
The flow conditions are the solution of the characteristic system of equations along the forward characteristics originating from the boundary (Step 4) and the negative characteristics intersecting the initial forward characteristic. It yields:

$$V(x, t) = \frac{1}{2}(V(x = 0, t_0) + 2C(x = 0, t_0) + V_0 - 2C_0)$$

$$C(x, t) = \frac{1}{4}(V(x = 0, t_0) + 2C(x = 0, t_0) - V_0 + 2C_0)$$

Discussion
It is important to remember the basic assumptions of the simple wave analysis. That is, a wave for which $S_o = S_f = 0$ with initially constant water depth d_o and flow velocity V_o. Further the front of the wave is assumed small enough for the initial forward characteristic trajectory to satisfy $dx/dt = V_o + C_o$.

In the solution of the simple wave problem, Step 2 is important to assess the extent of the influence of the disturbance. In Steps 3 and 5, the celerity C is a function of the flow depth: $C = \sqrt{gA/B}$ where the flow cross-section A and free-surface width B are both functions of the water depth.

Application
A long irrigation channel is controlled by a downstream gate. During a gate operation, the flow velocity, immediately upstream of the gate, increases linearly from 0 to 1 m/s in 2 min. Initially the flow is at rest and the water depth is 2 m. Neglecting bed slope and friction slope, calculate the water depths at the gate, at mid-distance and at the canal intake for the next 10 min.

Solution
Initially the water in the canal is at rest and the water depth is 2 m everywhere.

The prescribed boundary condition at the downstream end is the velocity. Let select a coordinate system with $x = 0$ at the downstream end and x positive in the upstream direction. The initial conditions are $V_o = 0$ and $C_o = 4.43$ m/s assuming a wide rectangular channel.

Then the flow conditions at the boundary ($x = 0$) are calculated at various times t_o:

$$C(x = 0, t_o) = 0.5(V(x = 0, t_o) - (V_o - 2C_o)) \qquad (12.12a)$$

$$d(x = 0, t_o) = \frac{C(x = 0, t_o)}{g}$$

Next the C1 characteristic trajectories can be plotted from the left boundary ($x = 0$). As the initial C1 characteristic is a straight line, all the C1 characteristics are straight lines with the slope:

$$\frac{dt}{dx} = \frac{1}{V(x, t) + C(x, t)} = \frac{1}{V(x = 0, t_o) + C(x = 0, t_o)}$$

Results are shown in Fig. 12.4 for the first few minutes. Figure 12.4(a) presents three C1 characteristics. Following the initial characteristic, the effects of the gate operation are felt as far upstream as $x = +1.59$ km at $t = 6$ min.

At $t = 6$ min, the flow property between $x = 0$ and $x = 1590$ m are calculated from:

$$V(x, t = 6\,\text{min}) + 2C(x, t = 6\,\text{min}) = V(x = 0, t_o) + 2C(x = 0, t_o)$$

Forward characteristic

$$V(x, t = 6\,\text{min}) - 2C(x, t = 6\,\text{min}) = V_o - 2C_o \qquad \text{Backward characteristic}$$

where the equation of the forward characteristic is:

$$t = t_o + \frac{x}{V(x = 0, t_o) + C(x = 0, t_o)} \qquad \text{Forward characteristic}$$

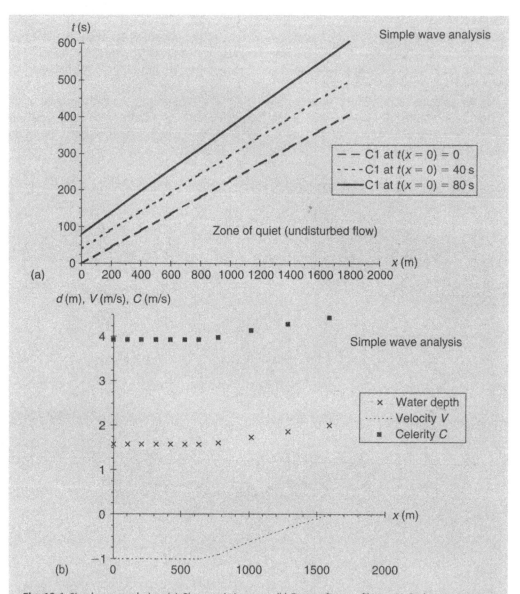

Fig. 12.4 Simple wave solution. (a) Characteristic curves. (b) Free-surface profile at $t = 6$ min.

These three equations are three unknowns: $V(x, t = 6\,\text{min})$, $C(x, t = 6\,\text{min})$ and $t_o = t(x = 0)$ for the C1 characteristics. The results in terms of the water depth, flow velocity and celerity at $t = 6$ min are presented in Fig. 12.4(b).

Remarks

1. Practically, the calculations of the free-surface profile are best performed by selecting the boundary point (i.e. the time t_o), calculating the flow properties at the boundary, and then drawing the C1 characteristics until they intersect the horizontal line $t = 6$ min. That is, the computations follow the order; select $t_o = t(x = 0)$, calculate $d(x = 0, t_o)$, compute

$V(x = 0, t_o)$, $C(x = 0, t_o)$, plot the C1 characteristics, calculate $x(t = 6\,min)$ along the C1 characteristics. Then compute $V(x, t = 6\,min)$ and $C(x, t = 6\,min)$:

$$V\left(x, t = 6\,min\right) = \frac{1}{2}\left(V\left(x = 0,\, t_o\right) + 2C\left(x = 0,\, t_o\right) + V_o - 2C_o\right)$$

$$C\left(x, t = 6\,min\right) = \frac{1}{4}\left(V\left(x = 0,\, t_o\right) + 2C\left(x = 0,\, t_o\right) - V_o + 2C_o\right)$$

Lastly the flow depth equals $d = C^2/g$.
2. In Fig. 12.4(b), note that the velocity is zero for $x > 1.59\,km$ at $t = 6\,min$.
3. Note that the C2 characteristics lines are not straight lines.
4. A relevant example is Henderson (1966, pp. 192–294).

12.4 Positive and negative surges

12.4.1 Presentation

A surge (or wave) induced by a rise in water depth is called a *positive surge*. A *negative surge* is associated with a reduction in water depth.

Considering a rise in water depth at the boundary, the increase in water depth induces an increase in small wave celerity C with time because $C = \sqrt{g/d}$ for a rectangular channel. As a result, the slope of the forward characteristics in the (x, t) plane decreases with increasing time along the boundary ($x = 0$) as:

$$\frac{dt}{dx} = \frac{1}{V + C}$$

It follows that the series of forward characteristics issuing from the boundary condition forms a network of converging lines (Fig. 12.5(a)).

Similarly, a negative surge is associated with a reduction in water depth, hence a decrease in wave celerity. The resulting forward characteristics issuing from the origin form a series of diverging straight lines (Fig. 12.5(b)). The negative wave is said to be dispersive. Locations of constant water depths move further apart as the wave moves outwards from the point of origin.

Remarks
1. A positive surge is also called bore, positive bore, moving hydraulic jump and positive wave. A positive surge of tidal origin is a tidal bore.
2. A negative surge is sometimes called negative wave.

12.4.2 Positive surge

Although the positive surge may be analysed using a quasi-steady flow analogy (see below), its inception and development is studied with the method of characteristics. When the water depth increases with time, the forward characteristics converge and eventually intersect (Fig. 12.5(a), points E2, F2 and G2). The intersection of two forward characteristics implies that the water

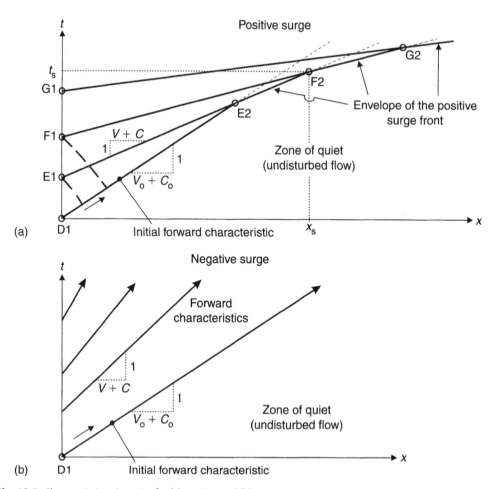

Fig. 12.5 Characteristic trajectories for (a) positive and (b) negative surges.

depth has two values at the same time. This anomaly corresponds to a wave front which becomes steeper and steeper until it forms an abrupt front: i.e. the positive surge front. For example, the forward characteristic issuing from E1 intersects the initial forward characteristic at E2 (Fig. 12.5(a)). For $x > x_{E2}$, the equation of the initial forward characteristic (D1–E2) is no longer valid as indicated by the thin dashed line in Fig. 12.5(a).

After the formation of the positive surge, some energy loss takes place across the surge front and the characteristic cannot be projected from one side of the positive surge to the other. The forward characteristic E2–F2 becomes the 'initial characteristic' for $x_{E2} < x < x_{F2}$. The points of intersections (e.g. points E2, F2, G2) form an envelope defining the zone of quiet (Fig. 12.5(a)).

Practically, *the positive surge forms at the first intersecting point*: i.e. the intersection point with the smallest time t (Fig. 12.6). Figure 12.6 illustrates the development of a positive surge from the onset of the surge (i.e. first intersection of forward characteristics) until the surge reaches its final form. After the first intersection of forward characteristics, *there is a water depth discontinuity along* the forward characteristics forming *the envelop of the surge front*: e.g. between points E2 and F2, and F2 and G2 in Fig. 12.5. The flow conditions across the surge front satisfy the continuity and momentum principles (see Discussion). Once fully developed, the surge front propagate as a stable bore (Figs 12.7 and 12.8).

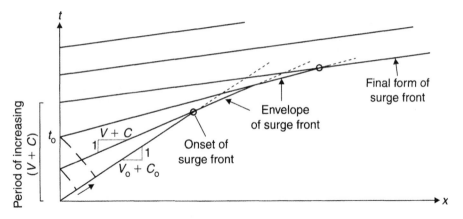

Fig. 12.6 Characteristic trajectories for the development of a positive surge.

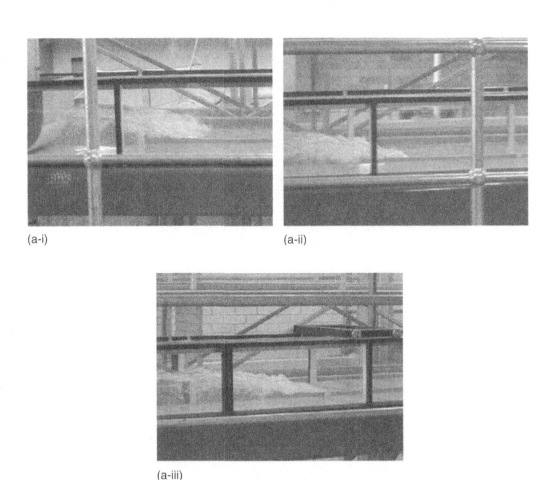

(a-i)

(a-ii)

(a-iii)

Fig. 12.7 Photographs of positive surges. (a) Advancing positive surge front in 0.5 m wide horizontal rectangular channel – surge propagation from a-i to a-iii after complete gate closure: $t = 1.5$, 5 and 8.7 s after start gate closure.

(b)

Fig. 12.7 (*Contd*) (b) Small tidal bore in the Baie du Mont Saint Michel (France), viewed from the Mont-Saint-Michel on 19 March 1996 (courtesy of Dr Pedro Lomonaco) – bore direction from right to left.

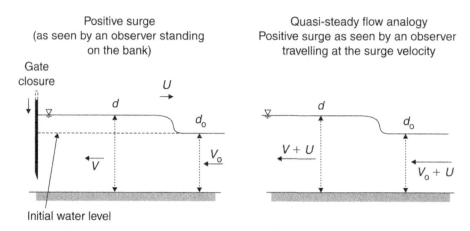

Fig. 12.8 Quasi-steady flow analogy for a fully developed positive surge.

Remarks

1. A positive surge has often a breaking front. It may also consist of a train of free-surface undulations if the surge Froude number is <1.35–1.4 (e.g. Chanson 1995b). Figure 12.7(a) show a breaking surge in a horizontal rectangular channel while Fig. 12.7(b) illustrates a non-breaking undular surge in a natural channel.
2. Henderson (1966, pp. 294–304) presented a comprehensive treatment of the surges using the method of characteristics.

Discussion: fully developed positive surge

Considering a fully developed positive surge in a rectangular channel (Fig. 12.8), the surge is a unsteady flow situation for an observer standing on the bank. But the surge is seen by an observer travelling at the surge speed U as a steady flow called a quasi-steady hydraulic jump.

For a smooth rectangular horizontal channel and considering a control volume across the front of the surge (Fig. 12.8), the continuity and momentum equations become:

$$(V_o + U)d_o = (V + U)d \qquad \text{Continuity equation}$$

$$\rho(V_o + U)\,d_o\,(V - V_o) = \frac{1}{2}\rho g d_o^2 - \frac{1}{2}\rho g d^2 \qquad \text{Momentum principle}$$

where U is the surge velocity as seen by an observer immobile on the channel bank, the subscript o refers to the initial flow conditions.

The continuity and momentum principles form a system of two equations with five variables (i.e. d_o, d, V_o, V, U). Usually the initial conditions V_o, d_o are known and the new flow rate Q/Q_o is determined by the rate of closure of the gate (e.g. complete closure: $Q = 0$, Fig. 12.7(a)).

The continuity and momentum equations yield:

$$\frac{d}{d_o} = \frac{1}{2}\left(\sqrt{1 + 8Fr_o^2} - 1\right)$$

$$Fr = \frac{2^{3/2}\,Fr_o}{\left(\sqrt{1 + 8Fr_1^2} - 1\right)^{3/2}}$$

where the surge Froude numbers Fr_o and Fr are defined as:

$$Fr_o = \frac{V_o + U}{\sqrt{gd_o}}$$

$$Fr = \frac{V + U}{\sqrt{gd}}$$

A positive surge travels faster than a small disturbance in front of it. The surge overtakes and absorbs any small disturbances that may exist at the free surface of the upstream water (i.e. in front of the surge). Similarly a positive surge travels more slowly than small disturbances behind it. Any small disturbance, downstream of the surge front and moving upstream toward the wave front, overtakes the surge and is absorbed into it. Basically the wave absorbs random disturbances on both sides of the surge and this makes the positive surge stable and self-perpetuating.

Simple wave calculations of a positive surge

For a simple wave, the apparition of the positive surge corresponds to the first intersecting point (point E2, Fig. 12.5(a)). The location and time of the wave front are deduced from the characteristic equation:

$$V_{E1} - 2C_{E1} = V_o - 2C_o \qquad \text{Backward characteristic to E1}$$

and from the trajectory equations:

$$x = (V_o + C_o)t \qquad \text{Initial forward characteristic}$$

$$x = (V_{E1} + C_{E1})(t - t_{E1}) \qquad \text{E1–E2 forward characteristic}$$

where V_o and C_o are the initial flow conditions. The first intersection of the two forward characteristics occurs at:

$$x_{E2} = \frac{(V_o + C_o)(V_o + 3C_{E1} - 2C_o)}{3\dfrac{C_{E1} - C_o}{t_{E1}}}$$

$$t_{E2} = \frac{x_{E2}}{V_o + C_o}$$

The reasoning may be extended to the intersection of two adjacent forward characteristics: e.g. the forward characteristics issuing from E1 and F1 and intersecting at F2, or the forward characteristics issuing from F1 and G1 and intersecting at G2. It yields a more general expression of the surge location:

$$x_s = \frac{\left(V_o + 3C(x = 0, t_o) - 2C_o\right)^2}{3\dfrac{\partial C(x = 0, t_o)}{\partial t}} \qquad (12.14\text{a})$$

$$t_s = t_o + \frac{x_s}{V_o + 3C(x = 0, t_o) - 2C_o} \qquad (12.14\text{b})$$

where $C(x = 0, t_o)$ is the specified disturbance celerity at the origin and x_s is the surge location at the time t_s (Fig. 12.5(a)). If the specified disturbance is the flow rate $Q(x = 0, t_o)$ or the flow velocity $V(x = 0, t_o)$, equation (12.13b) may be substituted into equation (12.14).

Equation (12.14) is the equation of the envelope delimiting the zone of quiet and defining the location of the positive surge front. The location where the surge first develops would correspond to point E2 in the (x, t) plane in Fig. 12.5(a).

Application

At the opening of a lock into a navigation canal of negligible slope, water flows into the canal with a linear increase between 0 and 30 s and a linear decrease between 30 and 60 s:

t (s)	0	30	60
Q (m³/s)	0	15	0

In the navigation canal, the water is initially at rest, the initial water depth is 1.8 m and the canal width is 22 m. Calculate the characteristics of the positive surge and its position with time.

Solution

Step 1: the initial conditions are $V_o = 0$ and $C_o = 4.2$ m/s.

Step 2: the trajectory of the initial forward characteristic is: $t = x/4.2$ where $x = 0$ at the water lock and x is positive in the downstream direction (i.e. toward the navigation canal).

Step 3: the boundary condition prescribes the flow rate. The velocity and celerity at the boundary satisfy:

$$Q(x = 0, t_o) = V(x = 0, t_o)\, d(x=0, t_o)\, W \qquad \text{Continuity equation}$$

$$V(x = 0, t_o) = V_o + 2(C(x = 0, t_o) - C_o) \qquad \text{C2 characteristic (12.12a)}$$

with $C(x = 0) = \sqrt{gd\,(x = 0)}$ It yields:

t_o (s)	0	5	10	15	20	25	30	40	50	60
$C(x = 0, t_o)$ (m/s)	1.8	4.23	4.26	4.29	4.32	4.35	4.37	4.32	4.26	4.20
$V(x = 0, t_o)$ (m/s)	4.2	0.06	0.12	0.18	0.24	0.29	0.35	0.24	0.12	0.0

Step 4: we draw a series of characteristics issuing from the boundary.

The characteristic curves for $0 < t_o < 30$ s are converging resulting in the formation of positive surge (Fig. 12.9). For $t_o > 30$ s, the reduction in flow rate is associated with the development of a negative surge (behind the positive surge) and a series of diverging characteristics. The position of the bore may be estimated from the first intersection of the converging lines with the initial characteristic or from equation (12.14). The results show a positive surge formation about 970 m downstream of the lock.

Remarks

1. The simple wave approximation neglects basically the bed slope and flow resistance (i.e. $S_o = S_f = 0$). In a frictionless horizontal channel, a rise in water depth must eventually form a surge with a steep front.
2. Barré de Saint-Venant predicted the development of tidal bore in estuaries (Barré de Saint-Venant 1871b, p. 240). He considered both cases of a simple wave and a surge propagating in uniform equilibrium flow (see below).

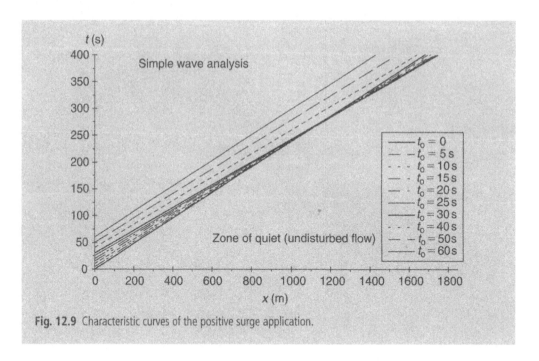

Fig. 12.9 Characteristic curves of the positive surge application.

Positive surge propagating in uniform equilibrium flow

For a wide rectangular channel, the friction slope equals:

$$S_f = \frac{f}{2}\frac{V^2}{gd}$$

At uniform equilibrium, the momentum principle states that the gravity force component in the flow direction equals exactly the flow resistance. It yields:

$$S_f = \frac{f}{2}\frac{V_o^2}{gd_o} = \sin\theta = S_o$$

The term $(S_f - S_o)$ may be rewritten as:

$$S_f - S_o = S_o\left(\frac{Fr^2}{Fr_o^2} - 1\right)$$

where Fr is the Froude number and Fr_o is the uniform equilibrium flow Froude number $(Fr_o = V_o/\sqrt{gd_o})$. Note that the celerity of a small disturbance is $C = \sqrt{gd}$ and the Froude number becomes: $Fr = V/C$.

Assuming that the initial flow conditions are uniform equilibrium (i.e. $S_o = S_f$), the characteristic system of equations becomes:

$$\frac{D}{Dt}(C(Fr + 2)) = -gS_o\left(\frac{Fr^2}{Fr_o^2} - 1\right) \qquad \text{Forward characteristic} \qquad (12.15a)$$

$$\frac{D}{Dt}(C(Fr - 2)) = -gS_0\left(\frac{Fr^2}{Fr_0^2} - 1\right) \qquad \text{Backward characteristic} \qquad (12.15b)$$

along:

$$\frac{dx}{dt} = C(Fr + 1) \qquad \text{Forward characteristic C1} \qquad (12.16a)$$

$$\frac{dx}{dt} = C(Fr - 1) \qquad \text{Backward characteristic C2} \qquad (12.16b)$$

Discussion

The initial flow conditions are at uniform equilibrium (i.e. $S_0 = S_f$). When a positive disturbance is introduced at one end of the channel (i.e. $x = 0$), the initial forward characteristic propagates in the uniform equilibrium flow ($S_0 = S_f$) and it is a straight line with a slope $dt/dx = 1/(V_0 + C_0)$ where V_0 is the uniform equilibrium flow depth.

Considering a backward characteristics upwards from the initial C1 characteristic, and at the intersection of the C2 characteristics with the initial forward characteristic, the following relationship holds:

$$\frac{D}{Dt}(C(Fr - 2)) = 0$$

Further upwards, the quantity $(S_f - S_0)$ may increase or decrease in response to a small change in Froude number ∂Fr:

$$\frac{\partial(S_f - S_0)}{\partial Fr} = \frac{2S_0}{Fr_0}$$

where the sign of both Fr_0 and S_0 is a function of the initial flow direction: i.e. $Fr_0 > 0$ and $S_0 > 0$ if $V_0 > 0$. For $t > 0$, the equations of the forward characteristics are:

$$x = (V_0 + C_0)(t - t_0)\frac{3(C(x = 0, t_0) - C_0)}{gS_0\dfrac{Fr_0 - 2}{2C_0Fr_0}}\left(\exp\left(gS_0\frac{Fr_0 - 2}{2C_0Fr_0}(t - t_0)\right) - 1\right)$$

Forward characteristic C1

where $t_0 = t(x = 0)$ (Henderson 1966). The above equation accounts for the effect of flow resistance on the surge formation and propagation. For $Fr_0 < 2$, the term in the exponential is negative. That is, the flow resistance delays the intersection of neighbouring forward characteristics and the positive surge formation. For $Fr_0 > 2$, surge formation may occur earlier than in absence of flow resistance (i.e. simple wave). Henderson (1966, pp. 297–304) showed that flow resistance makes positive waves more dispersive for uniform equilibrium flow conditions with Froude number <2.

Remarks

1. The above developments imply that the disturbance at the boundary is small enough to apply a linear theory and to neglect the terms of second order (Henderson 1966).
2. The friction slope has the same sign as the flow velocity. If there is a flow reversal ($V < 0$), the friction slope becomes negative to reflect that boundary friction opposes

the flow motion. In such a case, the friction slope must be rewritten as:

$$S_f = \frac{f}{2}\frac{V\,|V|}{gd} = \frac{f}{2}Fr\,|Fr|$$

where $|V|$ is the magnitude of the velocity.

3. A wave propagating upstream, against the flow, is sometimes called an *adverse wave*. A wave propagating downstream is named a *following wave*. Note that a small disturbance cannot travel upstream against a supercritical flow. That is, an adverse wave can exist only for $Fr_o < 1$.

12.4.3 Negative surge

A negative surge results from a reduction in water depth. It is an invasion of deeper waters by shallower waters. In still water, a negative surge propagates with a celerity $U = \sqrt{gd_o}$ for a rectangular channel, where d_o is the initial water depth. As the water depth is reduced, the celerity C decreases, the inverse slope of the forward characteristics increases and the family of forward characteristics forms a diverging lines.

A simple case is the rapid opening of a gate at $t = 0$ in a channel initially at rest (Fig. 12.10(a)). The coordinate system is selected with $x = 0$ at the gate and x positive in the upstream direction. The initial forward characteristic propagates upstream with a celerity

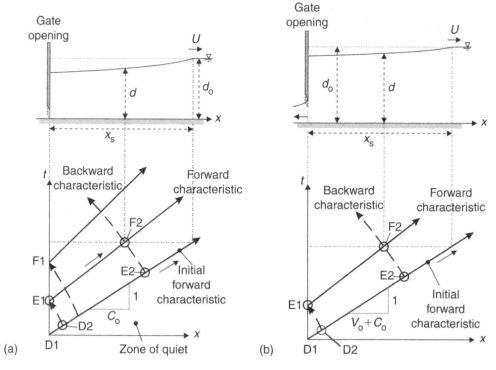

Fig. 12.10 Sketch of negative surges. (a) Negative surge in a channel initially at rest. (b) Sudden opening of a gate from a partially opened position.

$U = \sqrt{gd_o}$ where d_o is the initial water depth. The location of the leading edge of the negative surge is:

$$x_s = \sqrt{gd_o}\ t$$

Note that the gate opening induces a negative flow velocity.

A backward characteristic can be drawn issuing from the initial forward characteristic for $t > 0$ (e.g. point D2) and intersecting the boundary conditions at the gate (e.g. point E1). The backward characteristic satisfies:

$$V(x = 0, t_o) - 2C(x = 0, t_o) = -2C_o \qquad (12.17)$$

At the intersection of a forward characteristic issuing from the gate (e.g. at point E1) with a backwater characteristic issuing from the initial forward characteristic, the following conditions are satisfied:

$$V_{F2} + 2C_{F2} = V(x = 0, t_o) + 2C(x = 0, t_o) \qquad (12.18)$$
$$V_{F2} - 2C_{F2} = -2C_o \qquad (12.19)$$

where the point F2 is sketched in Fig. 12.10(a). Equations (12.17)–(12.18), plus the prescribed boundary condition, form a system of four equations with four unknowns V_{F2}, C_{F2}, $V(x = 0, t_o)$ and $C(x = 0, t_o)$. (In Fig. 12.10(a), $V_{E1} = V(x = 0, t_o)$ and $C_{E2} = C(x = 0, t_o)$.) In turn all the flow properties at the point F2 can be calculated.

In the particular case of a simple wave (paragraph 12.3), the forward characteristics are straight lines because the initial forward characteristic is a straight line (paragraph 12.3.1). The equation of forward characteristics issuing from the gate (e.g. from point E1) is:

$$\frac{dx}{dt} = V(x = 0, t_o) + C(x = 0, t_o) = V + C = 3C - 2C_o \qquad \text{Simple wave approximation}$$

The integration gives the water surface profile between the leading edge of the negative wave and the wave front:

$$\frac{x}{t - t_o} = 3\sqrt{gd} - 2\sqrt{gd_o} \quad \text{for}\ \ 0 \leqslant x \leqslant x_s \qquad (12.20)$$

assuming a rectangular channel. At a given time $t > t_o$, the free-surface profile (equation (12.20)) is a parabola.

The gate opening corresponds to an increase of flow rate beneath the gate. It is associated with both the propagation of a negative surge upstream of the gate and the propagation of the positive surge downstream of the gate.

Remarks

1. A negative surge is associated with a gradual decrease in water depth and the surge front is hardly discernible because the free-surface curvature is very shallow.
2. In the above application, the gate is partially opened from an initially closed position: i.e. $V_o = 0$.
3. In a particular case, experimental observations showed that the celerity of the negative surge was greater than the ideal value $U = -\sqrt{gd_o}$ (see Chapter 13). The difference was thought to be caused by streamline curvature effects at the leading edge of the negative surge.

Sudden complete opening

A limiting case of the above application is the sudden, complete opening of a gate from an initially closed position. The problem becomes a dam break wave and it is developed in Chapter 13.

Sudden partial opening

A more practical application is the sudden opening of a gate from a partially opened position (Fig. 12.10(b)). Using a coordinate system with $x = 0$ at the gate and x positive in the upstream direction, the initial velocity V_o must be negative. The initial forward characteristic propagates upstream with a celerity:

$$U = V_o + \sqrt{gd_o}$$

where d_o is the initial water depth and V_o is negative. Assuming a simple wave analysis, the development is nearly identical to the sudden opening of a gate from an initially closed position (equations (12.17)–(12.20)). It yields:

$$V(x = 0, t_o) - 2C(x = 0, t_o) = V_o - 2C_o \qquad \text{Characteristic D2–E1} \qquad (12.21)$$

$$V_{F2} + 2C_{F2} = V(x = 0, t_o) + 2C(x = 0, t_o) \qquad \text{Characteristic E1–F2} \qquad (12.22)$$

$$V_{F2} - 2C_{F2} = V_o - 2C_o \qquad (12.23)$$

where the point F2 is sketched in Fig. 12.10(b). Equations (12.21)–(12.23), plus the prescribed boundary condition, form a system of four equations with four unknowns. In turn all the flow properties at the point F2 can be calculated.

Remark

For the flow conditions sketched in Fig. 12.10(b), the location of the negative surge front is:

$$x_s = (V_o + \sqrt{gd_o})t$$

where $V_o < 0$.

Negative surge in a forebay

A particular application is the propagation of a negative surge in a forebay canal associated with a sudden increase in discharge into the penstock (Fig. 12.11). For example, the starting discharge of a hydropower plant supplied by a canal of small slope.

The problem may be solved for a simple wave in a horizontal forebay channel with initial flow conditions: $d = d_o$ and $V_o = 0$. Considering a point E1 located at $x = 0$ (i.e. penstock), the backward characteristic issuing from this point intersects the initial forward characteristic and it satisfies:

$$V_{E1} - 2C_{E1} = V_o - 2C_o \qquad (12.12a)$$

It yields:

$$V(x = 0) = 2(C(x = 0) - C_o)$$

(a-i)

(a-ii)

Fig. 12.11 Forebay canal. (a) Photographs of the forebay canal upstream of the inverted siphon on the Toyohashi-Tahara aqueduct (Japan) on 23 January 1999. The siphon characteristics are: length = 2780 m, diameter = 3.1 m, design flow = 17.1 m³/s. (a-i) Main canal immediately upstream the gates. (a-ii) Gates controlling the flow into the inverted siphon.

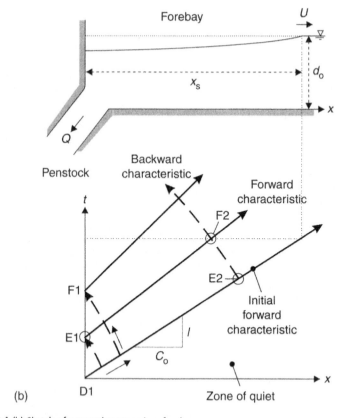

Fig. 12.11 (*Contd*) (b) Sketch of a negative surge in a forebay.

where V is negative with the sign convention used in Fig. 12.11. The discharge obtained from a rectangular forebay canal is:

$$Q(x = 0) = 2\,Wd(x = 0) = \left(\sqrt{gd(x = 0)} - \sqrt{gd_o}\right)$$

where W is the channel width. The discharge is maximum for $d = (4/9)d_o$ and the maximum flow rate equals:

$$Q_{max} = \frac{8}{27} d_o \sqrt{gd_o}\; W \qquad \text{Maximum discharge}$$

Practically, if the water demand exceeds this value, the forebay canal cannot supply the penstock and a deeper and wider canal must be designed.

Notes
1. A forebay is a canal or reservoir from which water is taken to operate a waterwheel or hydropower turbine, or to feed an inverted siphon (Fig. 12.11(a)).
2. A penstock is a gate for regulating the flow. The term is also used for a conduit leading water to a turbine.

3. The maximum flow rate for the simple wave approximation corresponds to $Q^2/(gW^2d^3) = 1$. That is, critical flow conditions. The flow rate equals the discharge at the dam break site for a dam break wave in an initially dry, horizontal channel bed (Chapter 13).
4. The operation of penstocks and inverted siphon systems is associated with both sudden water demand and stoppage. The latter creates a positive surge propagating upstream in the forebay channel (e.g. Fig. 12.11(a)).

12.5 The kinematic wave problem

12.5.1 Presentation

In a kinematic wave model, the differential form of the Saint-Venant equations is written as:

$$\frac{\partial A}{\partial t} + \frac{\partial Q}{\partial x} = 0 \qquad \text{Continuity equation} \tag{12.24}$$

$$S_f - S_o = 0 \qquad \text{Kinematic wave equation} \tag{12.25}$$

That is, the dynamic wave equation is simplified by neglecting the acceleration and inertial terms, and the free surface is assumed parallel to the channel bottom (paragraph 11.4.1, Chapter 11). The kinematic wave equation may be rewritten as:

$$Q = \sqrt{\frac{8g}{f}} A \sqrt{\frac{D_H}{4}} \sqrt{S_o}$$

where Q is the total discharge in the cross-section, f is the Darcy friction factor, A is the flow cross-sectional area and D_H is the hydraulic diameter. Equation (12.25) expresses an unique relationship between the flow rate Q and the water depth d, hence the cross-sectional area at a given location x. The differentiation of the flow rate with respect of time may be transformed:

$$\frac{\partial Q}{\partial t} = \left(\frac{\partial Q}{\partial A}\right)_{x=constant} \frac{\partial A}{\partial t}$$

The continuity equation becomes:

$$\frac{\partial Q}{\partial t} + \left(\frac{\partial Q}{\partial A}\right)_{x=constant} \frac{\partial Q}{\partial x} = 0 \qquad \text{Continuity equation} \tag{12.26}$$

It may be rewritten as:

$$\frac{DQ}{Dt} = \frac{\partial Q}{\partial t} + \frac{dx}{dt}\frac{\partial Q}{\partial x} = 0 \tag{12.27}$$

along:

$$\frac{dx}{dt} = \left(\frac{\partial Q}{\partial A}\right)_{x=constant} \tag{12.28}$$

That is, the discharge is constant along the characteristic trajectory defined by equation (12.28).

Note
1. The term

$$\frac{dx}{dt} = \left(\frac{\partial Q}{\partial A}\right)_{x=\text{constant}}$$

is sometimes called the speed of the kinematic wave.

12.5.2 Discussion

The trajectories defined by equation (12.28) are the characteristics of equations (12.26) and (12.27). There is only one family of characteristics which all propagate in the same direction: i.e. in the flow direction (Fig. 12.12). The solution of the continuity equation in terms of the flow rate $Q(x, t)$, in a river reach ($x_1 \leqslant x \leqslant x_2$) and for $t \geqslant 0$, requires one prescribed initial condition at $t = 0$ and $x_1 \leqslant x \leqslant x_2$, and one prescribed upstream condition ($x = x_1$) for $t > 0$. No prescribed downstream condition is necessary.

The kinematic wave model can only describe the downstream propagation of a disturbance. In comparison, the dynamic waves can propagate both upstream and downstream. Practically the kinematic wave equation is used to describe the translation of flood waves in some simple cases. One example is the monoclinal wave (paragraph 12.2.2). But the kinematic wave routing cannot predict the subsidence of flood wave and it has therefore limited applications.

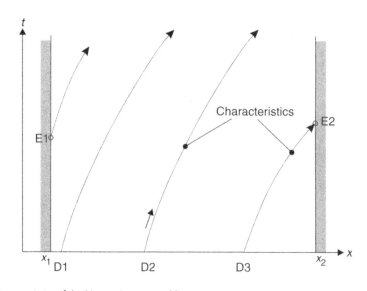

Fig. 12.12 Characteristics of the kinematic wave problem.

Notes

1. The *subsidence* of the flood wave is the trend to flatten out the flood peak discharge.
2. In summary, a kinematic wave model may predict the translation of a flood wave but not its subsidence (i.e. flattening out). The latter may be predicted with the diffusion wave model (paragraph 12.6).
3. An application of the kinematic wave equation will be developed for a dam break wave down sloping channel (Chapter 13).

12.6 The diffusion wave problem

12.6.1 Presentation

The diffusion wave equation is a simplification of the dynamic equation assuming that the acceleration and inertial terms are negligible. The differential form of the Saint-Venant equations becomes:

$$B\frac{\partial d}{\partial t} + \frac{\partial Q}{\partial x} = 0 \qquad \text{Continuity equation} \qquad (12.29\text{a})$$

$$\frac{\partial d}{\partial x} + S_\mathrm{f} - S_\mathrm{o} = 0 \qquad \text{Diffusion wave equation} \qquad (12.29\text{b})$$

The definition of the friction slope gives:

$$S_\mathrm{f} = \frac{Q|Q|}{\dfrac{8g}{f}A^2\,\dfrac{D_\mathrm{H}}{4}} \qquad (12.30)$$

where $|Q|$ is the magnitude of the flow rate.

Assuming a constant free-surface width B, the continuity equation (12.29a) is differentiated with respect to x, and the diffusion wave equation (12.29b) is differentiated with respect of the time t. It yields:

$$B\frac{\partial^2 d}{\partial x\,\partial t} + \frac{\partial^2 Q}{\partial x^2} = 0 \qquad (12.31\text{a})$$

$$\frac{\partial^2 d}{\partial t\,\partial x} + \frac{\partial}{\partial t}(S_\mathrm{f} - S_\mathrm{o}) = 0 \qquad (12.31\text{b})$$

The latter equation becomes:

$$\frac{\partial^2 d}{\partial t\,\partial x} + \frac{2|Q|}{\dfrac{8g}{f}A^2\,\dfrac{D_\mathrm{H}}{4}}\frac{\partial Q}{\partial t} - \frac{2Q\,|Q|}{\left(\dfrac{8g}{f}A^2\,\dfrac{D_\mathrm{H}}{4}\right)^{3/2}}\frac{\partial}{\partial t}\left(A\sqrt{\frac{8g}{f}\,\frac{D_\mathrm{H}}{4}}\right) = 0 \qquad (12.31\text{c})$$

Noting that:

$$\frac{\partial}{\partial t}\left(A\sqrt{\frac{8g}{f}\,\frac{D_\mathrm{H}}{4}}\right) = \frac{\partial}{\partial d}\left(A\sqrt{\frac{8g}{f}\,\frac{D_\mathrm{H}}{4}}\right)\frac{\partial d}{\partial t} = \frac{\partial}{\partial d}\left(A\sqrt{\frac{8g}{f}\,\frac{D_\mathrm{H}}{4}}\right)\left(-\frac{1}{B}\frac{\partial Q}{\partial x}\right)$$

and by eliminating $(\partial^2 d/\partial x \,\partial t)$ between equations (12.31a) and (12.31c), the diffusive wave equation becomes:

$$-\frac{1}{B}\frac{\partial^2 Q}{\partial x^2} + \frac{2|Q|}{\dfrac{8g}{f}A^2\dfrac{D_H}{4}}\frac{\partial Q}{\partial t}$$

$$+\frac{2Q|Q|}{B\left(\dfrac{8g}{f}A^2\dfrac{D_H}{4}\right)^{3/2}}\frac{\partial}{\partial d}\left(A\sqrt{\frac{8g}{f}\frac{D_H}{4}}\right)\frac{\partial Q}{\partial x}=0 \qquad (12.32a)$$

This equation may be rewritten as:

$$\frac{\partial Q}{\partial t} + \frac{Q}{B\sqrt{\dfrac{8g}{f}A^2\dfrac{D_H}{4}}}\frac{\partial}{\partial d}\left(A\sqrt{\frac{8g}{f}\frac{D_H}{4}}\right)\frac{\partial Q}{\partial x} = \frac{\dfrac{8g}{f}A^2\dfrac{D_H}{4}}{2B|Q|}\frac{\partial^2 Q}{\partial x^2} \qquad (12.32b)$$

Equation (12.32b) is an advective diffusion equation in terms of the discharge:

$$\frac{\partial Q}{\partial t} + U\frac{\partial Q}{\partial t} = D_t\frac{\partial^2 Q}{\partial x^2} \qquad (12.32c)$$

where U and D_t are respectively the celerity of the diffusion wave and the diffusion coefficient (Chapters 6 and 7). The advection velocity and diffusion coefficient are defined as:

$$U = \frac{Q}{B\sqrt{\dfrac{8g}{f}A^2\dfrac{D_H}{4}}}\frac{\partial}{\partial d}\left(A\sqrt{\frac{8g}{f}\frac{D_H}{4}}\right)$$

$$D_t = \frac{\dfrac{8g}{f}A^2\dfrac{D_H}{4}}{2B|Q|}$$

Considering a limited river reach ($x_1 \leqslant x \leqslant x_2$), the solution of the diffusion wave equation in terms of the flow rate $Q(x, t)$ requires the prescribed initial condition $Q(x, 0)$ for $t = 0$ and $x_1 \leqslant x \leqslant x_2$, and one prescribed condition at each boundary (i.e. $Q(x_1, t)$ and $Q(x_2, t)$) for $t > 0$.

Once the diffusion wave equation (12.32c) is solved in terms of the flow rates, the water depths are calculated from the continuity equation:

$$B\frac{\partial d}{\partial t} + \frac{\partial Q}{\partial x} = 0 \qquad \text{Continuity equation} \qquad (12.29a)$$

Notes
1. The diffusion wave problem is also called *diffusion routing*.
2. The term 'routing' or 'flow routing' refers to the tracking in space and time of a flood wave.

3. The above development was performed assuming a constant free-surface width B.
4. The diffusion coefficient may be rewritten as:

$$D_t = \frac{Q}{2BS_f}$$

Application

For a rectangular channel and assuming a constant Darcy friction factor, the following relationship holds:

$$\frac{\partial}{\partial d}\left(A\sqrt{\frac{8g}{f}\frac{D_H}{4}}\right) = \frac{3}{2}B\sqrt{\frac{8g}{f}\frac{A}{P_w}}\left(1 - \frac{2}{3}\frac{A}{BP_w}\right)$$

$$= \frac{3}{2}\frac{Q}{\sqrt{S_f d}}\left(1 - \frac{2}{3}\frac{A}{BP_w}\right) \qquad \text{Rectangular channel}$$

The speed of the diffusion wave equals:

$$U = \frac{3}{2}V\left(1 - \frac{2}{3}\frac{A}{BP_w}\right)$$

For a wide rectangular channel, the friction slope equals:

$$S_f = \frac{Q|Q|}{\dfrac{8g}{f}B^2d^3}$$

Assuming a constant Darcy friction factor f, the following relationship holds:

$$\frac{\partial}{\partial d}\left(A\sqrt{\frac{8g}{f}\frac{D_H}{4}}\right) = \frac{3}{2}B\sqrt{\frac{8g}{f}d} = \frac{3}{2}\frac{Q}{\sqrt{S_f d}}$$

Hence, for a wide rectangular channel and assuming a constant Darcy friction factor, the celerity of the diffusion wave equals:

$$U = \frac{3}{2}V$$

The celerity of the wave is equal to the celerity of the monoclinal wave. But the diffusion wave flattens out with longitudinal distance while the monoclinal wave has a constant shape (paragraph 12.2.2).

12.6.2 Discussion

For constant diffusion wave celerity U and diffusion coefficient D_t, equation (12.32c) may be solved analytically for a number of basic boundary conditions. Further, since equation (12.32c) is linear, the *theory of superposition* may be used to build-up solutions with more complex problems and boundary conditions (Chapters 5 and 6). Mathematical solutions of the

Table 12.1 Analytical solutions of the diffusion wave equation assuming constant diffusion wave celerity and diffusion coefficient

Problem (1)	Analytical solution (2)	Initial/boundary conditions (3)
Advective diffusion of a sharp front	$Q(x, t) = Q_o + \dfrac{\delta Q}{2}\left(1 - \mathrm{erf}\!\left(\dfrac{x - Ut}{\sqrt{4D_t t}}\right)\right)$	$Q(x, 0) = Q_o$ for $x > 0$ $Q(x, 0) = Q_o + \delta Q$ for $x < 0$
Initial volume slug introduced at $t = 0$ and $x = 0$ (i.e. sudden volume injection)	$Q(x, t) = Q_o + \dfrac{\delta Q\,\mathrm{Vol}}{A\sqrt{4\pi D_t t}}\exp\!\left(-\dfrac{(x - Ut)^2}{4D_t t}\right)$	$Q(x, 0) = Q_o$ for $t = 0$ Sudden injection of a water discharge (Vol) at a rate δQ at origin at $t = 0$
Sudden discharge injection in a river at a steady rate	$Q(x, t) = Q_o + \dfrac{\delta Q}{2}\left[1 - \mathrm{erf}\!\left(\dfrac{x - Ut}{\sqrt{4D_t t}}\right)\right.$ $\left. + \exp\!\left(\dfrac{Ux}{D_t}\right)\left(1 - \mathrm{erf}\!\left(\dfrac{x + Ut}{\sqrt{4D_t t}}\right)\right)\right]$	$Q(0, t) = Q_o + \delta Q$ for $0 < t < +\infty$ $Q(x, 0) = Q_o$ for $0 < x < +\infty$

Note: erf = Gaussian error function (Appendix A, Section 12.7).

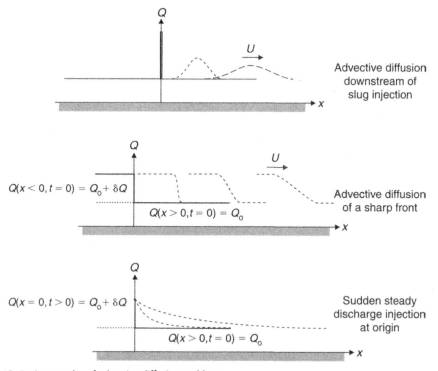

Fig. 12.13 Basic examples of advective diffusion problems.

diffusion and heat equations were addressed in two classical references (Crank 1956, Carslaw and Jaeger 1959). Simple analytical solutions of the diffusion wave equation are summarized in Table 12.1 assuming constant diffusion wave speed and diffusion coefficient. The flow situations are sketched in Fig. 12.13.

Note some basic limitations of such analytical solutions. First the diffusion coefficient D_t is a function of the flow rate. It may be assumed constant only if the change in discharge is small. Second the diffusion wave celerity is a function of the flow velocity and hence of the flow conditions.

12.6.3 The Cunge–Muskingum method

A simplification of the diffusion wave model is the Cunge–Muskingum method which was developed as an extension of the empirical Muskingum method.

Empiricism: the Muskingum method!?

The Muskingum method is based upon the continuity equation for a river reach $(x_1 \leqslant x \leqslant x_2)$:

$$\frac{\mathrm{dVol}}{\mathrm{d}t} = -(Q_2 - Q_1) \qquad \text{Continuity equation}$$

where Vol is the reach volume, $Q_1 = Q(x = x_1)$ is the inflow and $Q_2 = Q(x = x_2)$ is the outflow. The method *assumes* a relationship between the reach volume, and the inflow and outflow rates:

$$\mathrm{Vol} = K_M (X_M Q_1 + (1 - X_M) Q_2)$$

where K_M and X_M are empirical routing parameters of the river reach. By differentiating the above equation with respect to time and replacing into the continuity equation, it yields:

$$Q_2 + K_M (1 - X_M) \frac{\mathrm{d}Q_2}{\mathrm{d}t} = Q_1 - K_M X_M \frac{\mathrm{d}Q_1}{\mathrm{d}t}$$

assuming constant routing parameters K_M and X_M. If the inflow rate is a known function of time, the equation may be solved analytically or numerically in terms of the outflow Q_2.

Developed in the 1930s for flood control schemes in the Muskingum River catchment (USA), the Muskingum method is a purely empirical, intuitive method: '*the method uses the continuity equation together with an* empirical relation *linking the stored volume to the discharge*' (Montes 1998, p. 572); '*despite the popularity of the basic method,* it cannot be claimed that it is logically complete' (Henderson 1966, p. 364); '*although the conventional flood-routing methods* [including Muskingum] *may give good results under some conditions, they have the tendency to be* treacherous and unreliable' (Thomas in Miller and Cunge 1975, p. 246). Cunge *et al.* (1980) further stressed: if '*a forecasting system using the Muskingum method was built for a river basin and then a series of hydraulic works changed the river's flow characteristics,* the forecasting system might well become useless *and it might be necessary to wait 20 years in order to have enough new calibration data*' (Cunge *et al.* 1980, p. 355).

Discussion

The Muskingum method may be applied to specific situations to solve the downstream flow conditions as functions of the upstream flow conditions. Note that the downstream flow conditions have no influence on the upstream conditions, and the parameters K_M

and X_M must be calibrated and validated for each reach. The routing parameters are calculated from the inflow and outflow hydrographs of past floods in the reach.

Notes
1. The parameters K_M and X_M are *completely empirical*: i.e. they have not theoretical justification. The routing parameter K_M is homogeneous to a time while the parameter X_M is dimensionless between 0 and 1.
2. X_M must be between 0 and 0.5 for the method to be stable. For $X_M > 0.5$, the flood peak increases with distance.
3. For $X_M = 0$, the Muskingum method becomes a simple reservoir storage problem. For $X_M = 0.5$, the flood motion is a pure translation without attenuation and the Muskingum method approaches the kinematic wave solution.

Cunge–Muskingum method
The above method approaches a diffusion wave problem when:

$$U = \frac{x_2 - x_1}{K_M}$$

$$D_t = \left(\frac{1}{2} - X_M \right) U(x_2 - x_1)$$

where $(x_2 - x_1)$ is the river reach length, and K_M and X_M are the Muskingum routing parameters of the river reach. In turn, the routing parameters become functions of the flow rate:

$$X_M = \frac{1}{2} \left(1 - \frac{\left(A\sqrt{\frac{8g}{f} \frac{D_H}{4}} \right)^3}{(x_2 - x_1) Q|Q| \frac{\partial}{\partial d}\left(A\sqrt{\frac{8g}{f} \frac{D_H}{4}} \right)} \right)$$

$$K_M = \frac{x_2 - x_1}{\dfrac{Q}{B\left(\frac{8g}{f} A^2 \frac{D_H}{4} \right)^{1/2}} \frac{\partial}{\partial d}\left(A\sqrt{\frac{8g}{f} \frac{D_H}{4}} \right)}$$

This technique is called the Cunge–Muskingum method.

Discussion
The Cunge–Muskingum method may be applied to specific situations to solve the downstream flow conditions as functions of the upstream flow conditions. Limitations include: (1) the inertial terms must be negligible, (2) the downstream flow conditions have no influence on the upstream conditions and (3) the parameters K_M and X_M must be calibrated and validated for the reach.

In the Cunge–Muskingum method, the routing parameters K_M and X_M may be calculated from the physical characteristics of the channel reach, rather than on previous flood records only.

Notes
1. The extension of the Muskingum method was first presented by Cunge (1969).
2. Jean A. Cunge worked in France at Sogreah in Grenoble and he lectured at the Hydraulics and Mechanical Engineering School of Grenoble (ENSHMG).

12.7 Appendix A – Gaussian error functions

12.7.1 Gaussian error function

The Gaussian error function erf is defined as:

$$\text{erf}(u) = \frac{2}{\sqrt{\pi}} \int_0^u \exp(-\tau^2)\, d\tau \tag{12A.1}$$

Values of the Gaussian error function are summarized in Table 12A.1. Basic properties of the function are:

$$\text{erf}(0) = 0 \tag{12A.2}$$

$$\text{erf}(+\infty) = 1 \tag{12A.3}$$

$$\text{erf}(-u) = -\text{erf}(u) \tag{12A.4}$$

$$\text{erf}(u) = \frac{1}{\sqrt{\pi}}\left(u - \frac{u^3}{3 \times 1!} + \frac{u^5}{5 \times 2!} - \frac{u^7}{7 \times 3!} + \cdots \right) \tag{12A.5}$$

$$\text{erf}(u) \approx 1 - \frac{\exp(-u^2)}{\sqrt{\pi}\, u}\left(1 - \frac{1}{2 \times u^2} + \frac{1 \times 3}{(2 \times u^2)^2} - \frac{1 \times 3 \times 5}{(2 \times u^2)^3} + \cdots \right) \tag{12A.6}$$

where $n! = 1 \times 2 \times 3 \times \cdots \times n$.

Table 12A.1 Values of the error function erf

u	$\text{erf}(u)$	u	$\text{erf}(u)$
0	0	1	0.8427
0.1	0.1129	1.2	0.9103
0.2	0.2227	1.4	0.9523
0.3	0.3286	1.6	0.9763
0.4	0.4284	1.8	0.9891
0.5	0.5205	2	0.9953
0.6	0.6309	2.5	0.9996
0.7	0.6778	3	0.99998
0.8	0.7421	$+\infty$	1
0.9	0.7969		

Notes
In first approximation, the function erf(u) may be correlated by:

$$\text{erf}(u) \approx u \, (1.375511 - 0.61044u + 0.088439u^2) \qquad 0 \leqslant u < 2$$

$$\text{erf}(u) \approx \tanh(1.198787u) \qquad -\infty < u < +\infty$$

In many applications, the above correlations are *not accurate enough*, and equation (12A.5) and Table 12A.1 should be used.
For small values of u, the error function is about:

$$\text{erf}(u) = \frac{u}{\sqrt{\pi}} \qquad \text{Small values of } u$$

For large values of u, the erf function is about:

$$\text{erf}(u) \approx 1 - \frac{\exp(-u^2)}{u \sqrt{\pi}} \qquad \text{Large values of } u$$

12.7.2 Complementary error function

The complementary Gaussian error function erfc is defined as:

$$\text{erfc}(u) = 1 - \text{erf}(u) = \frac{2}{\sqrt{\pi}} \int_u^{+\infty} \exp(-\tau^2) \, d\tau \qquad (12\text{A}.7)$$

Basic properties of the complementary function include:

$$\text{erfc}(0) = 1 \qquad (12\text{A}.8)$$

$$\text{erf}(+\infty) = 0 \qquad (12\text{A}.9)$$

12.8 Exercises

1. List the key assumptions of the Saint-Venant equations.
2. What is the celerity of a small disturbance in (1) a rectangular channel, (2) a 90° V-shaped channel and (3) a channel of irregular cross-section? (4) Application: a flow cross-section of a flood plain has the following properties: hydraulic diameter = 5.14 m, maximum water depth = 2.9 m, wetted perimeter = 35 m, free-surface width = 30 m. Calculate the celerity of a small disturbance.
3. A 0.2 m high small wave propagates downstream in a horizontal channel with initial flow conditions $V = +0.1$ m/s and $d = 2.2$ m. Calculate the propagation speed of the small wave.
4. Uniform equilibrium flow conditions are achieved in a long rectangular channel ($W = 12.8$ m, concrete lined, $S_o = 0.0005$). The observed water depth is 1.75 m. Calculate the celerity of a small monoclinal wave propagating downstream. *Perform your calculations using the Darcy friction factor.*
5. The flow rate in a rectangular canal ($W = 3.4$ m, concrete lined, $S_o = 0.0007$) is 3.1 m^3/s and uniform equilibrium flow conditions are achieved. The discharge suddenly increases

to $5.9\,\text{m}^3/\text{s}$. Calculate the celerity of the monoclinal wave. How long will it take for the monoclinal wave to travel $20\,\text{km}$?

6. What is the basic definition of a simple wave? May the simple wave theory be applied to (1) a sloping, frictionless channel, (2) a horizontal, rough canal, (3) a positive surge in a horizontal, smooth channel with constant water depth and (4) a smooth, horizontal canal with an initially accelerating flow?

7. What is the 'zone of quiet'?

8. Another basic application is a river discharging into the sea and the upstream extent of tidal influence onto the free surface. Considering a stream discharging into the sea, the tidal range is $0.8\,\text{m}$ and the tidal period is $12\,\text{h}\ 25\,\text{min}$. The initial flow conditions are $V = 0.3\,\text{m/s}$, $d = 0.4\,\text{m}$ corresponding to low tide. Neglecting bed slope and flow resistance, and starting at a low tide, calculate how far upstream the river level will rise $3\,\text{h}$ after low tide and predict the free-surface profile at $t = 3\,\text{h}$.

9. Considering a long, horizontal rectangular channel ($W = 4.2\,\text{m}$), a gate operation, at one end of the canal, induces a sudden withdrawal of water resulting in a negative velocity. At the gate, the boundary conditions for $t > 0$ are: $V(x = 0, t) = -0.2\,\text{m/s}$. Calculate the extent of the gate operation influence in the canal at $t = 1\,\text{h}$. The initial conditions in the canal are: $V = 0$ and $d = 1.4\,\text{m}$.

10. Water flows in an irrigation canal at steady state ($V = 0.9\,\text{m/s}$, $d = 1.65\,\text{m}$). The flume is assumed smooth and horizontal. The flow is controlled by a downstream gate. At $t = 0$, the gate is very slowly raised and the water depth upstream of the gate decreases at a rate of $5\,\text{cm/min}$ until the water depth becomes $0.85\,\text{m}$. (1) Plot the free-surface profile at $t = 10\,\text{min}$. (2) Calculate the discharge per unit width at the gate at $t = 10\,\text{min}$.

11. A $200\,\text{km}$ long rectangular channel ($W = 3.2\,\text{m}$) has a reservoir at the upstream end and a gate at the downstream end. Initially the flow conditions in the canal are uniform: $V = 0.35\,\text{m/s}$, $d = 1.05\,\text{m}$. The water surface level in the reservoir begins to rise at a rate of $0.2\,\text{m/h}$ for $6\,\text{h}$. Calculate the flow conditions in the canal at $t = 2\,\text{h}$. *Assume $S_o = S_f = 0$.*

12. Waters flow in a horizontal, smooth rectangular channel. The initial flow conditions are $d = 2.1\,\text{m}$ and $V = +0.3\,\text{m/s}$. The flow rate is stopped by sudden gate closure at the downstream end of the canal. Using the quasi-steady flow analogy, calculate the new water depth and the speed of the fully developed surge front.

13. Let consider the same problem as above (i.e. a horizontal, smooth rectangular channel, $d = 2.1\,\text{m}$ and $V = 0.3\,\text{m/s}$) but the downstream gate is closed slowly at a rate corresponding to a linear decrease in flow rate from $0.63\,\text{m}^2/\text{s}$ down to 0 in $15\,\text{min}$. (1) Predict the surge front development. (2) Calculate the free-surface profile at $t = 1\,\text{h}$ after the start of gate closure.

14. Waters flow in a horizontal, smooth rectangular irrigation canal. The initial flow conditions are $d = 1.1\,\text{m}$ and $V = +0.35\,\text{m/s}$. Between $t = 0$ and $t = 10\,\text{min}$, the downstream gate is slowly raised at a rate implying a decrease in water depth of $0.05\,\text{m/min}$. For $t > 10\,\text{min}$, the gate position is maintained constant. Calculate and plot the free-surface profile in the canal at $t = 30\,\text{min}$. *Use a simple wave approximation.*

15. A $5\,\text{m}$ wide forebay canal supplies a penstock feeding a Pelton turbine. The initial conditions in the channel are $V = 0$ and $d = 2.5\,\text{m}$. (1) The turbine starts suddenly operating with $6\,\text{m}^3/\text{s}$. Predict the water depth at the downstream end of the forebay canal. (2) What is the maximum discharge that the forebay channel can supply? *Use a simple wave theory.*

16. Write the kinematic wave equation for a wide rectangular channel, in terms of the flow rate, bed slope and water depth. What is the speed of a kinematic wave? Does the kinematic wave routing predict subsidence?

17. A wide channel has a bed slope $S_o = 0.0003$ and the channel bed has an equivalent roughness height of 25 mm. The initial flow depth is 2.3 m and uniform equilibrium flow conditions are achieved. The water depth is abruptly increased to 2.4 m at the upstream end of the channel. Calculate the speed of the diffusion wave and the diffusion coefficient.

18. A 8 m wide rectangular canal (concrete lining) operates at uniform equilibrium flow conditions for a flow rate of 18 m^3/s resulting in a 1.8 m water depth. At the upstream end, the discharge is suddenly increased to 18.8 m^3/s. Calculate the flow rate in the canal 1 h later at a location $x = 15$ km. *Use diffusion routing.*

12.9 Exercise solutions

2. (2) $C = \sqrt{gd/2}$ (4) $C = 3.8$ m/s.
3. $U = +5.1$ m/s, $C = 4.64$ m/s.
4. $U = +3.0$ m/s (for $k_s = 1$ mm).
5. $d_2 = 1.23$ m, $U = +1.81$ m/s, $t = 11\,100$ s (3 h 5 min).
6. (1) No. (2) No. (3) Yes. (4) No.
8. The prescribed boundary condition at the river mouth is the water depth:

$$d(x = 0, t_o) = 0.4 + \frac{0.8}{2}\left(1 + \cos\left(\frac{2\pi}{T}t_o - \pi\right)\right)$$

where T is the tide period ($T = 44\,700$ s).

Let select a coordinate system with $x = 0$ at the river mouth and x positive in the upstream direction. The initial conditions are $V_o = -0.3$ m/s and $C_o = 1.98$ m/s assuming a wide rectangular channel. Then the flow conditions at the boundary ($x = 0$) are calculated at various times t_o:

$$C(x = 0, t_o) = \sqrt{gd(x = 0, t_o)}$$
$$V(x = 0, t_o) = 2C(x = 0, t_o) + V_o - 2C_o \qquad (12.12a)$$

The results are summarized below:

t_o (s) (1)	$d(x = 0, t_o)$ (m) (2)	$C(x = 0, t_o)$ (m/s) (3)	$V(x = 0, t_o)$ (m/s) (4)	$V + C$ (m/s) (5)	$V + 2C$ (m/s) (6)
0	0.40	1.98	-0.30	1.68	3.66
4470	0.48	2.16	0.06	2.22	4.38
8940	0.68	2.57	0.89	3.46	6.04
13410	0.92	3.01	1.76	4.77	7.77
17880	1.12	3.32	2.38	5.70	9.01
22350	1.20	3.43	2.60	6.03	9.46

Next the C1 characteristic trajectories can be plotted from the left boundary ($x = 0$). As the initial C1 characteristic is a straight line, all the C1 characteristics are straight lines with the slope:

$$\frac{dt}{dx} = \frac{1}{V(x, t) + C(x, t)} = \frac{1}{V(x = 0, t_o) + C(x = 0, t_o)}$$

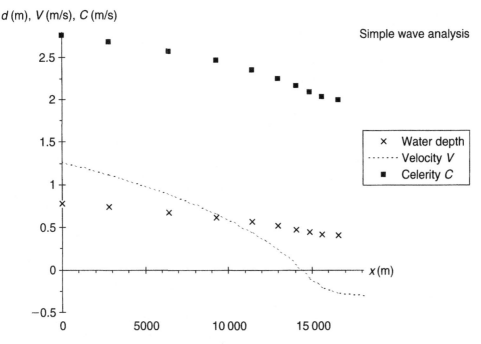

Fig. 12.14 Simple wave solution: flow properties at mid-tide ($t = 3$ h).

Results are shown in Fig. 12.5 for the first 3 h ($0 < t < 3$ h). Following the initial characteristic, the effects of the tide are felt as far upstream as $x = +18.1$ km at $t = 3$ h.

At $t = 3$ h, the flow property between $x = 0$ and $x = 18\,100$ m are calculated from:

$$V(x, t = 3\,\text{h}) + 2C(x, t = 3\,\text{h}) = V(x = 0, t_o) + 2C(x = 0, t_o) \qquad \text{Forward characteristic}$$

$$V(x, t = 3\,\text{h}) - 2C(x, t = 3\,\text{h}) = V_o - 2C_o \qquad \text{Backward characteristic}$$

where the equation of the forward characteristic is:

$$t = t_o + \frac{x}{V(x = 0, t_o) + C(x = 0, t_o)} \qquad \text{Forward characteristic}$$

These three equations are three unknowns: $V(x, t = 3\,\text{h})$, $C(x, t = 3\,\text{h})$ and $t_o = t(x = 0)$ for the C1 characteristics. The results in terms of the water depth, flow velocity and celerity at high tide are presented in Fig. 12.14.

Remarks

- Practically, the calculations of the free-surface profile are best performed by selecting the boundary point (i.e. the time t_o), calculating the flow properties at the boundary, and then drawing the C1 characteristics until they intersect the horizontal line $t = 3$ h. That is, the computations follow the order; select $t_o = t(x = 0)$, calculate $d(x = 0, t_o)$, compute $V(x = 0, t_o)$, $C(x = 0, t_o)$, plot the C1 characteristics, calculate $x(t = 3\,\text{h})$ along the C1 characteristics. Then compute $V(x, t = 3\,\text{h})$ and $C(x, t = 3\,\text{h})$:

$$V(x, t = 3\,\text{h}) = \frac{1}{2}(V(x = 0, t_o) + 2C(x = 0, t_o) + V_o - 2C_o)$$

$$C(x, t = 3\,\text{h}) = \frac{1}{4}(V(x = 0, t_o) + 2C(x = 0, t_o) - V_o + 2C_o)$$

Lastly the flow depth equals $d = C^2/g$.
- In Fig. 12.14, note that the velocity is positive (i.e. upriver flow) for $x < 14.4\,\text{km}$ at $t = 3\,\text{h}$.
- Note that the C2 characteristics lines are not straight lines.
- A relevant example is Henderson (1966, pp. 192–294).

9. The problem may be analysed as a simple wave. The initial flow conditions are: $V_o = 0$ and $C_o = 3.7\,\text{m/s}$. Let select a coordinate system with $x = 0$ at the gate and x positive in the upstream direction. In the (x, t) plane, the equation of the initial forward characteristics (issuing from $x = 0$ and $t = 0$) is given:

$$\frac{dt}{dx} = \frac{1}{V_o + C_o} = 0.27\,\text{s/m}$$

At $t = 1\,\text{h}$, the extent of the influence of the gate operation is 13.3 km.

10. The simple wave problem corresponds to a negative surge. In absence of further information, the flume is assumed wide rectangular.

Let select a coordinate system with $x = 0$ at the gate and x positive in the upstream direction. The initial flow conditions are: $V_0 = -0.9$ and $C_0 = 4.0\,\text{m/s}$. In the (x, t) plane, the equation of the initial forward characteristics (issuing from $x = 0$ and $t = 0$) is given:

$$\frac{dt}{dx} = \frac{1}{V_o + C_o} = 0.32\,\text{s/m}$$

At $t = 10\,\text{min}$, the maximum extent of the disturbance is $x = 1870\,\text{m}$. That is, the zone of quiet is defined as $x > 1.87\,\text{km}$.

At the gate $(x = 0)$, the boundary condition is: $d(x = 0, t_o \leqslant 0) = 1.65\,\text{m}$, $d(x = 0, t_o) = 1.65 - 8.33 \times 10^{-4}t_o$, for $0 < t_o < 960\,\text{s}$, and $d(x = 0, t_o \geqslant 960\,\text{s}) = 0.85\,\text{m}$. The second flow property is calculated using the backward characteristics issuing from the initial forward characteristics and intersecting the boundary at $t = t_o$:

$$V(x = 0, t_o) = V_o + 2(C(x = 0, t_o) - C_o) \qquad \text{Backward characteristics}$$

where $C(x, t_o) = \sqrt{gd(x = 0, t_o)}$.

At $t = 10\,\text{min}$, the flow property between $x = 0$ and $x = 1.87\,\text{km}$ are calculated from:

$$V(x, t = 600) + 2C(x, t = 600) = V(x = 0, t_o) + 2C(x = 0, t_o)$$
$$\text{Forward characteristics}$$

$$V(x, t = 600) - 2C(x, t = 600) = V_o - 2C_o \qquad \text{Backward characteristics}$$

where the equation of the forward characteristics is:

$$t = t_o + \frac{x}{V(x = 0, t_o) + C(x = 0, t_o)} \qquad \text{Forward characteristics}$$

These three equations are three unknowns: $V(x, t = 600)$, $C(x, t = 600)$ and $t_o = t(x = 0)$ for the C1 characteristics. The results of the calculation at $t = 12\,\text{min}$ are presented in Table 12.2 and Fig. 12.15.

Table 12.2 Negative surge calculations at $t = 10\,\text{min}$

$t_o(x = 0)$ C1 (1)	$d(x = 0)$ (2)	$C(x = 0)$ (3)	$V(x = 0)$ C2 (4)	$Fr\,(x = 0)$ (5)	x $t = 10\,\text{min}$ (6)	$V(x)$ $t = 10\,\text{min}$ (7)	$C(x)$ $t = 10\,\text{min}$ (8)	$d(x)$ $t = 10\,\text{min}$ (9)
0	1.65	4.02	−0.90	−0.22	1873	−0.90	4.02	1.65
60	1.6	3.96	−1.02	−0.26	1586	−1.02	3.96	1.60
120	1.55	3.90	−1.15	−0.29	1320	−1.15	3.90	1.55
180	1.5	3.83	−1.27	−0.33	1075	−1.27	3.83	1.50
240	1.45	3.77	−1.40	−0.37	852	−1.40	3.77	1.45
300	1.4	3.70	−1.53	−0.41	651	−1.53	3.70	1.40
360	1.35	3.64	−1.67	−0.46	473	−1.67	3.64	1.35
420	1.3	3.57	−1.80	−0.51	318	−1.80	3.57	1.30
480	1.25	3.50	−1.94	−0.55	187	−1.94	3.50	1.25
540	1.2	3.43	−2.08	−0.61	80.7	−2.08	3.43	1.20
600	1.15	3.36	−2.23	−0.66	0	−2.23	3.36	1.15

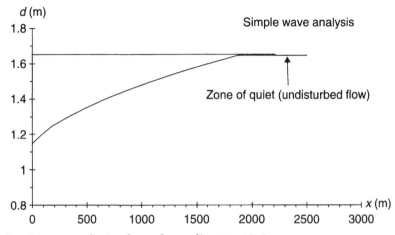

Fig. 12.15 Negative surge application: free-surface profile at $t = 10\,\text{min}$.

The flow rate at the gate is $-2.56\,\text{m}^2/\text{s}$ at $t = 600\,\text{s}$. The negative sign shows that the flow direction is in the negative x-direction.

12. $U = 4.46\,\text{m/s}$, $d_2 = 2.24\,\text{m}$ (gentle undular surge).

13. At $t = 1\,\text{h}$, the positive surge has not yet time to form (Fig. 12.16). That is, the forward characteristics do not intersect yet.

x (m)	V (m/s)	C (m/s)	d (m)
17 000	−0.30	4.54	2.10
15 251.5	−0.30	4.54	2.10
15 038.2	−0.27	4.55	2.11
14 941.1	−0.21	4.58	2.14
14 875.8	−0.15	4.61	2.17
14 763.4	−0.09	4.64	2.20
14 559.9	−0.04	4.67	2.22
14 262.4	−0.02	4.68	2.23
13 896.2	−0.01	4.68	2.24
13 492.8	0.00	4.69	2.24
8435.73	0.00	4.69	2.24
1405.96	0.00	4.69	2.24

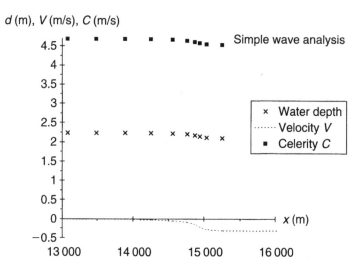

Fig. 12.16 Positive surge application: free-surface profile at $t = 1$ h.

15. (1) $d(x = 0) = 2.24$ m. (2) $Q = -18.3$ m³/s.
17. Initial uniform equilibrium flow conditions are: $d = 2.3$ m, $V = +1.42$ m/s, $f = 0.026$.
    Diffusion wave: $U = 2.1$ m/s, $D_t = 5.4 \times 10^3$ m²/s (for a wide rectangular channel).
18. Initially, uniform equilibrium flow conditions are: $V = +1.25$ m/s, $S_o = S_f = 0.00028$,
    $f = 0.017$.
    The celerity of the diffusion wave is:

$$U = \frac{Q}{B\left(\frac{8g}{f} A^2 \frac{D_H}{4}\right)^{1/2}} \frac{\partial}{\partial d}\left(A\sqrt{\frac{8g}{f}\frac{D_H}{4}}\right) = \frac{\sqrt{S_f}}{B}\frac{\partial}{\partial d}\left(A\sqrt{\frac{8g}{f}\frac{D_H}{4}}\right)$$

$$= \frac{\sqrt{S_f}}{B}\frac{3}{2}B\sqrt{\frac{8g}{f}\frac{A}{P_w}}\left(1 - \frac{2}{3}\frac{A}{BP_w}\right) = \frac{3}{2}V\left(1 - \frac{2}{3}\frac{A}{BP_w}\right) = 1.68 \text{ m/s}$$

The diffusion coefficient is:

$$D_t = \frac{\dfrac{8g}{f} A^2 \dfrac{D_H}{4}}{2B|Q|} = 4.02 \times 10^3 \text{ m}^2/\text{s}$$

The analytical solution of the diffusion wave equation yields:

$$Q(x, t) = Q_o + \frac{\delta Q}{2}\left(1 - \text{erf}\left(\frac{x - Ut}{\sqrt{4D_t t}}\right) + \exp\left(\frac{Ux}{D_t}\right)\left(1 - \text{erf}\left(\frac{x + Ut}{\sqrt{4D_t t}}\right)\right)\right)$$

assuming constant diffusion wave celerity and diffusion coefficient, where $Q_o = 18$ m³/s
and $\delta Q = 0.8$ m³/s. For $t = 1$ h and $x = 15$ km, $Q = 18.04$ m³/s.

13

Unsteady open channel flows: 3. Application to dam break wave

Summary

In this chapter the Saint-Venant equations for unsteady open channel flows are applied to the dam break wave problem. Basic applications include dam break wave in horizontal and sloping channels, flash floods and tsunamis.

13.1 Introduction

During the 19th Century, major dam break catastrophes included the failures of the Puentes Dam in 1802, Dale Dyke Dam in 1864, Habra Dam in 1881 and South Fork (Johnstown) Dam in 1889. In the 20th Century, three major accidents were the St Francis and Malpasset Dam failures in 1928 and 1959 respectively, and the overtopping of the Vajont Dam in 1963 (Fig. 13.1). Table 13.1 documents further accidents. These accidents yielded strong interest on dam break wave flows. A dam break is not only an engineering failure but, more importantly, a human tragedy often with some political implications.

Completed in 1926, the St Francis Dam was a 62.5 m high curved concrete gravity dam located about 72 km North of Los Angeles. The dam thickness was 4.9 m at the crest and 53.4 m at the lowest base elevation. The dam was equipped with 11 spillway openings (2.8 m^2 each) and five outlet pipes (diameter 0.7 m). The spillways discharged onto the downstream stepped face of the dam (0.4 m step height). The reservoir began filling in 1926 and was nearly full on the 5 March 1928. The volume of the reservoir was 46.9 Mm^3 at that date. Although the dam site was investigated on the morning of the 12 March 1928 and nothing was judged hazardous, the dam collapsed suddenly on the evening of the 12 March 1928 between 11:57 and 11:58 p.m. The peak discharge just below the dam reached 14 200 m^3/s. Four hundred and fifty people were killed by the flood wave. The dam failure was caused by a combination of a massive landslide on the left abutment and uplift pressure effects. Afterwards geological and geotechnical investigations revealed the very poor quality of the geological setting. During the collapse, a single erect monolith survived essentially unmoved from its original position (Fig. 13.1(a) and (b)). The remnant part was blasted on 23 May 1929.

(a)

(b)

Fig. 13.1 Photographs of dam break accidents. (a) Old photograph of the remnant part of the St Francis Dam after failure (courtesy of the Santa Clarita Valley Historical Society). Dam height: 62.5 m, failure on 12 March 1928. Eastern abutment showing the landslide area (viewed from the right abutment). (b) Old photograph of the St Francis Dam after failure, looking from inside the reservoir towards downstream (courtesy of the Santa Clarita Valley Historical Society).

Fig. 13.1 (c) Remains of the Malpasset Dam in 1981. Dam height: 60 m, failure on 2 December 1959. View from upstream looking downstream, with the right abutment on the top right of the photograph. (d) Left abutment of the Malpasset Dam in the late 1990s (courtesy of Didier Toulouze). (e) Möhne Dam shortly after the RAF raid on 16–17 May 1943. Almost 1300 people died in the floods following the dam buster campaign, mostly inmates of a Prisoner of War (POW) camp just below the dam.

Table 13.1 Examples of dam break failures and dam overtopping

Dam/reservoir (1)	Years of construction (2)	Date of accident (3)	Accident/failure (4)	Lives lost (5)
Dam break				
Puentes Dam, Spain	1785–1791	30 April 1802	Masonry dam break caused by a failure of wooden pile foundation.	608
Blackbrook Dam, UK	1795–1797	1799	Collapse caused by dam settlement and spillway inadequacy.	None
South Fork (Johnstown) Dam, USA	1839	May 1889	Overtopping and break of earth dam caused by spillway inadequacy.	Over 2000
Bilberry Dam, UK	1843	5 February 1852	Failure of earth dam caused by poor construction quality.	81
Dale Dyke Dam, UK	1863	11 March 1864	Earth embankment failure attributed to poor construction work. Surge wave volume $\sim 0.9\,\text{Mm}^3$.	150
Habra Dam, Algeria	1873	December 1881	Break of masonry gravity dam caused by inadequate spillway capacity leading to overturning. Note that the storm rainfall of 165 mm in one night lead to an estimated runoff of about three times the reservoir capacity.	209
Bouzey Dam, France	1878–1880	27 April 1895	Dam accident in 1884. Dam break in 1895.	85
Austin Dam, USA	1892	7 April 1900	Dam break during overflow caused by foundation failure.	–
Minneapolis Mill Dam, USA	1893–1894	1899	Dam break during a small spill (caused by cracks resulting from ice pressure on the dam).	–
Dolgarrog Dams, UK	1911/1910s	1925	Sequential failure of two earth dams following undermining of the upper structure.	25
Möhne Dam, Germany	1913	16–17 May 1943	RAF bombing of the dam to stop hydro-electricity production.	1300
Moyie River Dam, USA	1920s	1925	Failure of the left abutment and spillway (caused by undermining) during large flood. Arch wall still standing.	None
Lake Lanier Dam, USA	1925	21 January 1926	Failure by undermining and overturning of the left abutment. Arch wall still standing.	None
St Francis Dam, USA	1926	1928	Dam break (caused by foundation failures).	450
Malpasset Dam, France	1957–1958	2 December 1959	Arch dam break caused by uplift pressures.	Over 300
Belci Dam, Romania	1958–1962	1991	Dam overtopping and breach (caused by a failure of gate mechanism).	97
Teton Dam, USA	1976	5 June 1976	Dam failure caused by cracks and piping in the embankment near completion.	11
Tous Dam, Spain	1977	1982	Dam break (following an overtopping; collapse caused by an electrical failure).	None
Cité de la Jonquière Dam, Canada		July 1996	Right abutment overtopping during floods of the Rivière aux Sables, Saguenay region.	None
Lake Ha! Ha! Dam, Canada	–	July 1996	Dam overtopping caused by extreme rainfalls (18–22 July) in the Saguenay region.	None
Zeyzoun (or Zayaoun) Dam, Syria	1996	4 June 2002	Embankment dam cracks, releasing about $71\,\text{Mm}^3$ of water. A 3.3 m high wall of water rushed through the villages submerging over $80\,\text{km}^2$. The final breach was 80 m wide.	22

(Contd)

Table 13.1 (*Contd*)

Dam/reservoir (1)	Years of construction (2)	Date of accident (3)	Accident/failure (4)	Lives lost (5)
Glashütte Dam, Germany 1953		12 August 2002	Embankment dam overtopping during very large flood because of inadequate spillway capacity.	None
Dam overtopping				
Warren Dam, Australia	1916	1917	Dam overtopped (no damage).	None
Palagnedra Dam, Switzerland	1952	1978	Dam overtopping (caused by a combination of large flood and large volume of debris).	24
Vajont Dam, Italy	1956–1960	9 October 1963	Dam overtopping following a massive landslide into the $169 \times 10^6 \, \text{m}^3$ reservoir.	Over 2000

References: Smith (1971), Schnitter (1994), Chanson *et al.* (2000).

The Malpasset Dam was a double-curvature arch dam located on the Reyran River upstream of Fréjus, southern France. The town of Fréjus is on the site of an ancient naval base founded by Julius Caesar about BC 50, known originally as Forum Julii. Designed by André Coyne (1891–1960), the Malpasset Dam was 60 m high, the wall thickness was 6.0 and 1.5 m at base and at crest respectively, and the opening angle of the arch was 135°. The reservoir was an irrigation water supply. Completed in 1954, the reservoir was not full until late November 1959. On 2 December 1959 around 9:10 p.m., the dam wall collapsed completely. More than 300 people died in the catastrophe. Field observations showed that the surging waters formed a 40 m high wave at 340 m downstream of the dam site and the wave height was still about 7 m about 9 km downstream (Faure and Nahas 1965). The dam break wave took about 19 min to cover the first 9 km downstream of the dam site. (Relatively accurate time records were obtained from the destruction of a series of electrical stations located in the downstream valley.) The dam collapse was caused by uplift pressures in faults of the gneiss rock foundation which lead to the complete collapse of the left abutment (Fig. 13.1(d)).

Designed by Carlo Semenza (1893–1961), the Vajont Dam (Italy) is a double-curvature arch structure, built between 1956 and 1960. The 262 m high reinforced-concrete dam created a $169 \times 10^6 \, \text{m}^3$ reservoir which was filled up by a major landslide on 9 October 1963. The reservoir waters overtopped the dam and the flood wave devastated the downstream valley. More than 2000 people died in the catastrophe. Little damage was done however to the dam itself which is still standing and used despite a drastically reduced storage capacity. At the dam toe, the overtopping nappe reached velocities in excess of 70 m/s before impact. Although the bed friction slowed down the resulting wave, the warning time was too short.

Discussion: man-made dam failures

During armed conflicts, man-made flooding of an army or a city was carried out by building an upstream dam and destroying it. Historical examples include the Assyrians (Babylon, Iraq BC 689), the Spartans (Mantinea, Greece BC 385–384), the Chinese (Huai River, AD 514–515), the Russian army (Dnieprostroy Dam, 1941) (Dressler 1952, Smith 1971, Schnitter 1994). It may be added the aborted attempt to blow up Ordunte Dam, during the Spanish civil war, by the troops of General Franco, and the anticipation of German Dam destruction at the German–Swiss border to stop the crossing of the Rhine River by the Allied Forces in 1945 (Ré 1946). A related case is the air raid on the Möhne Dam in the Ruhr Basin conducted by the British, in 1943, during the dam buster campaign (Fig. 13.1(e)). In 1961, the Swiss army blew up with explosives a concrete arch dam to document the flooding of the downstream valley (Lauber 1997).

Dyke destruction and associated flooding played also a role in several wars. For example, the war between the cities of Lagash and Umma (Assyria) around BC 2500 was fought for the control of irrigation systems and dykes; in 1938, the Chinese army destroyed dykes along the Huang Ho River (Yellow River) to slow down the Japanese army.

13.2 Dam break wave in a horizontal channel

13.2.1 Dam break in a dry channel

Considering an ideal dam break surging over a dry river bed, the method of characteristics may be applied to solve completely the wave profile as first proposed by Ritter in 1892 (e.g. Henderson 1966, Montes 1998). Interestingly, Ritter's work was initiated by the South Fork Dam's (Johnstown) catastrophe (Table 13.1).

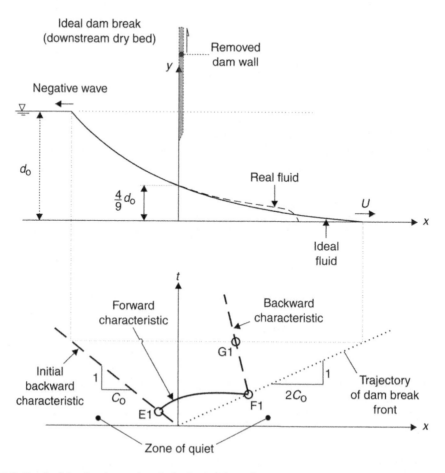

Fig. 13.2 Sketch of dam break wave in a dry horizontal channel.

The dam break may be idealized by a vertical wall that is suddenly removed (Fig. 13.2). After the removal of the wall, a negative wave propagates upstream and a dam break wave moves downstream. Although there is considerable vertical acceleration during the initial instants of fluid motion, such acceleration is not taken into account by the method of characteristics and the pressure distributions are assumed hydrostatic. For an ideal dam break over a dry horizontal channel, the basic equations are those of the simple wave (Chapter 12, Section 12.3.1):

$$\frac{D}{Dt}(V + 2C) = 0 \qquad \text{Forward characteristics} \qquad (13.1a)$$

$$\frac{D}{Dt}(V - 2C) = 0 \qquad \text{Backward characteristics} \qquad (13.1b)$$

along:

$$\frac{dx}{dt} = V + C \qquad \text{Forward characteristics C1} \qquad (13.2a)$$

$$\frac{dx}{dt} = V - C \qquad \text{Backward characteristics C2} \qquad (13.2b)$$

The instantaneous dam break creates a negative wave propagating upstream into a fluid at rest with known water depth d_o. In the (x, t) plane, the initial negative wave characteristic has a slope $dt/dx = -1/C_o$ where $C_o = \sqrt{gd_o}$ assuming a rectangular channel (Fig. 13.2).

The forward characteristics can be drawn issuing from the initial backward characteristic for $t > 0$ and intersecting the trajectory of the leading edge of the dam break wave front (Fig. 13.2, trajectory E1–F1). The forward characteristic satisfies:

$$V + 2C = V_o + 2C_o = 2C_o \qquad (13.3)$$

since $V = V_o = 0$ and $C = C_o = \sqrt{gd_o}$ at the point E1 (Fig. 13.2).

At the leading edge of the dam break wave front, the water depth is zero, hence $C = 0$, and the propagation speed of the dam break wave front equals:

$$U = 2C_o = 2\sqrt{gd_o} \qquad \text{Ideal dam break} \qquad (13.4)$$

Considering any backward characteristics issuing from the dam break wave front (Fig. 13.2, trajectory F1–G1), the C2 characteristics is a straight line because the initial backward characteristics is a straight line (Chapter 12, Section 12.3.1). The inverse slope of the backward characteristics is a constant:

$$\frac{dx}{dt} = V - C = 2C_o - 3C$$

using equation (13.3). The integration of the inverse slope gives the water surface profile at the intersection of the C2 characteristics with a horizontal line $t = $ constant (Fig. 13.2, point G1). That is, at a given time, the free-surface profile between the leading edge of the negative wave and the wave front is a parabola:

$$\frac{x}{t} = 2\sqrt{gd_o} - 3\sqrt{gd} \qquad \text{for } -\sqrt{gd_o} \leqslant \frac{x}{t} \leqslant +2\sqrt{gd_o} \qquad (13.5a)$$

At the origin ($x = 0$), equation (13.5a) predicts a constant water depth:

$$d(x = 0) = \frac{4}{9} d_o \qquad (13.6)$$

Similarly the velocity at the origin is deduced from equation (13.3):

$$V(x = 0) = \frac{2}{3} \sqrt{g d_o} \qquad (13.7)$$

After dam break, the flow depth and velocity at the origin are both constants, and the water discharge at $x = 0$ equals:

$$Q(x = 0) = \frac{8}{27} d_o \sqrt{g d_o}\, W \qquad (13.8)$$

where W is the channel width.

Discussion

Important contributions to the dam break wave problem in a dry horizontal channel include Ritter (1892), Schoklitsch (1917), Ré (1946), Dressler (1952, 1954) and Whitham (1955). The above development is sometimes called Ritter's theory. Calculations were performed assuming a smooth rectangular channel, an infinitely long reservoir and for a quasi-horizontal free surface. That is, bottom friction is zero and the pressure distribution is hydrostatic. Experimental results (e.g. Schoklitsch 1917, Faure and Nahas 1961, Lauber 1997) showed that the assumptions of hydrostatic pressure distributions and zero friction are reasonable, but for the initial instants and at the leading tip of the dam break wave.

Bottom friction affects significantly the propagation of the leading tip. Escande et al. (1961) investigated specifically the effects of bottom roughness on dam break wave in a natural valley. They showed that, with a very rough bottom, the wave celerity could be about 20–30% lower than for a smooth bed. The shape of a real fluid dam break wave front is sketched in Fig. 13.2 and experimental results are presented in Fig. 13.6 (see Section 13.3.1).

Notes

1. The above development, called Ritter's theory, was developed for a frictionless horizontal rectangular channel initially dry and an infinitely long reservoir. The free surface is also assumed to be quasi-horizontal.
2. The assumption of hydrostatic pressure distributions has been found to be reasonable, but for the initial instants: i.e. $t < 3\sqrt{d_o/g}$ (Lauber 1997).
3. Ritter's theory implies that the celerity of the initial negative wave is $U = -\sqrt{g d_o}$. Experimental observations suggested however that the real celerity is about $U = -\sqrt{2}\sqrt{g d_o}$ (Lauber 1997, Leal et al. 2001). It was suggested that the difference was caused by streamline curvature effects at the leading edge.

4. At the origin ($x = 0$), the flow conditions satisfy:

$$\frac{V(x=0)}{\sqrt{g\,d(x=0)}} = \frac{\frac{2}{3}\sqrt{gd_o}}{\sqrt{g\frac{4}{9}d_o}} = 1$$

That is, critical flow conditions take place at the origin (i.e. initial dam site) and the flow rate is a constant:

$$Q(x=0) = \sqrt{gd(x=0)^3} = \frac{8}{27}d_o\sqrt{gd_o}\ W$$

where W is the channel width. Importantly, the result is valid only within the assumptions of the Saint-Venant equations. The water free surface is quasi-horizontal and the pressure distribution is hydrostatic at the origin.

5. The above development was conducted for a semi-infinite reservoir. At a given location $x > 0$, equation (13.5a) predicts an increasing water depth with increasing time:

$$d = \frac{4}{9}d_o\left(1 - \frac{3}{2}\frac{x}{\sqrt{gd_o}\ t}\right)^2 \qquad (13.5b)$$

At any distance x from the dam site, the water depth d tends to $d = 4/9 \times d_o$ for $t = +\infty$ (for a semi-infinite reservoir).

6. Henderson (1966) analysed the dam break wave problem by considering the sudden horizontal displacement of a vertical plate behind which a known water depth is initially at rest. His challenging approach yields identical results (also Liggett 1994).

Remarks

For a dam break wave down a dry channel, the boundary conditions are *not* the vertical axis in the (x, t) plane. The two boundary conditions are: (1) the upstream edge of the initial negative wave where $d = d_o$ and (2) the dam break wave front where $d = 0$. At the upstream end of the negative surge, the boundary condition is $d = d_o$: i.e. this is the initial backward characteristic in the (x, t) plane (Fig. 13.2). The downstream boundary condition is at the leading edge of the dam break wave front where the water depth is zero.

Note that, in Chapter 12, paragraphs 12.3 and 12.4, most applications used a boundary condition at the origin: i.e. it was the vertical axis in the (x, t) plane. For a dam break wave on a horizontal smooth bed, the vertical axis is not a boundary.

13.2.2 Dam break in a horizontal channel initially filled with water

Presentation

The propagation of a dam break wave over still water with an initial depth $d_1 > 0$ is possibly a more practical application: e.g. sudden flood release downstream of a dam in a river. It is a different situation from an initially dry channel bed because the dam break wave is lead by a positive surge (Fig. 13.3).

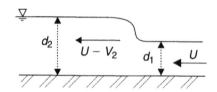

Quasi-steady flow analogy at the surge front
(as seen by an observer travelling at the surge velocity)

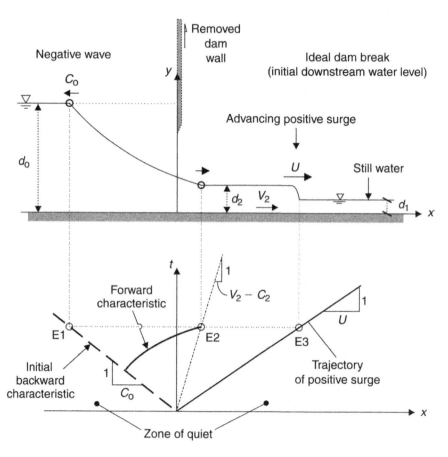

Fig. 13.3 Sketch of dam break wave in a horizontal channel initially filled with water for $d_1 < 0.1383 \times d_o$.

The basic flow equations are the characteristic system of equations (paragraph 13.2.1), and the continuity and momentum equations across the positive surge front. That is:

$$\frac{D}{Dt}(V + 2C) = 0 \quad \text{along} \quad \frac{dx}{dt} = V + C \qquad \text{Forward characteristics} \qquad (13.1a)$$

$$\frac{D}{Dt}(V - 2C) = 0 \quad \text{along} \quad \frac{dx}{dt} = V - C \qquad \text{Backward characteristics} \qquad (13.1b)$$

$$d_1 U = d_2(U - V_2) \qquad \text{Continuity equation} \qquad (13.9a)$$

$$d_2(U - V_2)^2 - d_1 U^2 = \frac{1}{2} g d_1^2 - \frac{1}{2} g d_2^2 \qquad \text{Momentum equation} \qquad (13.10a)$$

where U is the positive surge celerity for an observer fixed on the channel bank, and the subscripts 1 and 2 refer respectively to the flow conditions upstream and downstream of the positive surge front: i.e. initial flow conditions and new flow conditions behind the surge front (Fig. 13.3).

Immediately after the dam break, a negative surge propagates upstream into a fluid at rest with known water depth d_o. In the (x, t) plane, the initial negative wave characteristics has a slope $dt/dx = -1/C_o$ where $C_o = \sqrt{g d_o}$ assuming a rectangular channel (Fig. 13.3). The initial backward characteristics is a straight line, hence all the C2 characteristics are straight lines (Chapter 12, Section 3.3.1).

For $t > 0$, the forward characteristics issuing from the initial backward characteristics cannot intersect the downstream water level ($d = d_1$) because it would involve a discontinuity in velocity. The velocity is zero in still water ($V_1 = 0$) but, on the forward characteristics, it satisfies:

$$V + 2C = V_o + 2C_o = 2C_o \qquad (13.3)$$

Such a discontinuity can only take place as a positive surge which is sketched in Fig. 13.3.

Detailed solution

Considering a horizontal, rectangular channel, the water surface is horizontal upstream of the leading edge of the negative wave (Fig. 13.3, point E1). Between the leading edge of the negative wave (point E1) and the point E2, the free-surface profile is a parabola:

$$\frac{x}{t} = 2 \sqrt{g d_o} - 3 \sqrt{g d} \qquad x_{E1} \leq x \leq x_{E2} \qquad (13.5a)$$

Between the point E2 and the leading edge of the positive surge (point E3), the free-surface profile is horizontal. The flow depth d_2 and velocity V_2 satisfy equations (13.9a) and (13.10a), as well as the condition along the C_1 forward characteristics issuing from the initial negative characteristics and reaching point E2:

$$V_2 + 2 \sqrt{g d_2} = 2 \sqrt{g d_o} \qquad \text{Forward characteristics} \qquad (13.3)$$

Equations (13.3), (13.9a) and (13.10a) form a system of three equations with three unknowns V_2, d_2 and U. The system of equations may be solved graphically (Fig. 13.4). Figure 13.4 shows $U/\sqrt{g d_1}$, d_2/d_1 and $V_2/\sqrt{g d_2}$, (right axis), and $(V_2 - C_2)/\sqrt{g d_1}$ (left axis) as functions of the ratio d_o/d_1.

Once the positive surge forms (Fig. 13.3), the locations of the points E2 and E3 satisfy respectively:

$$x_{E2} = (V_2 - C_2)t$$

$$x_{E3} = U t$$

where $C_2 = \sqrt{g d_1}$ for a rectangular channel. Figure 13.4 shows that $U > (V_2 - C_2)$ for $t > 0$ and $0 < (d_1/d_o) < 1$. That is, the surge front (point E3) advances faster than the point E2.

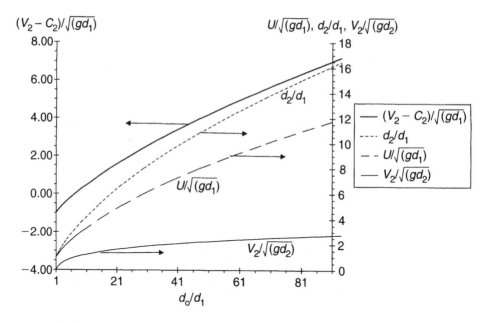

Fig. 13.4 Graphical solution of the flow conditions for a dam break wave in a horizontal channel initially filled with water.

Montes (1998) showed that the surge celerity satisfies:

$$\sqrt{\frac{d_o}{d_1}} = \frac{1}{2}\frac{U}{\sqrt{gd_1}}\left(1 - \frac{1}{X}\right) + \sqrt{X} \tag{13.11}$$

where

$$X = \frac{1}{2}\left(\sqrt{1 + 8\frac{U_2}{gd_1}} - 1\right)$$

The flow conditions behind the surge front are then deduced from the quasi-steady flow analogy (Chapter 12, Section 12.4.2).

Notes
1. The above equations were developed for initially still water ($V_1 = 0$). Experimental observations compared well with the above development and they showed that bottom friction is negligible.
2. Figure 13.4 is valid for $0 < (d_1/d_o) < 1$.
3. The flow depth downstream of (behind) the positive surge is deduced from the continuity and momentum equations. d_2 is independent of the time t and distance x. Practically, the flow depth d_2 is a function of the depths d_1 and d_o. It may be correlated by:

$$\frac{d_2}{d_o} = 0.931\,967\,1\left(\frac{d_1}{d_o}\right)^{0.371396}$$

Good agreement was observed between large-size experimental data and the above correlation (e.g. Chanson *et al.* 2000).

4. Equation (13.11) may be empirically correlated by:

$$\frac{U}{\sqrt{gd_1}} = \frac{0.635\,45 + 0.328\,6 \left(\dfrac{d_1}{d_o}\right)^{0.65167}}{0.002\,51 + \left(\dfrac{d_1}{d_o}\right)^{0.65167}}$$

5. Figure 13.3 illustrates the solution of the Saint-Venant equations. It shows some discontinuities in terms of free-surface slope and curvatures at the points E1 and E2. Such discontinuities are not observed experimentally and this highlights some limitation of the Saint-Venant equations.

Application

A 35 m high dam fails suddenly. The initial reservoir height was 31 m above the downstream channel invert and the downstream channel was filled with 1.8 m of water initially at rest. (1) Calculate the wave front celerity, and the surge front height. (2) Calculate the wave front location and free-surface profile 2 min after failure. (3) Predict the water depth 10 min after gate opening at two locations: $x = 2$ km and $x = 4$ km. *Assume an infinitely long reservoir and a horizontal, smooth channel.*

Solution

The downstream channel was initially filled with water at rest. The flow situation is sketched in Fig. 13.3. The x coordinate is zero ($x = 0$) at the dam site and positive in the direction of the downstream channel. The time origin is taken at the dam collapse.

First the characteristic system of equation, and the continuity and momentum principles at the wave front must be solved graphically using Fig. 13.4 or theoretically.

Using Fig. 13.4, the surge front celerity is: $U/\sqrt{gd_1} = 4.3$ and $U = 18.1$ m/s.

At the wave front, the continuity and momentum equations yield:

$$\frac{d_2}{d_1} = \frac{1}{2}\left(\sqrt{1 + 8\frac{U^2}{gd_1}} - 1\right) = 5.63 \qquad \text{Hence } d_2 = 10.14\,\text{m}$$

Equation (12.33) may be rewritten:

$$\frac{V_2}{\sqrt{gd_2}} = 2\left(\sqrt{\frac{d_o}{d_2}} - 1\right) = 1.5 \qquad \text{Hence } V_2 = 14.9\,\text{m/s}$$

Note that these results are independent of time.

At $t = 2$ min, the location of the points E1, E2 and E3 sketched in Fig. 13.3 is:

$$x_{E1} = -\sqrt{gd_o}\,t = -2.1\,\text{km}$$

$$x_{E2} = \left(V_2 - \sqrt{gd_2}\right)t = +0.59\,\text{km}$$

$$x_{E3} = Ut = +2.2\,\text{km}$$

Between the points E1 and E2, the free-surface profile is a parabola:

$$\frac{x}{t} = 2\sqrt{gd_o} - 3\sqrt{gd} \qquad x_{E1} \leqslant x \leqslant x_{E2} \qquad (13.5)$$

The free-surface profile at $t = 2\,\mathrm{min}$ is:

x (m)	−5000	−2100	−1000	0	595	2178	2178	5000
d (m)	31	31	21.1	13.7	10.14	10.14	1.8	1.8
Remark		Point E1			Point E2	Point E3		

Lastly $d(x = 2\,\mathrm{km}, t = 10\,\mathrm{min}) = 11.3\,\mathrm{m}$ and $d(x = 4\,\mathrm{km}, t = 10\,\mathrm{min}) = 10.1\,\mathrm{m}$.

Remarks
1. As $d_1/d_o < 0.138$, the water depth and flow rate at the dam site equal:

$$d(x = 0) = \frac{4}{9}d_o = 13.7\,\mathrm{m}$$

$$q(x = 0) = \frac{8}{27}\sqrt{gd_o^3} = 160\,\mathrm{m^2/s}$$

2. For a sudden dam break wave over an initially dry channel, the wave front celerity would be 35 m/s (equation (13.4), for an ideal dam break on a smooth frictionless channel).

Extension to non-zero initial flow velocity
This development may be easily extended for initial flow conditions with a non-zero, constant initial velocity V_1. Equations (13.3), (13.9a) and (13.10a) become:

$$V_2 + 2\sqrt{gd_2} = 2\sqrt{gd_o} \qquad \text{Forward characteristics} \qquad (13.3)$$

$$d_1(U - V_1) = d_2(U - V_2) \qquad \text{Continuity equation} \qquad (13.9b)$$

$$d_2(U - V_2)^2 - d_1(U - V_1)^2 = \frac{1}{2}gd_1^2 - \frac{1}{2}gd_2^2 \qquad \text{Momentum equation} \quad (13.10b)$$

Discussion
Figure 13.3 sketches a situation where the point E2 is located downstream of the initial dam wall. At the origin, the water depth and velocity are respectively (paragraph 13.2.1):

$$d(x = 0) = \frac{4}{9}d_o \qquad (13.6)$$

$$V(x = 0) = \frac{2}{3}\sqrt{gd_o} \qquad (13.7)$$

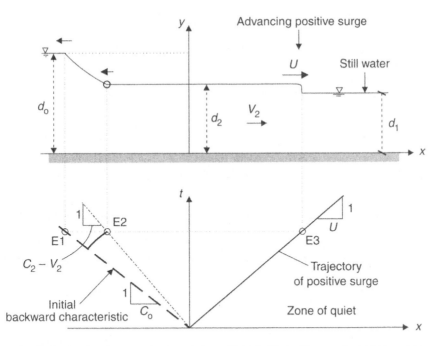

Fig. 13.5 Sketch of a dam break wave in horizontal channel initially filled with water for $0.1383 < d_1/d_o < 1$.

Another situation is sketched in Fig. 13.5 where the point E2 is located upstream of the dam wall. Between these two situations, the limiting case is a fixed point E2 at the origin. This occurs for

$$\frac{d_1}{d_o} = 0.1383$$

For $d_1/d_o < 0.1383$, the flow situation is sketched in Fig. 13.3. For $d_1/d_o > 0.1383$, the point E2 is always located upstream of the origin and the flow pattern is illustrated in Fig. 13.5.

Application

Demonstrates that the intersection of the forward characteristics issuing from the initial characteristics with the horizontal free-surface behind the positive surge (i.e. point E2, Fig. 13.3) is at the origin ($x = 0$) if and only if:

$$\frac{d_1}{d_o} = 0.1383$$

Solution

In the limiting case for which the point E2 is at the origin, the water depth and velocity at the origin are respectively:

$$d(x = 0) = \frac{4}{9} d_o = d_{E2} = d_2$$

$$V(x = 0) = \frac{2}{3} \sqrt{gd_o} = V_{E2} = V_2$$

where d_2 and the V_2 are the flow velocity behind (downstream of) the positive surge.
First the flow conditions at the origin satisfies $V_2/\sqrt{gd_2} = 1$
Second the continuity and momentum equations across the positive surge give:

$$\frac{d_2}{d_1} = \frac{1}{2}\left(\sqrt{1 + 8\frac{U^2}{gd_1}} - 1 \right) = \left(\frac{4}{9}\right)\left(\frac{d_o}{d_1}\right)$$

$$\frac{U - V_2}{\sqrt{gd_2}} = \frac{2^{3/2}\dfrac{U}{\sqrt{gd_1}}}{\left(\sqrt{1 + 8\dfrac{U^2}{gd_1}} - 1 \right)^{3/2}} = \frac{27}{8}\left(\frac{d_1}{d_o}\right)^{3/2} = \frac{U}{\sqrt{gd_1}}\sqrt{\frac{d_1}{d_2}} - 1$$

where U is the celerity of the positive surge (Chapter 12). One of the equations may be simplified into:

$$\frac{U}{\sqrt{gd_1}} = \frac{1}{\dfrac{3}{2}\left(\dfrac{d_1}{d_o}\right)^{1/2} - \dfrac{27}{8}\left(\dfrac{d_1}{d_o}\right)^{3/2}}$$

This system of non-linear equations may be solved by a graphical method and by test-and-trial. The point E2 is at the origin for:

$$\frac{d_1}{d_o} = 0.1382701411$$

Notes

1. For $d_1/d_o < 0.1383$, the discharge at the origin is a constant independent of time:

$$Q(x = 0) = \frac{8}{27} d_o \sqrt{gd_o}\, W \tag{13.8}$$

2. For $d_1/d_o > 0.1383$, the discharge at the origin is also a constant independent of the time. But the flow rate at $x = 0$ becomes a function of the flow depth d_2 and it is less than $8/27\sqrt{g}\ Wd_o^{3/2}$. At the limit $Q(x = 0) = 0$ for $d_1 = d_o$.

13.3 Effects of flow resistance

13.3.1 Flow resistance effect on dam break wave on horizontal channel

For a dam break wave in a dry horizontal channel, observations showed that Ritter's theory (paragraph 13.2.1) is valid but for the leading tip of the wave front. Experimental data indicated that the wave front has a rounded shape and its celerity U is less than $2C_o$ (e.g. Schoklitsch 1917, Dressler 1954, Faure and Nahas 1961). Escande *et al.* (1961) investigated specifically the effects of bottom roughness on dam break wave in a natural valley. They showed that, with a very rough bottom, the wave celerity could be about 20–30% lower than for a smooth bed. Faure and Nahas (1965) conducted both physical and numerical modelling of the catastrophe of the Malpasset Dam (Fig. 13.1(c) and (d)). In their study, field observations were best reproduced with a Gauckler–Manning coefficient $n_{Manning} = 0.025$–0.033 s/m$^{1/3}$.

Figure 13.6 compares the simple-wave theory (zero friction) with measured dam break wave profiles, and it illustrates the round shape of the leading edge. Dressler (1952) and Whitham (1955) proposed analytical solutions of the dam break wave that include the effects of bottom friction, assuming constant friction coefficient. Some results are summarized in Table 13.2 while their respective theories are discussed below.

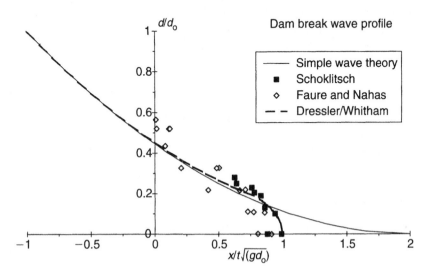

Fig. 13.6 Comparison of the dam break wave profile in an initially dry horizontal channel with and without bottom friction.

Table 13.2 Analytical solutions of dam break wave with and without bottom friction

Parameter	Simple wave Ritter's solution	Dressler (1952)	Whitham (1955)	Remarks
(1)	(2)	(3)	(4)	(5)
Wave front celerity U	$U = 2\sqrt{gd_o}$	–	$\dfrac{U}{\sqrt{gd_o}} = 2 - 3.452\sqrt{\dfrac{f}{8}} t_s \sqrt{\dfrac{g}{d_o}}$	For $t = t_s$ small
$\dfrac{Q(x=0)}{W} =$	$\dfrac{8}{27}\sqrt{gd_o^3}$	$\dfrac{8}{27}\sqrt{gd_o^3} \times \left(1 - 0.239\dfrac{f}{8}\sqrt{\dfrac{g}{d_o}} t\right)$	–	
Location of critical flow conditions	$x = 0$	$\dfrac{x}{d_o} = 0.395\dfrac{f}{8}\sqrt{\dfrac{g}{d_o}} t$	–	

Note: Theoretical results obtained for an initially dry horizontal channel.

> **Note**
> Faure and Nahas (1965) conducted detailed physical modelling of the Malpasset Dam failure using a Froude similitude. They used both undistorted and distorted scale models, as well as a numerical model. For the distorted model, the geometric scaling ratios were: $X_r = 1/1600$ and $Z_r = 1/400$. Field data were best reproduced in the undistorted physical model ($L_r = 1/400$). Both numerical results and distorted model data were less accurate.

Dam break wave calculations with flow resistance

The theoretical solutions of Dressler and Whitham give very close results and they are in reasonable agreement with experimental data (Fig. 13.6). Both methods yield results close to the simple-wave solution but next to the leading tip.

Dressler (1952) used a perturbation method. His first order correction for the flow resistance gives the velocity and celerity at any position:

$$\frac{V}{\sqrt{gd_o}} = \frac{2}{3}\left(1 + \frac{x}{t\sqrt{gd_o}}\right) + F_1 \frac{f}{8}\sqrt{\frac{g}{d_o}}\, t$$

$$\frac{C}{\sqrt{gd_o}} = \frac{1}{3}\left(2 - \frac{x}{t\sqrt{gd_o}}\right) + F_2 \frac{f}{8}\sqrt{\frac{g}{d_o}}\, t$$

where f is the Darcy friction factor. The functions F_1 and F_2 are respectively:

$$F_1 = -\frac{108}{7\left(2 - \dfrac{x}{t\sqrt{gd_o}}\right)^2} + \frac{12}{2 - \dfrac{x}{t\sqrt{gd_o}}} - \frac{8}{3} + \frac{8\sqrt{3}}{189}\left(2 - \frac{x}{t\sqrt{gd_o}}\right)^{3/2}$$

$$F_2 = \frac{6}{5\left(2 - \dfrac{x}{t\sqrt{gd_o}}\right)} - \frac{2}{3} + \frac{4\sqrt{3}}{135}\left(2 - \frac{x}{t\sqrt{gd_o}}\right)^{3/2}$$

Dressler compared successfully his results with the data of Schoklitch (1917) and his own data (Dressler 1954).

At $t = t_s$, the location x_s of the dam break wave front satisfies:

$$\frac{f}{8}\sqrt{\frac{g}{d_o}}\, t_s = \frac{1/6}{\left(\dfrac{54}{7\left(2 - \dfrac{x_s}{t_s\sqrt{gd_o}}\right)^3} - \dfrac{3}{\left(2 - \dfrac{x_s}{t_s\sqrt{gd_o}}\right)^2} + \dfrac{\sqrt{3}}{63}\left(2 - \dfrac{x_s}{t_s\sqrt{gd_o}}\right)^{1/2}\right)}$$

Further results are summarized in Table 13.2.

Whitham (1955) analysed the wave front as a form of boundary layer using an adaptation of the Pohlhausen method. For a horizontal dry channel, his estimate of the wave front celerity is best correlated by:

$$\frac{U}{\sqrt{gd_o}} = \frac{2}{\left(1 + 2.90724 \left(\frac{f}{8} \sqrt{\frac{gt^2}{d_o}}\right)\right)^{0.4255}}$$

where f is the Darcy friction factor. Whitham commented that his work was applicable only for $U/\sqrt{gd_o} > 2/3$. He further showed that the wave front shape would follow:

$$\frac{d}{d_o} = \sqrt{\frac{f}{4} \frac{x_s - x}{d_o}} \left(2 - 3.452 \left(\frac{f}{8} t \sqrt{\frac{g}{d_o}}\right)^{1/3}\right) \qquad \text{Wave front shape } (x < x_s)$$

for $\sqrt{g/d_o}\ tf/8$ small.

> **Notes**
> 1. For $f = 0$, Dressler's solution is Ritter's solution of the Saint-Venant equations.
> 2. In an earlier development, Whitham (1955) assumed the shape of the wave front to follow:
>
> $$d = \sqrt{\frac{f}{4g}}\ U \sqrt{x_s - x}$$
>
> where x_s is the dam break wave front location, but he discarded the result as inaccurate (?).

13.3.2 Dam break wave down a sloping channel

Basic theory
Considering the dam break wave down a sloping channel, the kinematic wave equation[1] may be solved analytically (Hunt 1982). Hunt's analysis gives:

$$\frac{V_H t}{L} = \frac{1 - \left(\dfrac{d_s}{H_{dam}}\right)^2}{\left(\dfrac{d_s}{H_{dam}}\right)^{3/2}} \tag{13.12}$$

$$\frac{x_s}{L} = \frac{\dfrac{3}{2}}{\dfrac{d_s}{H_{dam}}} - \frac{1}{2}\frac{d_s}{H_{dam}} - 1 \tag{13.13a}$$

$$\frac{U}{V_H} = -\frac{3}{4}\frac{V_H t}{L} + \sqrt{\frac{x_s + L}{L} + \left(\frac{3}{4}\frac{V_H t}{L}\right)^2} \tag{13.14}$$

[1] That is, the kinematic wave approximation of the Saint-Venant equations (Chapter 12, paragraph 12.5).

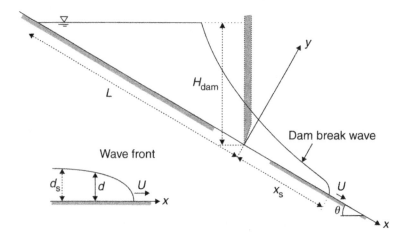

Fig. 13.7 Definition sketch of a dam break wave down a sloping channel.

where t is the time with $t = 0$ at dam break, d_s is the dam break wave front thickness, x_s is the dam break wave front position measured from the dam site, H_{dam} is the reservoir height at dam site, L is the reservoir length, and S_o is the bed slope, $S_o = H_{dam}/L$ (Fig. 13.7). The velocity V_H is the uniform equilibrium flow velocity for a water depth H_{dam}:

$$V_H = \sqrt{\frac{8g}{f} H_{dam} S_o} \tag{13.15}$$

where f is the Darcy friction factor which is assumed constant. Equations (13.12), (13.13a) and (13.14) are valid for $S_o = S_f$ where S_f is the friction slope, and when the free surface is parallel to the bottom of the sloping channel. Equation (13.13a) may be transformed:

$$\frac{d_s}{H_{dam}} = \frac{3}{2} \frac{L}{x_s - L} \qquad \text{For } x_s/L > 4 \tag{13.13b}$$

Hunt (1984, 1988) developed an analytical expression of the shock front shape:

$$\frac{x - x_s}{L} = \frac{d_s}{H_{dam}} \left(\frac{d}{d_s} + \ln\left(1 - \frac{d}{d_s}\right) + \frac{1}{2} \right) \tag{13.16}$$

where d is the depth (or thickness) measured normal to the bottom, and d_s and x_s are functions of time which may be calculated using equations (13.12) and (13.13a) respectively.

Discussion
The dynamic wave equation is simplified by neglecting acceleration and inertial terms, and the free surface is assumed parallel to the channel bottom ($S_o \approx S_f$). The kinematic wave approximation gives the relationship between the velocity and the water depth:

$$V = V_H \sqrt{\frac{d}{H_{dam}}}$$

Once the flood wave has travelled approximately four lengths of the reservoir downstream of the dam site, the free-surface profile of the dam break wave follows:

$$\frac{x+L}{L} = \frac{d}{H_{\text{dam}}} + \frac{3}{2}\frac{V_H\,t}{L}\sqrt{\frac{d}{H_{\text{dam}}}} \quad \text{For } x \leq x_s \text{ and } x_s/L > 4$$

Notes
1. The elegant development of Hunt (1982, 1984) was verified by several series of
   experiments (e.g. Hunt 1984, Nsom *et al.* 2000). It is valid however after the flood
   wave has covered approximately four reservoir lengths downstream of the dam site.
2. Hunt called equations (13.12)–(13.14) the *outer solution* of the dam break wave
   while equation (13.16) was called the *inner solution*. Note that Hunt's analysis
   accounts for bottom friction assuming a constant Darcy friction factor.
3. Bruce Hunt is a reader at the University of Canterbury, New Zealand.

Dam break wave down a sloping stepped chute
A particular type of dam break wave is the sudden release of water down a rough stepped
chute (Fig. 11.1(b)). Applications include the sudden release of water down a stepped spillway and flood runoff in stepped storm waterways during tropical storms. Figure 13.8 illustrates a stepped waterway that may be subjected to such sudden flood waves.

A review of basic experimental work (Table 13.3) demonstrated key features of dam break
wave propagation down a stepped profile. First visual observations showed a wave propagation
as a succession of free-jet, nappe impact on each step and quasi-horizontal runoff until the
downstream step edge. These observations highlighted also the chaotic nature of the flow
with strong aeration of the wave leading edge (Fig. 11.1(b)). For a 3.4° chute, wave front
location data were compared successfully with Hunt's (1982) theory assuming an equivalent

Fig. 13.8 A stepped waterways susceptible to be subjected to rapid flood wave: Robina stepped weir No. 1 along
the storm water diversion system around the Robina shopping town, Gold Coast, Australia on 2 April 1997. Design
flow: 50 m³/s, $\theta = 22°$, design storm concentration time: 30 min.

Table 13.3 Experimental conditions of dam break wave flow down a stepped chute

Experiment (1)	θ (degree) (2)	h (m) (3)	$Q(t = 0+)$ (m³/s) (4)	d_o (m) (5)	Remarks (6)
Chanson (2003, 2004d)	3.4	0.143	0.019–0.075	0.12–0.30	10 horizontal steps ($l = 2.4$ m). $W = 0.5$ m.
	3.4	0.0715	0.03–0.075	0.16–0.30	18 horizontal steps ($l = 1.2$ m). $W = 0.5$ m. Detailed air–water flow measurements.
Brushes Clough Dam (Baker 1994)	18.4	0.19	0.5	0.42	Inclined downward steps, trapezoidal channel (2 m bottom width).

Notes: h: step height; $Q(t = 0+)$: initial flow rate.

Darcy–Weisbach friction factor $f = 0.05$, irrespective of flow rate and chute configuration (Chanson 2003c, 2004d).

Second air–water flow measurements demonstrated quantitatively very strong flow aeration at the wave leading edge (see Discussion). The shape of the leading edge followed closely Hunt's (1984) inner solution. Unsteady air–water velocity profiles showed further the presence of an unsteady turbulent boundary layer next to the invert.

Notes

1. In steady stepped chute flows, three flow regimes may be observed depending upon the flow rate and step geometry: i.e. nappe flow, transition flow and skimming flow (e.g. Chanson 2001a). In a dam break wave flow down a stepped cascade, only one flow regime was observed: i.e. a nappe flow consisting of a succession of free-falling nappe, nappe impact and horizontal runoff (Fig. 11.1(c)).
2. Chanson (2001a, pp. 293–299) discussed other unsteady flow situations down stepped chutes, including roll wave phenomena and shock waves.

Discussion

Detailed unsteady air–water flow measurements were conducted in wave front leading edge (Chanson 2003, 2004d). At the front of the wave, the instantaneous vertical distributions of void fraction in the horizontal runoff had a roughly linear shape:

$$C = 0.9 \frac{y}{Y_{90}} \qquad 0.1 < t\sqrt{\frac{g}{d_o}} < 1.3$$

where Y_{90} is the height where $C = 0.90$ and t is the time measured from the first water detection by a reference probe. For larger times t, the distributions of air concentration were best described by the diffusion model:

$$C = 1 - \tanh^2\left(K' - \frac{\frac{y}{Y_{90}}}{2D_o} + \frac{\left(\frac{y}{Y_{90}} - \frac{1}{3}\right)^3}{3D_o} \right) \qquad t\sqrt{\frac{g}{d_o}} > 1.3$$

where K' and D_o are functions of the mean air content only (Chapter 17). The data highlighted a major change in void fraction distribution shape for $t\sqrt{g/d_o} \sim 1.3$

(Fig. 13.9). Possible explanations might include (a) a non-hydrostatic pressure field in the leading front of the wave, (b) some change in air–water flow structure between the leading edge and the main flow associated with a change in rheological fluid properties, (c) a gas–liquid flow regime change with some plug/slug flow at the leading edge and a homogenous bubbly flow region behind and (d) a change in boundary friction between the leading edge and the main flow behind. All these mechanisms would be consistent with high-shutter speed movies of leading edge highlighting very dynamic spray and splashing processes.

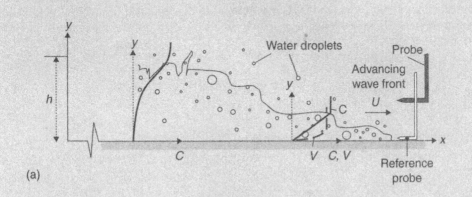

(a)

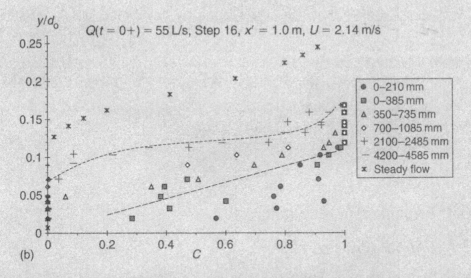

(b)

(mm)	0–210	0–385	350–735	700–1085	2100–2485	4200–4585
ΔX (m) =	0.210	0.385	0.385	0.385	0.385	0.385
$t\sqrt{g/d_o}$ =	0.313	0.573	1.615	2.657	6.826	13.08

Fig. 13.9 Void fraction distributions behind the leading edge of surge front. (a) Definition sketch. (b) Experimental data: $\theta = 3.4°$, $h = 0.07$ m, $U = 2.14$ m/s, Step 16, $x' = 1$ m (horizontal runoff).

13.3.3 Further dam break wave conditions

Additional dam break wave conditions were experimentally studied. Chanson *et al.* (2000) investigated the propagation of a dam break wave downstream of a free-falling nappe impact. They observed strong splashing and mixing at the nappe impact, and their data showed consistently a greater wave front celerity than for the classical dam break analysis for $x/d_o < 30$, where d_o is a function of the initial discharge. Khan *et al.* (2000) studied the effects of floating debris. The results showed an accumulation of debris near the wave front and a reduction of the front celerity both with and without initial water levels. Nsom *et al.* (2000) investigated dam break waves downstream of a finite reservoir in horizontal and sloping channels with very viscous fluids. The dam break wave propagation was first dominated by inertial forces and then by viscous processes. In the viscous regime, the wave front location followed:

$$x_s \propto \sqrt{\frac{\cos \theta \, t}{\mu}}$$

where θ is the bed slope and μ is the fluid viscosity.

> **DISCUSSION**
>
> Another related debris flow is the rockslide. Famous examples include the disastrous Vajont slide on 9 October 1963 (paragraph 13.1). Another major rockslide was the Mont Granier Cliff rockslide near Chambéry (France) in AD 1248. For both rockslides, it was proposed that mechanical energy dissipated in heat inside the slip zone led to vaporization of pore pressure, thus creating a cushion of zero or negligible friction (e.g. Vardoulakis 2000).
>
> In debris flow surges, large debris and big rocks are often observed 'rolling' and 'floating' at the wave front (e.g. Ancey 2000).

13.4 Embankment dam failures

13.4.1 Introduction

During the 19th Century, numerous embankment dams failed in Europe and North America. The two most common causes of failures were dam overtopping and cracking in the earthfill. The former was often caused by inadequate spillway facility. The latter resulted from a combination of bad understanding of basic soil mechanics, poor construction standards and piping at the connection between bottom outlet and earth material. Figure 13.10(a) presents the ruptured Dale Dyke Embankment Dam. The dam failure occurred rapidly as a result of piping in the embankment. Figure 13.10(b) shows a failed tailings dam. The failure was caused by overtopping of runoff waters. At the nearby township of Merriespruit, 17 people were killed when another tailings dam failed on the night of 22 February 1994. At Merriespruit, water

(a)

(b)

Fig. 13.10 Embankment dam failures. (a) Ruptured Dale Dyke Embankment Dam, UK: view from inside the reservoir few days after the disaster (courtesy of Michael Armitage). The embankment dam failed just before midnight on 11 March 1864; 150 people were killed. Although the spillway was operating at the time with a very small discharge, failure was attributed to poor construction standards and cracks in the embankment close to the culvert. (b) Failures of the Saaiplaas tailings facility, South Africa in 1993 (courtesy of Prof. Andre Fourie). The tailings dam is not far from the Merriespruit facility which collapsed on the night of 22 February 1994, killing 17 people.

had been stored on the tailings surface and was added to by 50 mm of rainfall falling in 1 h; 600 000 m^3 of tailings flowed into the town and travelled up to 4 km.

The overtopping of an embankment is a relatively slow process. It is not comparable to a sudden failure. For example, the failure of the 100 m high Teton Dam (earth fill dam) started around 11:00 a.m. on 5 June 1976 and the reservoir was drained by the evening. At its peak, the flow was estimated to be 28 000 m^3/s. During the failure of the Zeyzoun Dam (Syria, June 2002), the breach opened up to 6 m width about 3½h after the initial breach. In the township of Ziara, 2 km downstream of the dam, the water depth peaked at about 4 m and dropped down to 10 cm a few hours later.

Notes

1. Piping is the action of water cutting preferential channels in an embankment, following sometimes cracks and roots.
2. Tailings are mining residues.
3. In Sections 13.2 and 13.3, the dam break was assumed to be sudden. This assumption is untrue for most embankment dam overtopping.

Discussion

'Natural' lakes and reservoirs may be formed by landslides and rockslides. For example, during the Chi-Chi earthquake in Taiwan on 21 September 1999, the Chin-Shui and Ta-Chia Rivers, and the Tzao-Ling Valley were dammed by massive landslides (Hwang 1999). The Tzao-Ling Valley was previously dammed by record landslides in 1943 and 1974. In Tajikistan, Lake Sarez was formed by a massive rockslide (called Usoy Dam) which dammed the Murgab River Valley during a severe earthquake in 1911. The reservoir contains today $17 \times 10^9 \, \text{m}^3$ of water.

These landslide dams might become a hazard. In August 1191, a natural dam formed at Vaudaine (France) across the Romanche River. The reservoir, called Lac Saint-Laurent or Lac de l'Oisans, was located upstream of Bourg-d'Oisans. The natural dam was the result of massive landslides from the Belledonne range. The dam failed during the night of the 14–15 September 1219, the city of Grenoble suffered massive flooding and several thousand people were killed in the floods. In Taiwan, a 217 m high natural dam in the Tzao-Ling Valley was overtopped and failed in May 1951, and 154 people were killed in the subsequent floods. In Tajikistan, the 550 m high Usoy Dam is recognized as a potential hazard. The failure of the Usoy Dam would raise the level of the large Aral Sea by 1 m if all the water reached the Aral Sea (after causing massive destruction) (Waltham and Sholji 2001).

Glacier lake outburst floods (GLOF) is another form of dam break wave (e.g. Galay 1987). Galay described a GLOF in Nepal (Dig Cho glacier lake, 4 August 1985) in which a volume of 6–10 Mm^3 of water was drained: "*Local witnesses reported that the surge front advanced rather slowly down-valley as a huge 'black' mass of water full of debris. Trees and large boulders were dragged along and bounced around. The surge emitted a loud noise 'like many helicopters' and a foul mud-smell. The valley bottom was wreathed in misty clouds of water vapour; the river banks were trembling; houses were shaking; the sky was cloudless*" (Galay 1987, p. 2.36).

13.4.2 Embankment breach

There is some analogy between natural breach shape and the inlet designs of minimum energy loss (MEL) culvert and weirs. Photographs of breach profile illustrated a hourglass profile similar to MEL structures (Fig. 13.11). Figure 13.11 illustrates a natural breach shape which may be compared with Fig. 13.10(b). Experimental results show that the flow is near-critical in the breach (i.e. $0.5 < Fr < 1.8$) (Fig. 13.12). The total head remains constant throughout the breach inlet up to the throat (Fig. 13.12(b)). Head losses occurs downstream of the throat when the flow expands and separation takes place at the lateral boundaries. Separation is associated with form drag and head losses. Basically the movable boundary flow tends to an equilibrium that is associated with minimum energy conditions and maximum discharge per unit width for the available specific energy.

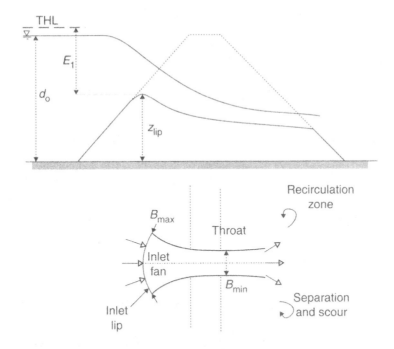

Fig. 13.11 Definition sketch of embankment breach for non-cohesive material. Cross-section through the breach centreline and view in elevation of breach flow. THL: total head line.

Discussion

The MEL culverts are designed with the concept of minimum head loss and nearly constant total head along the waterway (Apelt 1983, Chanson 1999a, 2003d). The flow in the approach channel is contracted through a streamlined inlet into the barrel where the channel width is minimum before being finally released in a streamlined outlet into the downstream natural channel. All the waterway must be streamlined to avoid significant form losses and flow separation, and the flow is critical from the inlet lip to the outlet lip.

A MEL inlet design is based basically upon a flow net analysis using irrotational flow theory (e.g. Vallentine 1969). The equipotential lines must be perpendicular to the flow direction (i.e. streamlines) everywhere. The flow net forms a network of converging 'quasi-square' elements. While the design theory is well understood for rectangular channels, the design of a natural channel is complicated by the irregular cross-sectional shape, but the inlet must be streamlined using a potential flow theory.

Remarks

1. Professor McKay suggested first an analogy between natural scour below a small bridge and the shape of MEL inlet design (McKay 1971).
2. In an MEL culvert, the outlet is streamlined to prevent flow separation and large head losses (e.g. Apelt 1983, Chanson 1999a, pp. 391–392).
3. For the data of Coleman *et al.* (2002), the breach geometry shows a dimensionless inlet length L_{inlet}/B_{max} of about 0.5–0.6: i.e. a result close to the minimum inlet length

recommended for MEL culvert design (Apelt 1983, p. 91). For shorter inlet length, separation would be observed in the throat.
4. A related form of embankment failure is a dyke breach and lagoon inlet breach. Gordon (1981) and Brodie (1988) observed lagoon breakouts at Dee Why, illustrating well the hourglass shape and some analogy with the inlet designs of MEL culvert and weirs, while Gordon (1990) compared breakout characteristics at three lagoons in South-East Australia. Visser *et al.* (1990) reported a prototype experiment with a 2.2 m high dyke breached during the rising tide. Their data showed that the breach width was about: $B_{min} \approx 2.8E_1$ that is consistent with a re-analysis of the data of Coleman *et al.* (2002).

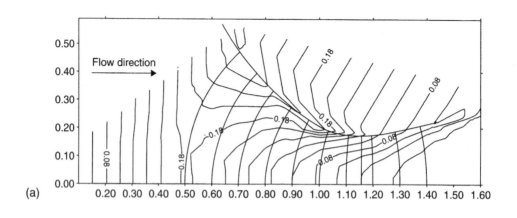

(a)

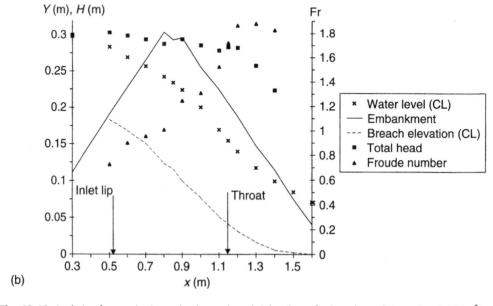

(b)

Fig. 13.12 Analysis of non-cohesive embankment breach inlet shape for breach conditions: $Q = 0.024 \, \text{m}^3/\text{s}$ at $t = 87 \, \text{s}$ for a 0.30 m high non-cohesive embankment (1.6 mm sand) (*data*: Coleman *et al.* 2002). (a) Flow net analysis for the 300 mm breach and contour lines of the breach. (b) Cross-section averaged Froude number and total head as functions of the longitudinal coordinate on the centreline ($Y = 0$).

Breach development

During breach development, the outflow rate equals:

$$Q = C_D \frac{2}{3} \sqrt{\frac{2}{3}} \, gE_1^3 \, B_{max} \qquad (13.17)$$

where E_1 is the upstream specific energy above centreline dam breach elevation, B_{max} is the free-surface width at the upper lip of the breach and C_D is a discharge coefficient ($C_D \sim 0.6 \, \text{m}^{1/2}/\text{s}$) (Fig. 13.11). During an overtopping event, the breach size increases with time resulting in the hydrograph of the breach. In equation (13.17), the breach free-surface width and specific energy are both functions of time, embankment properties and reservoir size. For an infinitely long reservoir, the re-analysis of embankment breach data suggests that:

$$\frac{z_{lip}}{d_o} = 1.08 \exp\left(-0.0013 \, t \sqrt{\frac{g}{d_o}}\right) \qquad \text{For } t \sqrt{\frac{g}{d_o}} < 1750 \qquad (13.18)$$

$$\frac{B_{max}}{d_o} = 2.73 \times 10^{-4} \left(t \sqrt{\frac{g}{d_o}}\right)^{1.4} \qquad \text{For } t \sqrt{\frac{g}{d_o}} < 1000 \qquad (13.19)$$

$$\frac{B_{min}}{d_o} = 4.01 \times 10^{-7} \left(t \sqrt{\frac{g}{d_o}}\right)^{2.28} \qquad \text{For } t \sqrt{\frac{g}{d_o}} < 1000 \qquad (13.20)$$

where z_{lip} is the inlet lip elevation on the breach centreline and B_{min} is the free-surface width at the breach throat (Fig. 13.11).

Notes
1. Equations (13.18)–(13.20) are based upon a re-analysis of the data of Coleman *et al.* (2002). The experiments were performed with 1V:2.7H embankment slopes and cohesionless materials ($0.9 < d_{50} < 2.4 \, \text{mm}$). The results were further obtained for an infinitely long reservoir.
2. Note that equations (13.19) and (13.20) were deduced from a very limited data set.
3. Importantly, equations (13.17)–(13.20) are valid during the development of the breach only. They do not account for overflow above the downstream slope of the embankment.

Application
The 9 m high Glashütte dam failed on Tuesday 12 August 2002. The dam was overtopped at 4:10 p.m. and failed rapidly. (1) Estimate the breach characteristics 5 min after failure. (2) Calculate the breach hydrograph for the first 10 min after overtopping. *Assume an infinitely long reservoir and cohesionless embankment.*

Solution
Using equations (13.17)–(13.20), the breach characteristics at $t = 5$ min are: $B_{max} = 7.7 \, \text{m}$, $B_{min} = 5.2 \, \text{m}$, $z_{lip} = 5.1 \, \text{m}$ and the flow through the breach is: $Q = 32 \, \text{m}^3/\text{s}$. At $t = 5$ min, the cumulative outflow volume is $24\,000 \, \text{m}^3$ of water.

Note that the upstream specific energy above breach inlet lip E_1 equals:

$$E_1 = d_o - z_{lip}$$

Discharge calculations yield a breach flow of nearly $210 \, \text{m}^3/\text{s}$ at about $10 \, \text{min}$ after the over-topping start, corresponding to an outflow volume of $34 \times 10^3 \, \text{m}^3$ of water (Fig. 13.13).

(a)

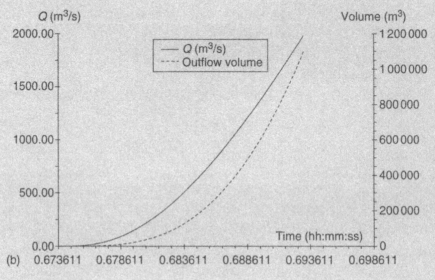

(b)

Fig. 13.13 Glashütte dam break accident. (a) Final breach (courtesy of Dr Antje Bornschein). Note outlet tunnel on right of photograph. (b) Breach outflow and outflow volume calculations at the Glashütte Dam.

Discussion

The Glashütte Dam was a small flood retention system completed in 1953. Located in the Elbe River, upstream of Dresden, the stepped spillway capacity became insufficient during a very heavy storm event in August 2002. Bornschein and Pohl (2003) presented a comprehensive study of the Glashütte Dam failure. The downstream slope of the dam was grasslined. Witness reports indicated that the dam was overtopped at 12:45 p.m. and that the wall failed completely within 30 min between 4:10 and 4:40 p.m.: i.e. more than 4 h after the overtopping start. Bornschein and Pohl's calculations suggested a maximum breach outflow of $120 \, m^3/s$.

The reservoir volume was only $50\,000 \, m^3$ of water. Present calculations assuming an infinitely long reservoir are inappropriate, but for the first few minutes. However they give some estimate of the reservoir drainage time (i.e. about 10 min) that is consistent with witness observations of dam failure in <30 min.

13.5 Related flow situations

Related cases of dam break wave flows include the flooding of a dry river bed, debris flow surges, wave runup in the swash zone and tsunami wave runup.

Ephemeral channels are usually not flowing above ground except during the rainy season. In Southern Africa, several large rivers may run dry during several months each year. When summer rainfalls take place in upper catchments, flood waves propagate downstream onto the dry river bed. Examples include the Upper Zambesi River, the Nata River in Zimbabwe Highlands, some rivers feeding the Okavango Swamps North of the Kalahari Desert, and the Molopo River in Southern Botswana. Figure 13.14 shows the Gascoyne River bed, in North-West Australia. The catchment area is about $67\,770 \, km^2$ and it extends 630 km inland. Average

Fig. 13.14 An ephemeral river: the Gascoyne River near Carnarvon WA (Australia) during a small flow (courtesy of Gascoyne Development Commission and Robert Panasiewicz).

annual rainfall is <250 mm throughout the basin. The river bed is generally wide, flat and sediment filled. There are typically one to two flow periods per year following seasonal rainfall or cyclone activity, but the river may fail to flow at all once every 5 or 6 years.

Notes
1. Ephemeral channels are also called arroyo, wadi, wash, dry wash, oued or coulee.
2. During the flooding of a dry river bed, infiltration may play a major role and the interactions between surface runoff and seepage cannot be ignored.
3. The Zambesi catchment lies near the Tropic of the Capricorn, and rainfalls take place predominantly during the summer months (November–April). At Maramba (formerly Livingstone), Zambia, the Zambezi River experiences its maximum flow in March–April. In October–November the discharge diminishes to <10% of the maximum.
4. The Molopo River starts in North-West, South Africa, and flows generally West for about 1000 km to join the Orange River. In its lower course the river passes through the Kalahari basin. Intermittent and usually dry, the Molopo marks part of the southern boundary between Botswana and South Africa.

Debris flow surges have been responsible for significant damages. Well-documented field studies were conducted in particular in China, Japan and Taiwan. Figure 13.15 shows a debris

Fig. 13.15 Osawa debris flow ravine (Japan) on 1 November 2001. Located at the foothill of Mt Fuji, the valley is subjected to major debris flows originating in the main (1 km wide) fault of Mt Fuji western slope. On 1 November 2001, the creek was dry and the last major debris flow event took place in Summer 2000 during a typhoon.

flow stream in Japan. (The rocks in left foreground are typically 0.5–2 m in size.) Debris flow control structures are built to protect developed areas. Figure 13.16 presents a range of examples including simple check dam and sophisticated tubular grid dam.

Capart and Young (1998) discussed specifically the wave propagation associated with sediment motion. They observed intense scouring of the bed at the leading edge of the bore. During the 1990s, further studies were conducted on dam break wave with sediment motion (e.g. Leal *et al.* 2001) while Chanson (2003) argued that strong air entrainment at the wave leading front affect sediment motion processes. For such studies, the Saint-Venant equations are not applicable (Chapter 12, Section 12.1).

(a)

(b)

Fig. 13.16 Debris flow control structures. (a) Check dam in the Hiakari-Gawa catchment (Japan) on 10 November 2001, located close to Toyota city, the Hiakari-Gawa sabo works include more than six checks dam (including two tubular grid dams) to protect a new 'Center for General Outdoor Activities'. (b) Sabo slit check dam and debris retention system on Inokubo stream (Japan) on 1 November 2001 The check dam is 104 m wide and 7.0 m high. Looking downstream at the six vertical slits.

(c)   (c) H. Okada

Fig. 13.16 (*Contd*) (c) Tubular grid dam on the Furan River (Japan) in 2002 (courtesy of Dr Marie Augendre and Prof. Okada).

Notes
1. Debris comprise mainly large boulders, rock fragments, gravel-sized to clay-sized material, tree and wood material that accumulate in creeks. The term debris flow is very broad. In its broad sense, it includes granular flows (and rockslides), mud flows (and paste flows) and wooden debris.
2. The Japanese word *sabo* means mountain protection system.

A related form of dam break wave is the wave runup in the swash zone. On the beach slope, waves run up after breaking like positive surges (Fig. 13.17). Figure 13.17 shows two examples of positive surges advancing against the gentle sandy slope. Note the significant amount of air bubble entrainment in the wave front (i.e. 'white waters') (Fig. 13.17(a)). Often the surge propagation takes place over retreating waters: i.e. against water run down. The swash zone wave runup is believed to be a significant factor in sediment processes in coastal zones (e.g. Longo *et al.* 2002, Elfrink and Baldock 2002).

(a)

(b)

Fig. 13.17 Surging waters on a sandy beach slope, Narrow Neck, Gold Coast, Australia on 18 February 2001. After wave breaking, the wave runup forms a series of positive surges at the upper end of the swash zone. (a) Strong positive surge advancing upslope from right to left and (b) weak positive surge, advancing from top left to bottom right.

Notes
1. In coastal engineering, the *swash* is the rush of water up a beach from the breaking waves.
2. The *swash line* is the upper limit of the active beach reached by highest sea level during big storms.

A tsunami is a long-period wave generated by ocean bottom motion during an earthquake. Occasionally it might be caused by another earth movement (e.g. underwater landslide, volcanic activity). The wave length is typically about 200–350 km and the tsunami behaves as a shallow-water wave, even in deep sea. Although the wave amplitude is moderate in the middle of the ocean (e.g. 0.5–1 m), the tsunami wave slows down and the wave height increases

Fig. 13.18 America (Peruvian Warship) beached and partially dismasted at Arica, Chile, following the 13 August 1868 tidal wave that washed her and other vessels ashore, photographed from the seaward side. The ship in the distance, beyond America's bow, is USS Wateree (courtesy of US Naval Historical Center, photograph NH 496 received from Captain Dudley W. Knox in 1934).

Fig. 13.19 Tsunami warning road sign post off Takatoyo Beach (Enshu Coast, Japan) on 27 March 1999). Note the drawings of surfers and fishermen.

near the shoreline, with periods ranging between 20 min and several hours typically. The wave runup height might reach several metres above the natural sea level. Major tsunami disasters were associated with well in excess of 140 000 losses of life (e.g. Yeh *et al.* 1996, Hebenstreit 1997, Chanson *et al.* 2000).

The tsunami wave runup may be a slow rise of water level or an advancing bore. The bore may be a surging bore resulting from the advance of an undular surge wave or a breaking wave associated with load noises. Visual observation suggest that the runup is a turbulent process characterized by significant scouring and sediment transport. When the coastline is flat, the abnormal rise of sea level associated with the tsunami wave may runup across flat lands, sweeping away buildings and carrying ships inland (e.g. Figs 13.18 and 13.19).

Notes

1. *Tsunami* is a Japanese word meaning 'harbour wave'. A tsunami is also called seismic sea wave. It is sometimes incorrectly termed 'tidal wave' but the process is not related to the tides.
2. One of the most devastating tsunami catastrophe occurred on 15 June 1896 along the Sanriku Coast which destroyed most of Yoshihama and Kamaishi townships (Japan). More than 26 000 people perished. Tsunami warning signs are often seen in Japan coastal zones (e.g. Fig. 13.19).
3. During the tsunami on 8 August 1868 in Peru and Northern Chile, the USS Wateree and the Peruvian warship America were carried about 1 km inland (Fig. 13.18).
4. A related case is the impulse wave generated by rockfalls, landslides, ice falls, glacier breakup or snow avalanches in lakes and man-made reservoirs. Some impulse waves might be induced by earthquake-generated falls. Vischer and Hager (1998) gave a thorough summary of the hydraulics of impulse waves.

13.6 Exercises

1. A 15 m high dam fails suddenly. The dam reservoir had a 13.5 m depth of water and the downstream channel was dry. (1) Calculate the wave front celerity, and the water depth at the origin. (2) Calculate the free-surface profile 2 min after failure. *Assume an infinitely long reservoir and use a simple-wave analysis ($S_o = S_f = 0$).*
2. A vertical sluice shut a trapezoidal channel (3 m bottom width, 1V:3H side slopes). The water depth was 4.2 m upstream of the gate and zero downstream (i.e. dry channel). The gate is suddenly removed. Calculate the negative wave celerity. Assuming an ideal dam break wave, compute the wave front celerity and the free-surface profile 1 min after gate removal.
3. A 5 m high spillway gate fails suddenly. The water depth upstream of the gate was 4.5 m depth and the downstream concrete channel was dry and horizontal. (1) Calculate the wave front location and velocity at $t = 3$ min. (2) Compute the discharge per unit width at the gate at $t = 3$ min. *Use Dressler's theory assuming f = 0.01 for new concrete lining.* (3) Calculate the wave front celerity at $t = 3$ min using Whitham's theory.
4. A horizontal, rectangular canal is shut by a vertical sluice. There is no flow motion on either side of the gate. The water depth is 3.2 m upstream of the gate and 1.2 m downstream. The gate is suddenly lifted. (1) Calculate the wave front celerity, and the surge front height. (2) Compute the water depth at the gate. Is it a function of time?

5. The 40 m high Zayzoun Dam failed on Tuesday 4 June 2002. The dam impounded a 35 m depth of water and failed suddenly. The depth of water in the downstream channel was 0.5 m. (1) Estimate the free-surface profile 7 min after the failure. (2) Calculate the time at which the wave will reach a point 10 km downstream of the dam and the surge front height. *Assume an infinitely long reservoir and a horizontal, smooth channel and use a simple-wave analysis* ($S_o = S_f = 0$).

6. A narrow valley is closed by a tall arch dam. The average bed slope is 0.08 and the valley cross-section is about rectangular (25 m width). At full reservoir level, the water depth immediately upstream of the dam is 42 m. In a dam break wave situation, (a) predict the wave front propagation up to 45 km downstream of the dam and (b) calculate the wave front celerity and depth as it reaches a location 25 km downstream of the dam site. *Use Hunt's theory assuming f = 0.06.*

7. The Cercey reservoir is a an artificial water supply reservoir held by a 18-m high embankment. A breach develops in the embankment. In first approximation, the reservoir level may be assumed constant. Estimate the breach characteristics 3 min after failure. Calculate the breach hydrograph for the first 15 min after overtopping. *Assume a cohesionless embankment. Notes*: Completed in 1836, the Cercey Reservoir (France) is part of the water supply of the Burgundy canal. The earthfill embankment was subjected to several slips prior to 1842. The embankment height is 14 m and its length is 1 km.

8. A 32 m high embankment dam is overtopped and fails rapidly. Estimate the breach characteristics for the first 20 min after failure, including the outflow volume. *Assume an infinitely long reservoir and cohesionless embankment.*

13.7 Exercise solutions

1. $U = 23$ m/s, $d(x = 0) = 6$ m.
2. The assumption of hydrostatic pressure distribution is valid for $t < 3\sqrt{d_o/g} \sim 2$ s. Hence the Saint-Venant equations may be applied for $t = 60$ s. Note the non-rectangular channel cross-section. For a trapezoidal channel, the celerity of a small disturbance is:

$$C = \sqrt{g\frac{A}{B}} = \sqrt{g\frac{d(W + d\cot\delta)}{W + 2d\cot\delta}}$$

where W is the bottom width and δ is the angle with the horizontal (i.e. $\cot\delta = 3$).

The method of characteristics predicts that the celerity of the negative wave is: $-C_o = -4.7$ m/s. The celerity of the wave front is $U = +2 \times C_o = +9.6$ m/s. Considering a backward characteristics issuing from the dam break wave front, the inverse slope of the C2 characteristics is a constant:

$$\frac{dx}{dt} = V - C = 2C_o - 3C$$

The integration gives the free-surface profile equation at a given time t:

$$\frac{x}{t} = 2\sqrt{g\frac{d_o(W + d_o\cot\delta)}{W + 2d_o\cot\delta}} - 3\sqrt{g\frac{d(W + d\cot\delta)}{W + 2d\cot\delta}}$$

At $t = 60$ s, the free-surface profile between the leading edge of the wave front and the negative wave most upstream location is:

d (m)	4.2	3	2	1.725	1	0.5	0
x (m)	-282	-160	-38.4	0	119	231	564

3. $x_s = 330$ m, $U = V(x = x_s) = 4.05$ m/s, $q(x = 0) = 0.21$ m²/s   (Dressler's   theory). $U = 4.7$ m/s (Whitham's theory).

4. (1) $d_1/d_0 = 0.375$, $U = 5.25$ m/s, $d(x = 0) = 2.07$ m. (2) $d_2 - d_1 = 0.87$ m.

5. $d_1/d_0 = 0.014$, $U = 22.3$ m/s, $d_2 - d_1 = 6.4$ m.
The free-surface profile at $t = 7$ min is:

x (m):	$-10\,000$	-7783	-5318	-2084	$+1894$	5124	5225	9353	9353	12 000
d (m):	35	35	28	20	12	7	6.86	6.86	0.5	0.5
Remark:		Point E1					Point E2	Point E3		

At a point located 10 km downstream of the dam site, the wave front arrives at $t = 449$ s (7 min 29 s). The height of the surge front is $\Delta d = d_2 - d_1 = 6.4$ m.

6. (b) $d_s = 1.3$ m, $U = 11.6$ m/s at $x_s = 25$ km (i.e. more than four reservoir lengths downstream of the dam site).

14

Numerical modelling of unsteady open channel flows

Summary

The numerical modelling of unsteady open channel flows is described. The development is based upon the Saint-Venant equations and the method of characteristics. Simple examples of explicit and implicit methods are presented.

14.1 Introduction

This chapter deals with the numerical integration of the Saint-Venant equations and the solutions of unsteady open channel flows. The Saint-Venant equations were developed for one-dimensional flows, hydrostatic pressure distributions, small bed slopes, constant water density and assuming that the flow resistance is the same as for a steady uniform flow for the same depth and velocity. The differential form of the equations is:

$$\frac{\partial A}{\partial t} + \frac{\partial Q}{\partial x} = 0 \qquad \text{Continuity equation} \tag{14.1}$$

$$\frac{\partial Q}{\partial t} + \frac{\partial}{\partial x}(V^2 A) + gA\frac{\partial d}{\partial x} = gA\,(S_o - S_f) \qquad \text{Momentum equation} \tag{14.2}$$

where A is the flow cross-section, Q is the flow rate, V is the flow velocity, g is the gravity acceleration, S_o is the bed slope and S_f is the friction slope (Chapter 11).

It is nearly impossible to achieve an exact solution of the Saint-Venant equations, or of the characteristic system of equations, because of the *non-linear terms* $V(\partial V/\partial x)$ and S_f, and because of the *complexity* of several functions: e.g. $A(d)$, $B(d)$. As a result, a numerical integration is required.

The basic form of the characteristic system of equations is:

$$\frac{\mathrm{D}}{\mathrm{D}t}(V + 2C) = -g(S_f - S_o) \qquad \text{Forward characteristics} \tag{14.3a}$$

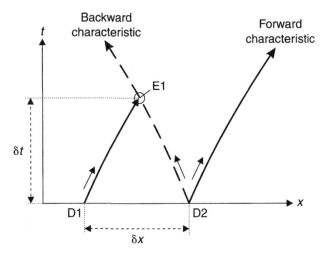

Fig. 14.1 Sketch of the characteristic trajectories.

$$\frac{D}{Dt}(V - 2C) = -g(S_f - S_o) \qquad \text{Backward characteristics} \qquad (14.3b)$$

along respectively:

$$\frac{dx}{dt} = V + C \qquad \text{Forward characteristics C1} \qquad (14.4a)$$

$$\frac{dx}{dt} = V - C \qquad \text{Backward characteristics C2} \qquad (14.4b)$$

Considering the characteristics C1 and C2 through the points D1 and D2 (Fig. 14.1), the characteristic trajectories intersect at point E1 (Fig. 14.1). The integration of the characteristic system of equations gives:

$$V_{E1} + 2C_{E1} = V_{D1} + 2C_{D1} - \int_{t_{D1}}^{t_{E1}} g(S_f - S_o)\,dt \qquad \text{Forward characteristics} \qquad (14.5a)$$

$$V_{E1} - 2C_{E1} = V_{D2} - 2C_{D2} - \int_{t_{D2}}^{t_{E1}} g(S_f - S_o)\,dt \qquad \text{Backward characteristics} \qquad (14.5b)$$

$$x_{E1} = x_{D1} + \int_{t_{D1}}^{t_{E1}} (V + C)\,dt \qquad \text{Forward characteristics C1} \qquad (14.6a)$$

$$x_{E1} = x_{D2} + \int_{t_{D2}}^{t_{E1}} (V - C)\,dt \qquad \text{Backward characteristics C2} \qquad (14.6b)$$

If the initial conditions at points D1 and D2 are known, equations (14.5a), (14.5b), (14.6a) and (14.6b) form a system of four equations with four unknowns: x_{E1}, t_{E1}, d_{E1} and V_{E1}. They are exact *but* non-linear equations because:

$$C_{E1} = \sqrt{g\frac{A_{E1}}{B_{E1}}} \quad A_{E1} = A(d_{E1}) \quad B_{E1} = B(d_{E1}) \quad S_{f_{E1}} \approx \frac{f_{E1}V_{E1}^2}{8gd_{E1}} \qquad \text{for a wide channel}$$

A *first approximation* consists in performing a trapezoidal integration of the characteristic system of equations:

$$V_{E1} + 2C_{E1} = V_{D1} + 2C_{D1} - g(t_{E1} - t_{D1})\left(\frac{S_{f_{E1}} + S_{f_{D1}}}{2} - \frac{S_{o_{E1}} + S_{o_{D1}}}{2}\right) \qquad (14.7a)$$

$$V_{E1} - 2C_{E1} = V_{D2} - 2C_{D2} - g(t_{E1} - t_{D2})\left(\frac{S_{f_{E1}} + S_{f_{D2}}}{2} - \frac{S_{o_{E1}} + S_{o_{D2}}}{2}\right) \qquad (14.7b)$$

$$x_{E1} = x_{D1} + (t_{E1} - t_{D1})\left(\frac{V_{E1} + V_{D1}}{2} + \frac{C_{E1} + C_{D1}}{2}\right) \qquad (14.8a)$$

$$x_{E1} = x_{D2} + (t_{E1} - t_{D2})\left(\frac{V_{E1} + V_{D2}}{2} - \frac{C_{E1} + C_{D2}}{2}\right) \qquad (14.8b)$$

The error on the flow conditions at the point E1 is a function of two parameters: i.e. the error introduced by the trapezoidal integration and the number of iterations. It can be shown mathematically that the accuracy of the numerical integration is greatly enhanced by increasing the number of iterations and decreasing the distance $\delta x = x_{D1} - x_{D2}$, and that it converges to the exact solution.

> ### Note
> The trapezoidal integration is the most accurate method of integration technique between two points without additional points.

Finite differences methods

One of the most frequently used methods of obtaining approximate solutions of partial differential equations is the method of finite differences, which consists essentially in replacing each partial derivative by a ratio of differences between two immediate values:

$$\frac{\partial V}{\partial t} \approx \frac{\delta V}{\delta t} \qquad (14.9)$$

where δV is the increase in the function V during the time step δt.

Although finite difference techniques are simple to program, including with a spreadsheet, there are a number of difficulties in particular associated with numerical instabilities. For example, the following explicit finite difference scheme is unstable:

$$\frac{\partial V_i^{n+1}}{\partial t} = \frac{V_i^{n+1} - V_i^n}{\delta t} \qquad \text{Unstable scheme} \qquad (14.10a)$$

$$\frac{\partial V_i^{n+1}}{\partial x} = \frac{V_{i+1}^n - V_{i-1}^n}{2\delta x} \qquad \text{Unstable scheme} \qquad (14.10b)$$

where the subscript i and superscript n refer respectively to the x-direction and t-axis (Fig. 14.2). The numerical scheme (equation (14.10)) is unstable for any value of δx and δt.

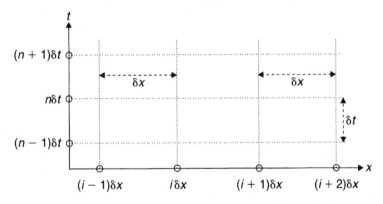

Fig. 14.2 Numerical integration of the characteristic system of equations: definition sketch.

The following explicit scheme is, under appropriate conditions, stable:

$$\frac{\partial V_i^{n+1}}{\partial t} = \frac{V_i^{n+1} - \dfrac{V_{i+1}^n + V_{i-1}^n}{2}}{\delta t} \qquad \text{Stable scheme} \qquad (14.11a)$$

$$\frac{\partial V_i^{n+1}}{\partial x} = \frac{V_{i+1}^n - V_{i-1}^n}{2\delta x} \qquad \text{Stable scheme} \qquad (14.11b)$$

This technique (equation (14.11)) is a diffusive scheme that is stable for:

$$\frac{U\,\delta t}{\delta x} \leqslant 1 \qquad (14.12)$$

where U is the velocity along the trajectory: i.e. $U = |V| + C$ where $|V|$ is the magnitude of the flow velocity. The ratio $U(\delta t/\delta x)$ is called the Courant number denoted by Cr. For $Cr > 1$, the numerical solution does not converge toward the exact solution.

Notes

1. There are several finite difference schemes. Classical expressions include:

$$\frac{\partial V_i^n}{\partial x} = \frac{V_{i+1}^n - V_i^n}{\delta x} \qquad \text{Forward difference}$$

$$\frac{\partial V_i^n}{\partial x} = \frac{V_i^n - V_{i-1}^n}{\delta x} \qquad \text{Backward difference}$$

$$\frac{\partial V_i^n}{\partial x} = \frac{V_{i+1}^n - V_{i-1}^n}{2\,\delta x} \qquad \text{Central difference}$$

and similar expressions apply to the time derivatives.

2. The Courant number was named after Richard Courant (1888–1972), American mathematician, born in Germany, who worked at New York University from 1934 until his retirement in 1958.

14.2 Explicit finite difference methods

Two well-known explicit finite difference methods are the Lax diffusive scheme and the leap-frog method (Table 14.1, Fig. 14.3) (Liggett and Cunge 1975). Considering the (x, t) plane sketched in Fig. 14.2, grid points are identified by the subscript i and by the superscript n to

Table 14.1 Finite difference methods for unsteady open channel flows

Scheme	$\dfrac{\partial V_i^{n+1}}{\partial t}$	$\dfrac{\partial V_i^{n+1}}{\partial x}$	Remarks
(1)	(2)	(3)	(4)
Explicit schemes Lax diffusive scheme	$\dfrac{V_i^{n+1} - \left(\alpha V_i^n + (1 - \alpha)\dfrac{V_{i+1}^n + V_{i-1}^n}{2} \right)}{\delta t}$	$\dfrac{V_{i+1}^n - V_{i-1}^n}{2\,\delta x}$	Stability: $0 \leqslant \alpha < 1$ and $\dfrac{\delta t\,(\lvert V \rvert + C)}{\delta x} \leqslant 1$
Leap-frog method	$\dfrac{V_i^{n+1} - V_i^{n-1}}{2\,\delta t}$	$\dfrac{V_{i+1}^n - V_{i-1}^n}{2\,\delta x}$	Stability: $\dfrac{\delta t\,(\lvert V \rvert + C)}{\delta x} \leqslant 1$
Implicit scheme Preissmann–Cunge scheme	$\dfrac{1}{2}\left(\dfrac{V_{i+1}^{n+1} - V_{i+1}^n}{\delta t} + \dfrac{V_i^{n+1} - V_i^n}{\delta t} \right)$	$\theta \dfrac{V_{i+1}^{n+1} - V_i^{n+1}}{\delta x}$ $+ (1 - \theta)\dfrac{V_{i+1}^n - V_i^n}{\delta x}$	Stability: $0.5 \leqslant \theta \leqslant 1$

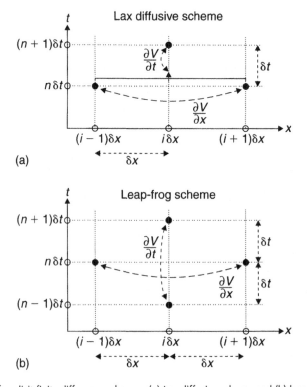

Fig. 14.3 Sketch of explicit finite difference schemes: (a) Lax diffusive scheme and (b) leap-frog scheme.

indicate spatial intervals and time steps respectively. For example, V_i^n is the discrete value of the flow velocity at a distance $x = i\delta x$ from the left boundary where $x = 0$ and $i = 0$, and at time $t = n\delta t$ from the initial conditions where $t = 0$ and $n = 0$, assuming constant spatial and time intervals δx and δt.

Discussion
The numerical integration of the characteristic system of equations does not require constant, uniform spatial and time steps. In a general case, V_i^n becomes the discrete value of the velocity at a distance x from left boundary and a time t from the origin that satisfy:

$$x = \sum_{j=1}^{i} \delta x_j$$

$$t = \sum_{k=1}^{n} \delta t_k$$

Note
The term 'explicit' implies that the flow properties at $x = x_1 > 0$ and $t = t_1 > 0$ can be calculated as functions of the flow conditions at $t < t_1$ and that they are independent of the flow properties for $t > t_1$.

14.2.1 Lax diffusive method

The continuity and momentum equations (14.1) and (14.2) may be rewritten in terms of the flow rate and free-surface elevation:

$$\frac{\partial Y}{\partial t} + \frac{1}{B}\frac{\partial Q}{\partial x} = 0 \qquad \text{Continuity equation} \qquad (14.13)$$

$$\frac{\partial Q}{\partial t} + \frac{\partial}{\partial x}\left(\frac{Q^2}{A}\right) + gA\frac{\partial Y}{\partial x} + gAS_f = 0 \qquad \text{Dynamic equation} \qquad (14.14)$$

where the friction slope S_f equals:

$$S_f = \frac{Q|Q|}{\frac{8g}{f}A^2\frac{D_H}{4}}$$

The Lax diffusive scheme is shown in Table 14.1. It is similar to equation (14.11), with the difference of a coefficient α satisfying $0 \leq \alpha < 1$. For $\alpha = 0$, the Lax method equals equation (14.11) which is sometimes called a diffusive scheme. For $\alpha = 1$, the Lax method is unstable (equation (14.10)). Using the Lax diffusive scheme, the discretization of the continuity and dynamic equations yields:

$$\frac{Y_i^{n+1} - \left(\alpha\,Y_i^n + (1-\alpha)\dfrac{Y_{i+1}^n + Y_{i-1}^n}{2}\right)}{\delta t} + \frac{1}{B}\frac{Q_{i+1}^n - Q_{i-1}^n}{2\,\delta x} = 0$$

Continuity equation (14.15)

$$\frac{Q_i^{n+1} - \left(\alpha Q_i^n + (1-\alpha)\dfrac{Q_{i+1}^n + Q_{i-1}^n}{2}\right)}{\delta t} + \frac{\left(\dfrac{Q^2}{A}\right)_{i+1}^n - \left(\dfrac{Q^2}{A}\right)_{i-1}^n}{2\,\delta x}$$

$$+ gA\frac{Y_{i+1}^n - Y_{i-1}^n}{2\,\delta x} + gAS_f = 0 \qquad \text{Dynamic equation} \qquad (14.16)$$

In the continuity equation, the term B is not explicitly defined. A simple assumption is: $B = B_i^n$. Similarly the terms A and S_f in the dynamic equation (14.16) may be assumed to be equal to A_i^n and S_{fi}^n respectively.

The continuity and dynamic equations form a system of two equations with two unknowns Y_i^{n+1} and Q_i^{n+1}. It may be solved explicitly:

$$Y_i^{n+1} = \left(\alpha Y_i^n + (1-\alpha)\frac{Y_{i+1}^n + Y_{i-1}^n}{2}\right) + \frac{\delta t}{B_i^n}\frac{Q_{i+1}^n - Q_{i-1}^n}{2\,\delta x} \qquad (14.17)$$

$$Q_i^{n+1} = \left(\alpha Q_i^n + (1-\alpha)\frac{Q_{i+1}^n + Q_{i-1}^n}{2}\right) - \frac{\delta t}{2\,\delta x}\left(\left(\frac{Q^2}{A}\right)_{i+1}^n - \left(\frac{Q^2}{A}\right)_{i-1}^n\right)$$

$$- \frac{\delta t}{2\,\delta x}gA_i^n(Y_{i+1}^n - Y_{i-1}^n) - \delta t\, gA_i^n S_{fi}^n \qquad (14.18)$$

The Lax diffusive method is characterized by some numerical diffusion for $\alpha \geqslant 0$ that introduces numerical inaccuracies. At the limit, for a steady flow, the Lax diffusive scheme predicts a horizontal free surface which is untrue. Basically the method is stable numerically for $\alpha < 1$ and:

$$\frac{\delta t(|V| + C)}{\delta x} \leqslant 1 \qquad (14.19)$$

Practically equation (14.19) imposes an upper limit to the time step δt. For example, considering a large river with a mean water depth of 2 m and mean velocity of 1 m/s, the time step δt must satisfy: $\delta t \leqslant \delta x/4.4$. If the spatial interval δx equals 1 km, the time step must be less than 226 s. A natural flood in such a river system is likely to spread over several days, possibly few weeks, and complete unsteady flow calculations may require several thousand iterations.

Notes

1. The continuity and momentum equations (14.1) and (14.2) are:

$$\frac{\partial A}{\partial t} + \frac{\partial Q}{\partial x} = 0 \qquad \text{Continuity equation} \qquad (14.1)$$

$$\frac{\partial Q}{\partial t} + \frac{\partial}{\partial x}(V^2 A) + gA\frac{\partial d}{\partial x} = gA\,(S_o - S_f) \qquad \text{Momentum equation} \qquad (14.2)$$

They may be rewritten in terms of the discharge Q and free-surface elevation Y as:

$$\frac{\partial Y}{\partial t} + \frac{1}{B}\frac{\partial Q}{\partial x} = 0 \qquad \text{Continuity equation} \qquad (14.13)$$

$$\frac{\partial Q}{\partial t} + \frac{\partial}{\partial x}\left(\frac{Q^2}{A}\right) + gA\frac{\partial Y}{\partial x} + gAS_f = 0 \qquad \text{Dynamic equation} \qquad (14.14)$$

where

$$S_f = \frac{Q|Q|}{\frac{8g}{f}A^2\frac{D_H}{4}}$$

2. The momentum equation (14.2) is commonly called the dynamic equation.
3. The Lax diffusive scheme was advocated by J.J. Stoker although it is often attributed to J. Keller and P. Lax (Montes 1998, p. 502). Isaacson *et al.* (1958) developed an initial discretization scheme assuming $\alpha = 0.5$ while Liggett and Cunge (1975) detailed the present scheme commonly called Lax diffusive scheme.

Boundary conditions

The free-surface elevation Y_i^{n+1} can be computed, using equation (14.17), at the time step $t = (n+1)\delta t$ and at the locations $x = i\delta x$ for $i = 1, 2, ..., N-1$, but at the boundaries (i.e. Y_0^{n+1} and Y_N^{n+1}). Equation (14.18) allows the computations of the flow rate Q_i^{n+1} at the time $t = (n+1)\delta t$ and at the locations $x = i\delta x$ for $i = 1, 2, ..., N-1$, but at the boundaries (i.e. Q_0^{n+1} and Q_N^{n+1}).

The implementation of boundary conditions has been discussed in Chapter 11 (Section 11.3.2 and Table 11.1). Basically, for a limited reach with subcritical flow conditions, one flow condition must be prescribed at each boundary for $t > 0$. In supercritical flows, two flow properties are required at the upstream boundary.

Considering the left boundary sketched in Fig. 14.4, at the time step $t = n\delta t$, the flow conditions at the previous time step are known: i.e. Y_0^n and Q_0^n, Y_1^n and Q_1^n. One boundary condition may be deduced from the method of characteristics. For the subcritical flow sketched in Fig. 14.4, the equations of the characteristics C2 through the left boundary give:

$$(V_{i=0}^{n+1} - 2)C_{i=0}^{n+1} = V_S - 2C_S - \int_{n\delta t}^{(n+1)\delta t} g(S_f - S_o)\,dt \qquad \text{Backward characteristics} \qquad (14.5b)$$

$$x_{i=0} = x_S + \int_{n\delta t}^{(n+1)\delta t}(V - C)\,dt = 0 \qquad \text{Backward characteristics C2} \qquad (14.6b)$$

where the point S is located at the intersection of the backward characteristics with the time step $t = n\delta t$ and assuming $x_{i=0} = 0$ at the left boundary. Equations (14.5b) and (14.6b) form a system of two equations with two unknowns $V_{i=0}^{n+1}$ and x_S (or $Q_{i=0}^{n+1}$ and x_S). A similar reasoning may be conducted at the right boundary ($i = N$) using a forward characteristic trajectory.

This treatment of the boundary conditions is not specific to the Lax diffusive method. It is general to all explicit schemes including the leap-frog method (Section 14.2.2).

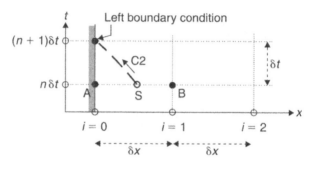

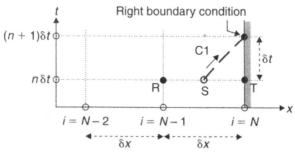

Fig. 14.4 Boundary conditions.

Application
Develop the characteristic equations for subcritical flows.

Solution (1): Left boundary
Considering the left boundary sketched in Fig. 14.4, the equations of the characteristics C2 through the left boundary give:

$$(V_0^{n+1} - 2) C_0^{n+1} = V_S - 2C_S - \int_{n\delta t}^{(n+1)\delta t} g(S_f - S_o) \, dt \qquad (14.5b)$$

$$x_0 = x_S + \int_{n\delta t}^{(n+1)\delta t} (V - C) \, dt = 0 \qquad (14.6b)$$

The flow properties at the point S may be interpolated with those at adjacent points at the same time level: i.e. points A and B in Fig. 14.4. With linear interpolations, the velocity V and celerity C are determined as:

$$V_S = \frac{x_S - x_0}{\delta x} (V_1^n - V_0^n) + V_0^n$$

$$C_S = \frac{x_S - x_0}{\delta x} (C_1^n - C_0^n) + C_0^n$$

Using the parabolic procedure, it yields:

$$V_S = \left(\frac{x_S - x_0}{\delta x}\right)^2 \frac{V_0^n - 2V_1^n + V_2^n}{2} + \left(\frac{x_S - x_0}{\delta x}\right) \frac{-3V_0^n + 4V_1^n - V_2^n}{2} + V_0^n$$

$$C_S = \left(\frac{x_S - x_0}{\delta x}\right)^2 \frac{C_0^n - 2C_1^n + C_2^n}{2} + \left(\frac{x_S - x_0}{\delta x}\right) \frac{-3C_0^n + 4C_1^n - C_2^n}{2} + C_0^n$$

The parabolic interpolation is more precise than the linear one.

Altogether this gives a system of four equations with four unknowns x_S, V_S, C_S and V_0^{n+1} (or Q_0^{n+1}).

Solution (2): Right boundary
Considering the right boundary sketched in Fig. 14.4, the equations of the characteristics C1 give:

$$V_N^{n+1} + 2C_N^{n+1} = V_S + 2C_S - \int_{n\delta t}^{(n+1)\delta t} g(S_f - S_o)\, dt \qquad (14.5a)$$

$$x_N = x_S + \int_{n\delta t}^{(n+1)\delta t} (V + C)\, dt = 0 \qquad (14.6a)$$

Using a linear interpolation between points R and T (Fig. 14.4), the flow properties at the point S are:

$$V_S = \frac{x_S - x_{N-1}}{\delta x}(V_N^n - V_{N-1}^n) + V_{N-1}^n$$

$$C_S = \frac{x_S - x_{N-1}}{\delta x}(C_N^n - C_{N-1}^n) + C_{N-1}^n$$

14.2.2 Leap-frog scheme

The leap-frog method uses centred differences for time and space (Table 14.1, Fig. 14.3(b)). The discretization of the continuity and dynamic equations yields:

$$\frac{Y_i^{n+1} - Y_i^{n-1}}{2\,\delta t} + \frac{1}{B_i^n}\frac{Q_{i+1}^n - Q_{i-1}^n}{2\,\delta x} = 0 \qquad \text{Continuity equation} \qquad (14.20)$$

$$\frac{Q_i^{n+1} - Q_i^{n-1}}{2\,\delta t} + \frac{\left(\dfrac{Q^2}{A}\right)_{i+1}^n - \left(\dfrac{Q^2}{A}\right)_{i-1}^n}{2\,\delta x}$$
$$+ gA_i^n\frac{Y_{i+1}^n - Y_{i-1}^n}{2\,\delta x} + gA_i^n S_{f\,i}^n = 0 \qquad \text{Dynamic equation} \qquad (14.21)$$

The continuity and dynamic equations form a system of two equations which may be solved explicitly:

$$Y_i^{n+1} = Y_i^{n-1} + \frac{\delta t}{\delta x}\frac{1}{B_i^n}(Q_{i+1}^n - Q_{i-1}^n) \qquad (14.22)$$

$$Q_i^{n+1} = Q_i^{n-1} - \frac{\delta t}{\delta x}\left(\left(\frac{Q^2}{A}\right)_{i+1}^n - \left(\frac{Q^2}{A}\right)_{i-1}^n\right) - \frac{\delta t}{\delta x}gA_i^n(Y_{i+1}^n - Y_{i-1}^n)$$
$$- 2\,\delta t\, gA_i^n S_{f\,i}^n \qquad (14.23)$$

Equations (14.22) and (14.23) give the free-surface elevation Y_i^{n+1} and flow rate Q_i^{n+1} at the $(n + 1)$ time step for $i = 1, 2, ..., N - 1$, but at the boundaries (i.e. $i = 0$ and N).

The leap-frog scheme is stable for:

$$\frac{\delta t\,(|V| + C)}{\delta x} \leqslant 1 \qquad (14.19)$$

Notes

1. The leap-frog scheme is called 'schéma en quinconce' in French. It is one of the earliest numerical schemes and was used by Lewis Fry Richardson in his 1910 paper on stress calculation in masonry dams (Montes 1998, p. 509).
2. Lewis Fry Richardson (1881–1953) was a British meteorologist. He pioneered mathematical weather forecasting. He made contributions to the theory of calculus and the study of diffusion.
3. Note that the calculations of free-surface elevation Y_i^{n+1} and flow rate Q_i^{n+1} are basically independent of Y_i^n and Q_i^n (Fig. 14.3(b)). Basically the points in the (x, t) space form two independent grids. This may result in saw-tooth distributions of flow rate and free-surface elevation.

14.2.3 Discussion

Practically the leap-frog scheme introduces less numerical errors than the Lax diffusive method. The leap-frog scheme is of second order rather than first order, and it is non-dissipative. There is no numerical diffusion terms like in the Lax diffusive method. Liggett and Cunge (1975) added on the leap-frog method: '*mass conservation [...] is very good because the approximation of the continuity equation is of second order for B = constant. Unfortunately the solution obtained is a saw-tooth line since the points where the dependent variables are computed are alternately odd or even*' (p. 121).

14.3 Implicit finite difference methods

Explicit schemes are very simple and they are well suited for interior grid points. Two major disadvantages are the treatment of the boundary conditions and the stability criterion (equation (14.19)). Modern numerical models of unsteady open channel flows are based upon some implicit finite difference methods.

The most common implicit method for unsteady open channel flows is the Preissmann–Cunge scheme (Fig. 14.5). The dependent variables and their derivatives are discretized as:

$$V = \frac{\theta}{2}\left(V_{i+1}^{n+1} + V_i^{n+1}\right) + \frac{1 - \theta}{2}\left(V_{i+1}^n + V_i^n\right) \qquad (14.24)$$

$$\frac{\partial V_i^{n+1}}{\partial t} = \frac{1}{2}\left(\frac{V_{i+1}^{n+1} - V_{i+1}^n}{\delta t} + \frac{V_i^{n+1} - V_i^n}{\delta t}\right) \qquad (14.25)$$

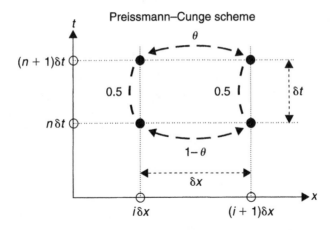

Fig. 14.5 Sketch of the Preissmann–Cunge implicit finite difference scheme.

$$\frac{\partial V_i^{n+1}}{\partial x} = \theta \frac{V_{i+1}^{n+1} - V_i^{n+1}}{\delta x} + (1 - \theta) \frac{V_{i+1}^n - V_i^n}{\delta x} \tag{14.26}$$

where θ is a weighting parameter. It can be shown that unconditional stability occurs only in the range:

$$0.5 \le \theta \le 1$$

The discretization of the continuity and dynamic equations yields:

$$\frac{1}{2}\left(\frac{Y_{i+1}^{n+1} - Y_{i+1}^n}{\delta t} + \frac{Y_i^{n+1} - Y_i^n}{\delta t} \right)$$

$$+ \frac{\theta \dfrac{Q_{i+1}^{n+1} - Q_i^{n+1}}{\delta x} + (1 - \theta) \dfrac{Q_{i+1}^n - Q_i^n}{\delta x}}{\dfrac{\theta}{2}\left(B_{i+1}^{n+1} + B_i^{n+1}\right) + \dfrac{1 - \theta}{2}\left(B_{i+1}^n + B_i^n\right)} = 0 \quad \text{Continuity equation} \tag{14.27}$$

$$\frac{1}{2}\left(\frac{Q_{i+1}^{n+1} - Q_{i+1}^n}{\delta t} + \frac{Q_i^{n+1} - Q_i^n}{\delta t} \right)$$

$$+ \left[\theta \frac{\left(\dfrac{Q^2}{A}\right)_{i+1}^{n+1} - \left(\dfrac{Q^2}{A}\right)_i^{n+1}}{\delta x} + (1 - \theta)\frac{\left(\dfrac{Q^2}{A}\right)_{i+1}^{n} - \left(\dfrac{Q^2}{A}\right)_i^{n}}{\delta x} \right.$$

$$+ g\left(\frac{\theta}{2}\left(A_{i+1}^{n+1} + A_i^{n+1}\right) + \frac{1 - \theta}{2}\left(A_{i+1}^n + A_i^n\right)\right)\left(\theta \frac{Y_{i+1}^{n+1} - Y_i^{n+1}}{\delta x} + (1 - \theta)\frac{Y_{i+1}^n - Y_i^n}{\delta x}\right)$$

$$+ g\left(\frac{\theta}{2}\left(A_{i+1}^{n+1} + A_i^{n+1}\right) + \frac{1 - \theta}{2}\left(A_{i+1}^n + A_i^n\right)\right)$$

$$\left. \times \left(\frac{\theta}{2}\left(S_{f\,i+1}^{n+1} + S_{f\,i}^{n+1}\right) + \frac{1 - \theta}{2}\left(S_{f\,i+1}^n + S_{f\,i}^n\right)\right)\right] = 0 \quad \text{Dynamic equation} \tag{14.28}$$

Note that any function of water depth, e.g. the free-surface width B, satisfies:

$$B_i^{n+1} - B_i^n = \left(\frac{\partial B}{\partial d}\right)_i^n (d_i^{n+1} - d_i^n) = \left(\frac{\partial B}{\partial Y}\right)_i^n (Y_i^{n+1} - Y_i^n)$$

Considering a river reach ($i = 0, 1, ..., N$), there are $2(N + 1)$ unknowns at the time step ($n + 1$) while equations (14.27) and (14.28) provide $2N$ equations between two adjacent points $\{i\}$ and $\{i - 1\}$. Including the two boundaries conditions at $i = 0$ and N, it yields a system of $2N + 2$ equations with $2(N + 1)$ unknowns.

Notes

1. In an 'implicit' scheme, the flow properties must be calculated simultaneously at once. For example, this may be achieved by a matrix inversion operation. In other words, the flow properties at $x = x_1 > 0$ and $t = t_1 > 0$ are functions of the flow conditions at $t = t_2 > t_1$.
2. The Preissmann–Cunge method was first developed in 1960 and extended in the following years (Preissmann 1960, Preissmann and Cunge 1961, Cunge and Berthier 1962, Cunge and Wegner 1964).
3. The Preissmann–Cunge scheme is also called the four-point implicit method, box model, Preissmann implicit scheme or Sogreah implicit method.
4. Alexandre Preissmann (1916–1990) was born and educated in Switzerland. From 1958, he worked on the development of mathematical models at Sogreah in Grenoble.
5. Born and educated in Poland, Jean A. Cunge worked in France at Sogreah in Grenoble and he lectured at the Hydraulics and Mechanical Engineering School of Grenoble (ENSHMG).

Discussion

The Preissmann–Cunge method is an implicit scheme. Although more complicated than the explicit methods, its major advantages include the stability of the scheme, the computation of discontinuities (e.g. hydraulic jump), the calculations of the boundary conditions and the variable spatial interval δx.

The Preissmann–Cunge scheme is unconditionally stable for $0.5 \leqslant \theta \leqslant 1$: i.e. the stability is neither a function of the time step nor Courant number. Practically Liggett and Cunge (1975) recommended to select a weighting parameter satisfying:

$$0.6 \leqslant \theta \leqslant 1$$

For $\theta < 0.5$, the scheme is always unstable while parasite oscillations were found for $0.5 \leqslant \theta < 0.66$ (Liggett and Cunge 1975, p. 163). Further studies suggested an optimum value θ for $0.6 \leqslant \theta \leqslant 0.75$.

The boundary conditions (i.e. $i = 0$ and N) may be introduced simply. For example, if the free-surface elevation $Y(t)$ is a known function of the time at the left boundary, this gives the following additional relationships:

$$Y_0^{n+1} = f(t_{n+1}) \qquad \text{Boundary condition}$$

$$\delta Y_0^n = Y_0^{n+1} - Y_0^n = f(t_{n+1}) - f(t_n)$$

If the discharge is a known function of time at the right boundary, it gives:

$$Q_N^{n+1} = f(t_{n+1}) \qquad \text{Boundary condition}$$

$$\left(\frac{\partial Q}{\partial Y}\right)_N^n \left(Y_N^{n+1} - Y_N^n\right) = Q_N^{n+1} - Q_N^n = f(t_{n+1}) - f(t_n)$$

If the discharge is a known function of the free-surface elevation time at the right boundary, this gives:

$$Q_N^{n+1} = f(Y_N^{n+1}) \qquad \text{Boundary condition}$$

$$Q_N^{n+1} - Q_N^n = \left(\frac{\partial Q}{\partial Y}\right)_N^n (Y_N^{n+1} - Y_N^n) = \left(\frac{df}{dY}\right)_N^n (Y_N^{n+1} - Y_N^n)$$

where df/dY is the derivative of the stage–discharge relationship at the boundary.

In the Preissmann–Cunge scheme (equations (14.27) and (14.28)), only the locations $\{i\}$ and $\{i + 1\}$ are linked: i.e. over one spatial interval δx (Fig. 14.5). As a result, it is possible to change the longitudinal increment δx without affecting the stability nor accuracy of the results.

Cunge (1975) showed that hydraulic discontinuities can be calculated using a variable weighting factor θ in the Preissmann–Cunge method.

Note
Although the stability of the Preissmann–Cunge method is independent of the Courant number, the accuracy of the results decreases with increasing Courant number for $Cr > 1$.

14.4 Exercises

1. Why the Saint-Venant equations cannot usually be solved with analytical equations?
2. Using a sketch, explain the basic differences between forward, backward and central difference methods.
3. Write the Courant number for a finite difference scheme.
4. What are the basic differences between an explicit and implicit numerical scheme?
5. Is the Lax diffusive scheme always stable? Detail your answer.
6. Is the leap-frog scheme an implicit method? What is(are) the unusual feature(s) of the leap-frog scheme?
7. Is the Preissmann–Cunge scheme an implicit method? For what range of the Courant number is it stable?

Part 3 Revision exercises

Summary

In this section, applications of unsteady open channel flows are reviewed. Simple exercises are developed for different types of applications.

Revision exercise no. 1

A long irrigation channel is controlled by a downstream gate. During a gate operation, the flow velocity, immediately upstream of the gate, increases linearly from 0 to 1 m/s in 2 min. Initially the flow is at rest and the water depth is 2 m. Neglecting bed slope and friction slope, calculate the water depths at the gate, at mid-distance and at the canal intake at $t = 6$ min. Compare the simple wave solution with the Hartree method.

Solution

The simple wave solution was developed in Chapter 12. Results at $t = 6$ min are presented in Fig. R.3, where the coordinate system is $x = 0$ at the downstream end and x is positive in the upstream direction.

The simple wave solution is an exact, theoretical solution of the Saint-Venant equations. The Hartree method is only a numerical approximation, and discrepancies between the exact solution and the numerical integration are seen.

Revision exercise no. 2

The Virdoule River flows in Southern France. It is known for its flash floods associated with extreme rainfall events, locally called 'épisode cévenol'. These are related to large masses of air reaching the Cévennes Mountain range. The Virdoule River is about 50 m wide. Initially the water depth is 0.5 m and the flow rate is 25 m³/s. During an extreme rainfall event, the discharge at Quissac increases from 25 to 1200 m³/s in 6 h. (Assume a linear increase of flow rate with time.) The flow rate remains at 1200 m³/s for an hour and then decreases down to 400 m³/s in the following 12 h:

(a) Compute the celerity of the monoclinal wave for the flash flood (i.e. increase in flow rate from 25 to 1200 m³/s). *Assume uniform equilibrium flows and $S_o = 0.001$.*

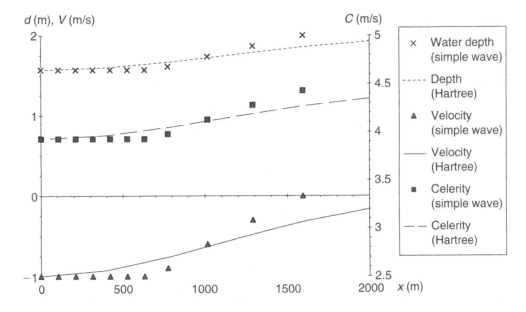

Fig. R.3 Free-surface profiles at $t = 6$ min. Comparison between simple wave solution and Hartree numerical integration ($\delta x = 400$ m, $\delta t = 20$ s).

(b) Approximate the flash flood with a dam break wave such that the flow rate at the origin (i.e. Quissac) is 1200 m^3/s after dam break.

(c) Calculate the water depth in the village of Sommières, located 21 km downstream of Quissac, for the first 24 h. *Use a simple wave theory.*

(d) Compare the results between the three methods at Sommières.

Remarks

Such an extreme hydrological event took place between Sunday 8 September and Monday 9 September 2002 in Southern France. More than 37 people died. At Sommières, the water depth of the Virdoule River reached up to 7 m. Interestingly, the old house in the ancient town of Sommières had no ground floor because of known floods of the Virdoule River.

Solution

(a) For a monoclinal wave, uniform equilibrium flow calculations are assumed upstream and downstream of the flood wave. The celerity of the monoclinal wave is:

$$U = \frac{Q_2 - Q_1}{A_2 - A_1}$$

Assuming an identical Darcy friction factor for both initial and new flow conditions, the new flow conditions yield $d_2 = 7.1$ m at uniform equilibrium and $U = 3.54$ m/s. The monoclinal wave will reach Sommières 1.6 h after passing Quissac.

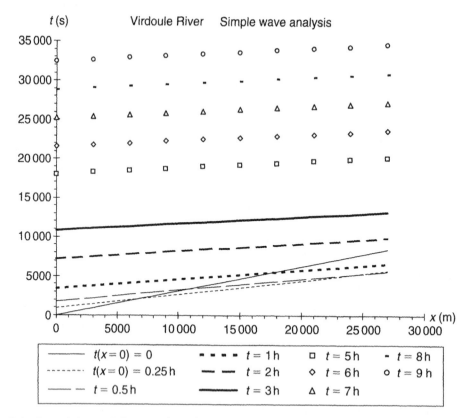

Fig. R.4   Forward characteristics curves for Virdoule River flash flood.

(b) For a dam break wave, the flow rate at the dam site ($x = 0$) is $1200 \, \text{m}^3/\text{s}$:

$$Q(x = 0) = \frac{8}{27} d_o \sqrt{g d_o} \, W$$

It yields $d_o = 8.74 \, \text{m}$. Note that $d_1/d_o = 0.057$ and it is reasonable to assume critical flow conditions at $x = 0$. Downstream of the dam, the initial flow conditions are $d_1 = 0.5 \, \text{m}$ and $V_1 = +1 \, \text{m/s}$. Complete calculations show that $U = 10.66 \, \text{m/s}$, $d_2 = 2.46 \, \text{m}$ and $V_2 - C_2 = 3.79 \, \text{m/s}$.

The positive surge leading to the dam break wave will reach Sommières 33 min after the dam break at Quissac and the wave front height will be 1.96 m. The water depth at Sommières will remain as $d = d_2$ until $t = 1.53 \, \text{h}$. Afterwards, the water depth will gradually increase. (Ultimately it would reach 3.9 m assuming an infinitely long reservoir.)

(c) For a simple wave analysis, the boundary conditions correspond to the formation of a positive surge between $t = 0$ and 6 h, and the formation of a negative surge between $t = 7$ and 19 h. The forward characteristics are plotted in Fig. R.4. A positive surge will form rapidly downstream of Quissac and the surge front will reach the township of Sommières at about $t = 1.3 \, \text{h}$.

(d) Discussion: Between the three calculations, the last one (i.e. simple wave calculations) would be the most accurate. Practically, an important result is the arrival time of the flash flood at Sommières. If the flood is detected at Quissac, the travel time between Quissac and Sommières is basically the warning time. The simple calculation predicts 1.3 h. The monoclinal wave calculations predict 1.6 h and the dam break wave theory gives 33 min. Interestingly all the results are of the same order of magnitude.

Revision exercise no. 3

A senior coastal engineer wants to study sediment motion in the swash zone. For 0.5 m high breaking waves, the resulting swash is somehow similar to a dam break wave running over retreating waters. (1) Assuming an initial reservoir water depth of 0.5 m, an initial water depth $d_1 = 0.07$ m and an initial flow velocity $V_1 = -0.4$ m/s, calculate the surge front celerity and height. *Assume a simple wave (S$_o$ = S$_f$ = 0).* (2) Calculate the bed shear stress immediately behind the surge front. The beach is made of fine sand ($d_{50} = 0.3$ mm, $d_{90} = 0.8$ mm). *Assume k$_s$ = 2d$_{90}$ (Chanson 1999, Table 12.2). For sea water, $\rho = 1024$ kg/m^3 and $\mu = 1.22 \times 10^{-3}$ Pa s.* (3) Predict the occurrence of bed load motion and sediment suspension. (4) During a storm event, breaking waves near the shore may be 3–5 m high. For a 2 m high breaking wave, calculate the surge front height and bed shear stress behind the surge front assuming an initial reservoir water depth of 2 m, an initial water depth $d_1 = 0.15$ m and an initial flow velocity $V_1 = -1$ m/s.

Solution
Select a positive x-direction toward the shore. The dam break wave ($d_o = 0.5$ m) propagates in a channel initially filled with water ($d_1 = 0.045$ m) with an opposing flow velocity ($V_1 = -0.4$ m/s). The x coordinate is zero ($x = 0$) at wave breaking (i.e. pseudo-dam site) and the time origin is taken at the start of wave breaking.

The characteristic system of equation, and the continuity and momentum principles at the wave front must be solved theoretically. The free-surface profile is horizontal between the leading edge of the positive surge (point E3) and the intersection with the C_1 forward characteristics issuing from the initial negative characteristics. The flow depth d_2 and velocity V_2 behind the surge front satisfy the continuity and momentum equations as well as the condition along the C_1 forward characteristics:

$$d_1(U - V_1) = d_2(U - V_2) \qquad \text{Continuity equation}$$

$$d_2(U - V_2)^2 - d_1(U - V_1)^2 = \frac{1}{2}gd_1^2 - \frac{1}{2}gd_2^2 \qquad \text{Momentum equation}$$

$$V_2 + 2\sqrt{gd_2} = 2\sqrt{gd_o} \qquad \text{Forward characteristics}$$

These three equations form a system of three non-linear equations with three unknowns V_2, d_2 and U. An iterative calculation shows that the surge front celerity is: $U/\sqrt{gd_1} = 3.2$ and $U = 2.12$ m/s. At the wave front, the continuity and momentum equations yield:

$$\frac{d_2}{d_1} = \frac{1}{2}\left(\sqrt{1 + 8\frac{(U - V_1)^2}{gd_1}} - 1\right) = 4.6 \qquad \text{Hence } d_2 = 0.21 \text{ m}$$

The momentum equation may be rewritten as:

$$\frac{V_2}{\sqrt{gd_2}} = 2\left(\sqrt{\frac{d_o}{d_2}} - 1\right) = 1.1 \qquad \text{Hence } V_2 = 1.58 \,\text{m/s}$$

Behind the surge front, the boundary shear stress equals:

$$\tau_o = \frac{f}{8}\rho V_2^2 = 7.7 \,\text{Pa}$$

The Shields parameter τ_* equals 1.55 which is almost one order of magnitude greater than the critical Shields parameter for bed load motion $(\tau_*)_c \sim 0.035$ (Graf 1971). For a 0.3 mm sand particle, the settling velocity is 0.034 m/s. The ratio V_*/w_o equals 2.5 implying sediment suspension (Chanson 1999, Chapter 9, Section 9.2).

For a 2 m high breaking wave during a storm event, the surge front height equals: $d_2 - d_1 = 0.69$ m. The boundary shear stress behind the surge front equals: $\tau_o = 21$ Pa.

Remarks

1. The above development has a number of limitations. The reservoir is assumed infinite although a breaking wave has a finite volume, assuming the beach slope to be frictionless and horizontal.
2. Note that the calculations of U, V_2 and d_2 are independent of time.

Revision exercise no. 4

The Qiantang River discharges into the Hangzhou Wan in East China Sea. Between Laoyancang and Jianshan the river is 4 km wide, the bed slope corresponds to a 5 m bed elevation drop over the 50 km reach. At the river mouth (Ganpu, located 30 km downstream of Jianshan), the tidal range at Jianshan is 7 m and the tidal period is 12 h 25 min. At low tide, the river flows at uniform equilibrium (2.4 m water depth). Discuss the propagation of the tidal bore.

Assume $S_o = S_f$, $f = 0.015$ and a wide rectangular prismatic channel between Laoyancang and Ganpu. (This is an approximation off course. Between Ganpu and Babao (upstream of Jianshan), the channel width contracts from 20 down to 4 km while the channel bed rises gently. The resulting funnel shape amplifies the tide in the estuary.) Sea water density and dynamic viscosity are respectively 1024 kg/m³ and 1.22×10^{-3} Pa s.

Solution

Uniform equilibrium flow calculations are conducted in the Qiantang River at low tide. It yields: normal velocity = 1.12 m/s (positive in the downstream direction), $C_o = 4.85$ m/s for $d_o = 2.4$ m, $f = 0.015$ and $S_o = 1 \times 10^{-4}$.

Select a coordinate system with $x = 0$ at the river mouth (i.e. at Ganpu) and x positive in the upstream direction. The initial conditions are $V_o = -1.12$ m/s and $C_o = 4.85$ m/s. The prescribed boundary condition at the river mouth (Ganpu) is the water depth given as:

$$d(x = 0,\, t_o) = 3.6 + \frac{7}{2}\left(1 + \cos\left(\frac{2\pi}{T}t - \pi\right)\right)$$

where $t_o = t(x = 0)$ and T is the tide period ($T = 44\,700$ s).

Fig. R.5   Photograph of the Hangzhou Tidal Bore (China), looking at the incoming bore. Close view from the left bank in 1997 (courtesy of Dr Eric Jones).

The initial forward characteristic trajectory is a straight line:

$$t = \frac{1}{V_o + C_o} x \qquad \text{Initial forward characteristic}$$

Preliminary calculations are conducted assuming a simple wave for $0 < x < 70$ km (i.e. Laoyancang). The results showed formation of the tidal bore at $x = 48$ km and $t = 12\,950$ s, corresponding to the intersection of the initial characteristic with the C1 characteristic issuing from $t_o \sim 5400$ s. (Simple wave calculations would show that the bore does not reach its final form until $x > 300$ km.)

Discussion

The Qiantang Bore, also called Hangchow or Hangzhou Bore, is one of the world's most powerful tidal bores with the Amazon River Bore (*pororoca*) (Fig. R.5). As the tide rises into the funnel-shaped Hangzhou Bay, the tidal bore develops and its effects may be felt more than 60 km upstream. Relevant references include Dai and Chaosheng (1987), Chen *et al.* (1990) and Chyan and Zhou (1993).

In the Hangzhou Bay, a large sand bar between Jianshan and Babao divides the tidal flow forming the East Bore and the South Bore propagating East–North–East and North–North–East respectively. At the end of the sand bar, the intersection of the tidal bores can be very spectacular: e.g. water splashing were seen to reach heights in excess of 10 m!

PART 4

Interactions between Flowing Water and its Surroundings

Air entrainment at Chinchilla Weir (Australia) in November 1997. Note self-aeration down the chute and the hydraulic jump in foreground right. The three-phase mixing (air, water and sediment) gave a beige appearance to the flow.

15

Interactions between flowing water and its surroundings: introduction

15.1 Presentation

Open channel hydraulics is possibly the most complicated field in fluid mechanics. First the basic principles form a complex system of non-linear equations (Chapters 2 and 11). For a given flow rate, there is an infinity of solutions depending upon the bed slope, boundary friction and channel cross-sectional shape (e.g. Henderson 1966, Chanson 1999a, 2004b). River flow rates may further range from extreme values: the hydraulics of droughts and floods are both important for the survival and development of a country. Second there are strong interactions between the flowing waters and the surrounding environment. This covers the transport of solids, the dispersion of chemicals, the mixing of air and water at the free surfaces. Comprehensive reviews include Graf (1971), Rutherford (1994) and Chanson (1997).

Natural channels have the ability to scour channel bed and banks, to carry sediment materials and to deposit sediment load. For example, during a flood event, the Yellow River eroded its bed by 7 m in <60 h at Longmen (Middle reach of Yellow River, July 1966, peak discharge: $7460 \, m^3/s$) (Wan and Wang 1994). Figure 15.1(a) shows a river in flood. The dark colour of the waters highlights significant sediment suspension. Figure 15.1(b) and (c) illustrates intense sediment transport leading to massive scour (Fig. 15.1(b)) and extreme sediment deposition (Fig. 15.1(c)). This phenomenon (i.e. sediment transport) is of great economical importance and numerous failures have resulted from the inability of engineers to predict sediment motion (e.g. Chanson and James 1998). Traditional fixed-boundary hydraulics cannot predict the morphology changes of natural streams because of the numerous interactions with the catchment, its hydrology and the sediment transport processes. It is now recognized that sediment motion is characterized by strong interactive processes between rainfall intensity and duration, water runoff, soil erosion resistance, topography of the stream and catchment, and stream discharge.

(a)

(b)

Fig. 15.1 Examples of sediment transport. (a) Sediment suspension in the Mur River (Graz, Austria) during a flood on 21 August 1999. Looking downstream, note the large eddies at the free surface and the dark colour of the waters caused by sediment suspension. (b) Rutherford Bridge (British Columbia, Canada) destroyed by a flood in October 2003 (courtesy of Acres International). Flow velocities in the vicinity of concrete bridge pier were about 7–8 m/s. The entire downstream river course was altered by some 20–30 m to the right.

(c)

Fig. 15.1 (c) Delta of the fully silted Saignon Reservoir (France) in June 1998. Looking downstream at the delta and fully silted reservoir with dam wall in far background. Dam height: 14.5 m (completed in 1961), reservoir capacity: 1.4×10^5 m^3, catchment area: 3.5 km^2. Fully silted reservoir after 2 years of operation.

In Nature, air–water flows are commonly encountered at waterfalls, in mountain torrents and at wave breaking (Fig. 15.2). They are also observed in aesthetical fountains and in hydraulic structures (e.g. Plumptre 1993, Chanson 1997) (Figs 1.2(d) and 15.2(c)). One of the first scientific accounts was made by Leonardo da Vinci (AD 1452–1519). In supercritical flows, air bubbles may be entrained when the turbulent kinetic energy is large enough to overcome both surface tension and gravity effects. The process is also called 'white waters'. Free-surface aeration is caused by a combination of wave instabilities and turbulence fluctuations acting next to the air–water free surface. Through this interface, there are continuous exchanges of both mass and momentum between water and atmosphere. Figure 15.2(a) and (b) illustrates strong air entrainment in natural streams, while Fig. 15.2(c) shows 'white waters' on a stepped spillway chute.

The air–water mixing is an important re-oxygenation process, because the air bubble entrainment increases drastically the air–water interface area, hence the air–water transfer rate (Chanson 1997). In spillway design, free-surface aeration may affect the thickness of flowing water and hence sidewall design. The presence of air bubbles in shear flows may reduce the shear stress between flow layers and induce some drag reduction. It may also prevent or reduce the damage caused by cavitation (Wood 1991, Chanson 1997).

Interactions between river waters and aquatic life are even more complex. Fish and aquatic species respond dynamically to river flows and changes in discharges and vegetation. Aquatic life must account for extreme flow events ranging from long drought periods (i.e. several years) to very large floods (e.g. during cyclones). Flooding and riverine migration are essential for the survival of many species and biodiversity of aquatic systems, but this is an interactive process. For example, floods have significant negative impacts on fish habitats. In-stream vegetation can be removed by scouring, hence heightening competition and increasing susceptibility to predation. In response to these pressures, the population may decline markedly. But fish microhabitats may reduce the impacts of floods: e.g. fish using in-stream vegetation to mitigate the negative impacts of excessive stream flows, for predator avoidance or reproduction. In many cases, migratory and reproductive behaviour can be triggered by

(a)

(b)

Fig. 15.2 Examples of air entrainment in open channel flows. (a) Nishizawa-keikoku River, Yamanishi prefecture, Japan in November 1998, looking upstream. Note the 'white waters' in the waterfall in foreground, and in the pool in the background. (b) White waters at Le Grand Remous (Québec, Canada) on 13 July 2002. View from the Pont Savoyard, looking downstream at the cascading waters and plunge point.

slight changes in water level, in flow velocity and sediment load (e.g. Dorava *et al.* 2001). While these natural processes are not yet fully understood, it is acknowledged that many hydraulic structures act as barriers to fish migration. Figure 15.3 shows a weir and a man-made fish pass next to the left bank. In that case, however, the writer noted that, on the downstream,

(c)

Fig. 15.2 (c) Air entrainment on the Opuha Dam stepped spillway (courtesy of Tonkin and Taylor). Dam height: 47 m (completed in 1999), crest length: 100 m, reservoir capacity: 85 Mm³.

Fig. 15.3 Interactions between hydraulic structures and aquatic life: small weir on the Joyou-gawa River, Japan on 1 December 2001, looking upstream. Note the fishway on the left bank.

the fish were attracted near the right bank, away from the fish pass, because of downstream flow concentrations caused by the differential settlement of concrete blocks. The example illustrates the difficult interactions between civil, hydraulic and ecological engineering (e.g. Yasuda *et al.* 2002, Nikora *et al.* 2003).

15.2 Terminology

Classical (clear-water) hydraulics is sometimes referred to as *fixed-boundary hydraulics*. It can be applied to most man-made channels (e.g. concrete-lined chute) and to some extent to grassed waterways. In natural streams, the channel boundaries are movable. *Movable boundary hydraulics* applies to streams with gravel or sand beds, estuaries (i.e. silt or sand beds), sandy coastlines and man-made canals in earth, sand or gravel. Movable boundary hydraulics is characterized by variable boundary roughness and variable channel dimensions. Strong interactive processes between the water flow and the bed form changes take place. *Accretion*, or *deposition*, refers to an increase of channel bed elevation resulting from the accumulation of sediment deposits. *Scour* or *erosion* is the removal of bed material caused by the eroding power of the flow. *Sediment transport* is the general term used for the transport of material (e.g. silt, sand, gravel, boulder) in rivers and streams. The transported material is called the *sediment load*. Distinction is made between the *bed load* and the *suspended load*. The bed load characterizes grains rolling along the bed while suspended load refers to grains maintained in suspension by turbulence. The distinction is however sometimes arbitrary when both loads are of the same material.

White water is the non-technical term used to design free-surface aerated flows. The refraction of light by the entrained air bubbles gives the 'whitish' appearance to the free surface of the flow. Natural aeration occurring at the free surface of high-velocity flows is referred to as *free-surface aeration* or *self-aeration*.

15.3 Structure of this section

In Part 4, sediment processes are discussed in Chapter 16. Free-surface aeration and 'white waters' are dealt with in Chapter 17. Both chapters present introductory material that must be complemented by expert references.

16

Interaction between flowing water and solid boundaries: sediment processes

Summary

This chapter deals with sediment motion in open channel flows primarily. After some basic definitions, both bed-load and suspension processes are discussed. Later complete calculations are developed.

16.1 Introduction

Numerous failures resulted from the inability of engineers to predict sediment motion: e.g. bridge collapse (pier foundation erosion), formation of sand bars in estuaries and navigable rivers, destruction of banks and levees (Fig. 15.1). In most cases, river and stream flows behave as quasi-steady gradually varied flows and the flow conditions are very close to uniform equilibrium flow conditions. Application of the momentum equation provides an expression of the mean flow velocity V at equilibrium:

$$V = \sqrt{\frac{8g}{f}} \sqrt{\frac{D_{\mathrm{H}}}{4} \sin\theta} \qquad \text{Uniform equilibrium flow} \qquad (16.1)$$

where f is the Darcy friction factor, θ is the bed slope and D_{H} is the hydraulic diameter.

The knowledge of the velocity profile, more specifically of the velocity next to the channel bed, is further required to predict accurately the occurrence of sediment motion. For an alluvial stream, the bed roughness effect might be substantial (e.g. in a gravel-bed stream) and the velocity profile is affected by the ratio of the sediment size to the inner wall region thickness: i.e. $d_{\mathrm{s}}/(10(\nu/V_*))$ where d_{s} is the sediment size, ν is the water kinematic viscosity and V_* is the shear velocity. If the sediment size is small compared to the sub-layer thickness (i.e. $V_*(d_{\mathrm{s}}/\nu) < 4$–5), the flow is *smooth turbulent*. If the sediment size is much larger than the

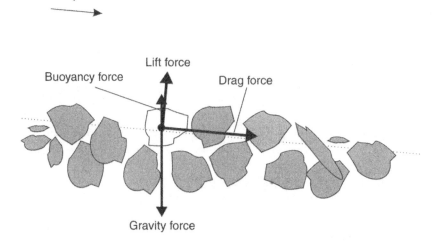

Fig. 16.1 Forces acting on a sediment particle; the inter-granular forces are not shown for clarity.

sub-layer thickness (i.e. $V_*(d_s/\nu) > 75$–100), the flow is called *fully rough turbulent*. In first approximation, the velocity distribution may be approximated by a power law function:

$$\frac{V}{V_{max}} = \left(\frac{y}{d}\right)^{1/N} \tag{16.2}$$

where V_{max} is the free-surface velocity, d is the water depth, y is the distance normal to the bed and N is a function of the boundary roughness: $N = K\sqrt{8/f}$ where f is the Darcy friction factor and K is a constant (K = 0.4).

> ### Notes
> 1. The hydraulic diamater is defined as: $D_H = 4(A/P_w)$ where A is the flow cross-sectional area and P_w is the wetted perimeter. D_H is also called the equivalent pipe diameter.
> 2. The constant K is called the von Karman constant after Theodore von Karman. It is a constant of proportionality between the Prandtl mixing length and the distance from the boundary. Experimental results give: K = 0.40.
> 3. Theodore von Karman (or von Kármán) (1881–1963) was a Hungarian fluid dynamicist and aerodynamicist who worked in Germany (1906–1929) and later in USA. He was a student of Ludwig Prandtl in Germany. He gave his name to the vortex shedding behind a cylinder (Karman vortex street).

Forces acting on a sediment particle

In open channel flows, the forces acting on each sediment particle are the gravity and buoyancy forces, the drag and lift forces, and the reaction forces of the surrounding grains (Fig. 16.1). The gravity force and the buoyancy force act both in the vertical direction. The drag force acts in the flow direction while the lift force in the direction perpendicular to the flow direction. The inter-granular forces are related to the grain disposition and packing.

16.2 Physical properties of sediments

16.2.1 Introduction

Distinction is made between two kinds of sediment: cohesive material (e.g. clay, silt) and non-cohesive material (e.g. sand, gravel). In this chapter, we will consider primarily the non-cohesive materials.

The density of quartz and clay minerals is typically: $\rho_s = 2650\,kg/m^3$. Most natural sediments have densities similar to that of quartz. The relative density of sediment particle equals: $s = \rho_s/\rho$ where ρ is the fluid density. For a quartz particle in water and air, $s = 2.65$ and 2200, respectively.

A key property of sediment particle is its characteristic size called diameter or sediment size d_s. Large particles are harder to move than small ones. Natural sediment particles are not spherical but exhibit irregular shapes, and there are several definitions for the sediment size: e.g. sieve diameter, sedimentation diameter and nominal diameter. A typical sediment classification is shown as follows:

Name	Size range (mm)
Clay	$d_s < 0.002$–0.004
Silt	0.002–$0.004 < d_s < 0.06$
Sand	$0.06 < d_s < 2.0$
Gravel	$2.0 < d_s < 64$
Cobble	$64 < d_s < 256$
Boulder	$256 < d_s$

Notes
1. The sieve diameter is the size of particle which passes through a square mesh sieve of given size but not through the next smallest size sieve.
2. The sedimentation diameter is the size of a quartz sphere which settles down (in the same fluid) with the same settling velocity as the real sediment particle.
3. The nominal diameter is the size of the sphere of same density and same mass as the actual particle.

Considering a single particle on a horizontal bed, the threshold condition for motion is achieved when the centre of gravity of the particle is vertically above the point of contact. The critical angle at which motion occurs is called the angle of repose ϕ_s. For sediment particles, the angle of repose ranges usually from 26° to 42° while it is typically between 26° and 34° for sands.

In a river bed, the density of wet sediment is: $(\rho_{sed})_{wet} = Po\rho + (1 - Po)\rho_s$ where ρ is the water density and Po is the porosity factor typically about 0.36–0.40.

Note
The density of sediments may be expressed also as a function of the void ratio: i.e. the ratio of volume of voids (or pores) to volume of solids. The void ratio is related to the porosity as:

Void ratio $= Po/(1 - Po)$

16.2.2 Particle fall velocity

In a fluid at rest, a suspended particle heavier than water has a downward motion. The *terminal fall velocity* of the particle is its velocity when the sum of the gravity force, buoyancy force and fluid drag force equals zero. For a spherical particle settling in a still fluid, the terminal fall velocity w_o equals:

$$w_o = -\sqrt{\frac{4gd_s}{3C_d}}\,(s-1) \tag{16.3}$$

where d_s is the particle diameter, C_d is the drag coefficient and $s = \rho_s/\rho$. The negative sign indicates a downward motion for $s > 1$. A re-analysis of numerous experimental data with spherical particles that were unaffected by sidewall effects yielded:

$$C_d = \frac{24}{Re}(1 + 0.150\,Re^{0.681}) + \frac{0.407}{1 + \dfrac{8710}{Re}} \qquad \text{Spherical particles } (Re < 2 \times 10^5)$$

where Re is the particle Reynolds numbers: $Re = w_o(d_s/\nu)$ and ν is the fluid kinematic viscosity (Brown and Lawler 2003).

For natural sand and gravel particles, experimental values of drag coefficient were best fitted by (Cheng 1997):

$$C_d = \left(\left(\frac{24}{Re}\right)^{2/3} + 1\right)^{3/2} \qquad \text{Natural sediment particles } (Re < 1 \times 10^4) \tag{16.4}$$

The terminal fall velocity of a sediment particle may be estimated by combining equations (16.3) and (16.4). Computed values for natural sediment particles were compared favourably with experimental data (e.g. Engelund and Hansen 1972), and typical values are reported below.

d_s (m)	Re	C_d	w_o (m/s)
0.0001	7.6×10^{-1}	36.2	0.008
0.0002	4.6×10^{0}	8.0	0.023
0.0005	3.3×10^{1}	2.4	0.067
0.001	1.2×10^{2}	1.6	0.117
0.002	3.7×10^{2}	1.3	0.186
0.005	1.6×10^{3}	1.1	0.314
0.01	4.5×10^{3}	1.0	0.454
0.02	1.3×10^{4}	1.0	0.650
0.05	5.1×10^{4}	1.0	1.034
0.1	1.5×10^{5}	1.0	1.466
0.2	4.1×10^{5}	1.0	2.075

Notes: $s = 2.65$; w_o: terminal fall velocity of single particle in water at 20°C.

Discussion

Large-size particles fall faster than small particles. At the limits, the relationship between fall velocity and particle size must satisfy:

$$w_o \propto d_s^2 \qquad \text{Laminar flow motion } (w_o(d_s/\nu) \ll 1)$$

$$w_o \propto \sqrt{d_s} \qquad \text{Turbulent flow motion } (w_o(d_s/\nu) > 1000)$$

However the settling velocity of a single particle may be affected by the presence of surrounding particles. Experiments showed that thick homogeneous suspensions have a slower fall velocity than that of a single particle. This effect, called *hindered settling*, results from the interaction between the downward fluid motion induced by each particle on the surrounding fluid and the return flow (i.e. upward fluid motion) following the passage of a particle. As a particle settles down, a volume of fluid equal to the particle volume is displaced upwards.

16.3. Threshold of sediment bed motion

16.3.1 Introduction

The term *threshold of sediment motion* describes the flow and boundary conditions for which the transport of sediment starts to occur. The threshold of sediment motion cannot be defined with an exact (absolute) precision but most experimental observations provide reasonably consistent results.

The inception of sediment motion is related to the boundary shear stress τ_o. Considering a given channel and bed material, no sediment motion is observed at very low bed shear stress until the shear stress τ_o exceeds a critical value $(\tau_o)_c$. For τ_o larger than the critical value, *bedload motion* takes place (e.g. Fig. 16.2). The grain motion along the bed is not smooth, and some particles bounce and jump over the others. With increasing shear velocities, the number of particles bouncing and rebounding increases until the cloud of particles becomes a *suspension*.

Notes

The average shear stress on the wetted surface or boundary shear stress equals:

$$\tau_o = \frac{f}{8} \rho V^2$$

where f is the Darcy friction factor and V is the mean flow velocity. The shear velocity V_* is defined as:

$$V_* = \sqrt{\frac{\tau_o}{\rho}}$$

where τ_o is the boundary shear stress and ρ is the density of the flowing fluid. The shear velocity is a measure of shear stress and velocity gradient near the boundary.

Fig. 16.2 Bed-load material in the Hayagawa Catchment (Japan) in November 1998; the Haya River is a tributary of the Fuji River.

16.3.2 Threshold of bed-load motion

Particle movement occurs when the moments of the destabilizing forces (i.e. drag, lift, buoyancy) with respect to the point of contact become larger than the stabilizing moment of the weight force. Experimental observations highlighted the importance of the Shields parameter τ_* which may be derived from dimensional analysis:

$$\tau_* = \frac{\tau_0}{\rho(\mathbf{s} - 1)\, g d_s} \tag{16.5}$$

A critical value of the Shields stability parameter may be defined at the inception of bed-load motion: i.e. $\tau_* = (\tau_*)_c$. Bed-load motion occurs for: $\tau_* > (\tau_*)_c$. Basically bed-load transport occurs when the boundary shear stress τ_0 is larger a critical value: $(\tau_0)_c = \rho(\mathbf{s} - 1) g d_s(\tau_*)_c$. Experimental observations showed that the critical Shields parameter $(\tau_*)_c$ is primarily a function of the shear Reynolds number $(d_s(V_*/\nu))$ (Fig. 16.3).

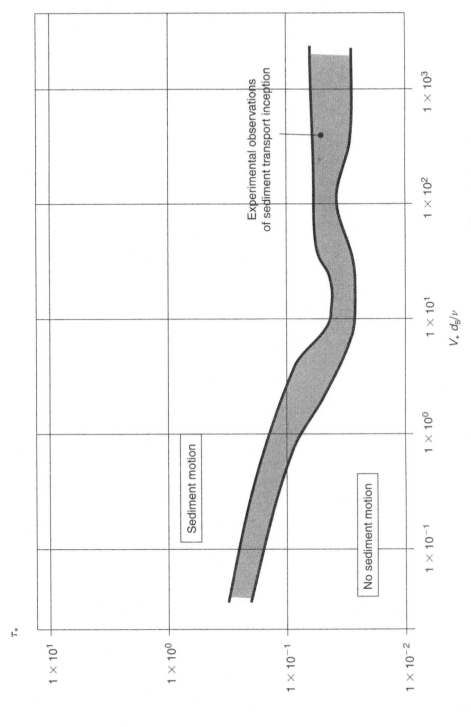

Fig. 16.3 Threshold of bed-load motion (Shields diagram), Shields parameter as a function of the particle Reynolds number for sediment in water.

Notes

1. The stability parameter τ_* is called commonly the *Shields parameter* after A. Shields who introduced it first. $(\tau_*)_c$ is commonly called the critical Shields parameter.
2. The stability parameter may be rewritten as:

$$\tau_* = \frac{V_*^2}{(s-1)gd_s} = \frac{\tau_o}{\rho(s-1)gd_s}$$

Discussion

For given fluid and sediment properties, and given boundary shear stress, the Shields parameter τ_* decreases with increasing sediment size: i.e. $\tau_* \propto 1/d_s$. For given flow conditions, sediment motion may occur for small particle sizes while no particle motion occurs for large grain sizes. The particle size distribution has an effect when the size range is wide. After an initial erosion of the fine particles, the coarser particles will form an armour layer preventing further erosion. This process is called *bed armouring*. Basically the fine particles become shielded by the larger particles (i.e. bed armour).

On steep channels, the bed slope assists in destabilizing the particles and bed motion occurs at lower bed shear stresses than in flat channels. At the limit, when the bed slope becomes larger than the repose angle, the grains roll even in absence of flow, i.e. the bed slope is unstable.

For clay and silty sediment beds, the cohesive forces between sediment particles may become important. This causes a substantial increase of the bed resistance to scouring.

16.3.3 Initiation of sediment suspension

Considering a particle in suspension, the particle motion in the direction normal to the bed is related to the balance between the particle fall velocity component ($w_o \cos \theta$) and the turbulent velocity fluctuation in the direction normal to the bed. The latter is of the same order of magnitude as the shear velocity V_*. In Fig. 16.4, some suspended sediment load is highlighted by the brownish colour of the flow.

Based upon the experimental observations, a simple criterion for the initiation of suspension is:

$$\frac{V_*}{w_o} > 0.2\text{--}2 \qquad \text{Sediment suspension} \qquad (16.6)$$

The flow conditions at onset of sediment suspension are summarized in Fig. 16.5. This modified Shields diagram presents the Shields parameter τ_* as a function of a dimensionless particle parameter $d_* = d_s \sqrt[3]{(s-1)g/\nu^2}$ (Fig. 16.5). The critical Shields parameter for initiation of bed-load motion is also plotted in solid line.

Fig. 16.4 Sediment suspension in Moggill Creek, Brisbane (Australia) at Rafting Ground Reserve on 20 June 2002, looking downstream at the muddy/brownish water, muddy bottom and banks at low tide.

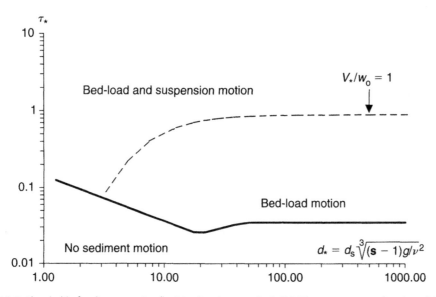

Fig. 16.5 Threshold of sediment motion (bed-load and suspension), Shields parameter as a function of the dimensionless particle parameter $d_* = d_s \sqrt[3]{(s-1)g/\nu^2}$

16.4 Sediment transport

16.4.1 Bed-load transport rate

When the bed shear stress exceeds a critical value, sediments are transported in the form of bed load and suspended load. For bed-load transport, the basic modes of particle motion are rolling motion, sliding motion and saltation motion.

Bed-load transport is closely associated with inter-granular forces. It takes place in a thin region of fluid close to the bed called *bed-load layer*. The bed-load transport rate per unit width may be defined as:

$$q_s = C_s \delta_s V_s \qquad \text{Bed-load transport} \qquad (16.7)$$

where C_s is the mean sediment concentration in the bed-load layer, δ_s is its thickness, and V_s is the average sediment velocity in the bed-load layer. A simple model yields (Nielsen 1992):

$$C_s = 0.65$$

$$\frac{\delta_s}{d_s} = 2.5 \, (\tau_* - (\tau_*)_c)$$

$$\frac{V_s}{V_*} = 4.8$$

Discussion

Several researchers proposed formulae to estimate the characteristics of the bed-load layer. Overall the results are not consistent and there is great uncertainty.

The prediction of bed-load transport rate is *not* an accurate prediction. One researcher stated explicitly that: "*the overall inaccuracy [...] may not be less than a factor 2*" (van Rijn 1984).

16.4.2 Suspension transport rate

Sediment suspension can be described as the motion of sediment particles during which the particles are surrounded by fluid. The grains are maintained within the mass of fluid by turbulent agitation without (frequent) bed contact. The amount of particles transported by suspension is called the *suspended load*.

In a stream with particles heavier than water, the sediment concentration is larger next to the bottom and turbulent mixing induces an upward migration of the grains to region of lower concentrations. A time-averaged balance between settling and diffusive flux derives from the continuity equation for sediment matter:

$$D_s \frac{dc_s}{dy} = -w_o \cos\theta \, c_s \qquad (16.8)$$

where c_s is the local sediment concentration at a distance y measured normal to the channel bed, D_s is the sediment diffusivity and w_o is the particle settling velocity. In natural (flowing) streams, the turbulence is generated by boundary friction: it is stronger close to the channel bed than near the free surface.

Assuming the sediment diffusivity to be nearly equal to the momentum exchange coefficient (i.e. 'eddy viscosity'), the sediment diffusivity D_s may be estimated as:

$$D_s \approx K V_* (d - y) \frac{y}{d}$$

where d is the flow depth, V_* is the shear velocity and K is the von Karman constant (K = 0.4). The integration of equation (16.8) gives the distribution of sediment concentration across the flow depth:

$$C_s = C_s \left(\frac{\frac{d}{y} - 1}{\frac{d}{\delta_s} - 1} \right)^{w_o \cos\theta / (KV_*)} \qquad \delta_s < y < d \qquad (16.9)$$

where C_s is the reference sediment concentration in the bed-load layer ($y < \delta_s$) (paragraph 16.4.1).

DISCUSSION

Equation (16.9) was first developed by Rouse (1937) and it was successfully verified with laboratory and field data. Suspension takes place above the bed-load layer. Hence a logical choice for the limiting conditions of the integration of equation (16.8) is the outer edge of the bed-load layer.

Although equation (16.9) was successfully compared with numerous data, its derivations implies a parabolic distribution of the sediment diffusivity. The re-analysis of model and field data (e.g. Anderson 1942, Coleman 1970) shows that the sediment diffusivity distribution is better estimated by a semi-parabolic profile:

$$D_s \approx KV_* (d - y) \frac{y}{d} \qquad y/d < 0.5$$

$$D_s \approx 0.1 V_* d \qquad y/d > 0.5$$

The suspended-load transport rate equals:

$$q_s = \int_{\delta_s}^{d} C_s v \, dy \qquad \text{Sediment load} \qquad (16.10)$$

where q_s is the volumetric suspended-load transport rate per unit width, c_s is the sediment concentration (equation (16.9)), v is the local velocity at a distance y measured normal to the channel bed (equation (16.2)), d is the flow depth and δ_s is the bed-load layer thickness.

Remarks

When both bed-load motion and suspension take place, the total sediment transport rate may be calculated using equations (16.7) and (16.10). The result is called the sediment transport capacity. It is not always equal to the observed sediment motion (see next section).

16.5 Total sediment transport rate

16.5.1 Presentation

The total sediment discharge is the total volume of sediment particles in motion per unit time. It includes the sediment transport by bed-load motion and by suspension. In Section 16.4, the *sediment transport capacity* of a known bed sediment mixture was estimated:

$$q_s = C_s \delta_s V_s + \int_{\delta_s}^{d} C_s v \, dy \qquad \text{Sediment transport capacity} \qquad (16.11)$$

This represents the maximum sediment transport rate. It does not take into account the sediment inflow nor erosion and accretion. Further equation (16.11) is valid for a given channel configuration with a flat movable bed. Although reasonable predictions might be obtained for straight prismatic channels with relatively wide cross-section formed with uniform bed material, these calculations are often not valid in natural streams because of the non-uniformity of the flow, channel bends and irregularities, formation of bars, presence of bed forms, but also any change of flow regime associated with change in bed slope.

DISCUSSION: BED FORMS IN RIVERS AND STREAMS

In natural streams, the sediments behave as a non-cohesive material and the river flow can distort the bed into various shapes. Bed forms result from the drag force exerted by the bed on the fluid flow as well as the sediment motion induced by the flow onto the sediment material. At low velocities, the bed does not move. With increasing flow velocities, the inception of bed movement is reached and the sediment bed begins to move. The basic bed forms which may be encountered are the ripples (usually of heights $> 0.1\,\text{m}$), dunes, flat bed, standing waves (for $0.5 \leqslant Fr \leqslant 1$ typically), and antidunes for $Fr > 1$. At high flow velocities (e.g. mountain streams, torrents), chutes and step-pools may form.

Note that ripples and dunes move in the downstream direction. Antidunes and step-pools are observed with supercritical flows and they migrate in the upstream flow direction. Typical bed forms are illustrated in Fig. 16.6.

(a)

Fig. 16.6 Photographs of bed forms. (a) Ripples at Cudgera Creek River mouth around on 15 June 2003 at low tide, note the small rock in the middle of the photograph.

Fig. 16.6 (b) Dune bed forms in a small stream, flow from right to left. (c) Standing wave flow with standing wave bed forms at Serizawa Beach (Japan) in March 1999, looking upstream.

(d)

Fig. 16.6 (*Contd*) (d) Supercritical flow with large antidune bed forms at Serizawa Beach (Japan) in a rip channel leading into breaking waves; Terasawa Beach (Japan) on 13 October 2001, looking downstream.

16.5.2 Flow resistance in natural systems

In alluvial streams the mean boundary shear stress τ_o may be expressed as:

$$\tau_o = \tau'_o + \tau''_o \tag{16.12}$$

where τ'_o is the skin friction shear stress and τ''_o is the form-related shear stress. The skin friction shear stress equals:

$$\tau'_o = \frac{f}{8}\rho V^2$$

where ρ is the fluid density, V is the mean flow velocity and f is the Darcy–Weisbach friction factor. The bed-form shear stress τ''_o is related to the fluid pressure distribution on the bed form and to the form loss. The form loss may be crudely analysed as a sudden expansion downstream of the bed-form crest. For a two-dimensional bed-form element, it yields:

$$\tau''_o = \frac{1}{2}\rho V^2 \frac{h^2}{ld}$$

where h and l are respectively the bed-form height and length (e.g. Chanson 1999a, 2004b).

Importantly the bed-load transport must be related to the effective shear stress (skin friction shear stress) only and not to the form roughness. In natural rivers, the Shields parameter and bed-load layer characteristics must be calculated using the skin friction bed shear stress. The onset of sediment motion and bed-load transport rate are predicted using the Shields parameter defined as:

$$\tau_* = \frac{\tau_o'}{\rho g \, (s-1) \, d_s}$$

For the suspended material, the sediment concentration and velocity distribution properties are related of the total bed shear stress τ_0. That is, the Rouse number ($w_o \cos \theta/(K V_*)$), the shear velocity V_* and the mean flow velocity V are calculated in terms of the mean boundary shear stress $\tau_0 = \tau_o' + \tau_o''$.

16.5.3 Design calculations

The interactions between sediment transport, bed-form and flow properties are extremely complicated. For complete calculations, Engelund and Hansen (1967) developed a simple design chart which regroups the relevant parameters: i.e. the Froude number $(Vd)/\sqrt{g(s-1)d_s^3}$, the dimensionless sediment transport rate $q_s/\sqrt{g(s-1)d_s^3}$, the bed slope $S_o = \sin \theta$, the dimensionless flow depth d/d_s, and the bed form (Fig. 16.7). This design chart was developed for fully rough turbulent flows and it was validated with experimental data.

Notes
1. Figure 16.7 takes into account the effect of bed forms and it may be used also to predict the type of bed form.
2. In the design chart, the sediment size is the median grain size $d_s = d_{50}$ (Fig. 16.7).
3. Calculations are not valid for ripple bed forms with $d_s V_*/\nu < 12$.

Application
The most important variables in designing alluvial channels are the water discharge Q, the sediment transport rate Q_s and the sediment size d_s. Calculations of flow properties and sediment transport are deduced by an iterative process. In a first stage, simplified design charts (e.g. Engelund and Hansen 1967, Fig. 16.7) may be used to 'guess' the type of bed form. In a second stage, complete calculations must be developed to predict the hydraulic flow conditions (V, d), the type of bed forms and the sediment transport capacity. Then the continuity equation for sediment material may be used to assess the rate of erosion (or accretion).

For a prismatic section of alluvial channel, complete calculations include several successive steps:

1. Determination of the channel characteristics (bed slope, cross-sectional shape, movable bed properties).
2. Selection of inflow conditions (discharge Q, sediment inflow).
3. Calculations of sediment-laden flow properties.

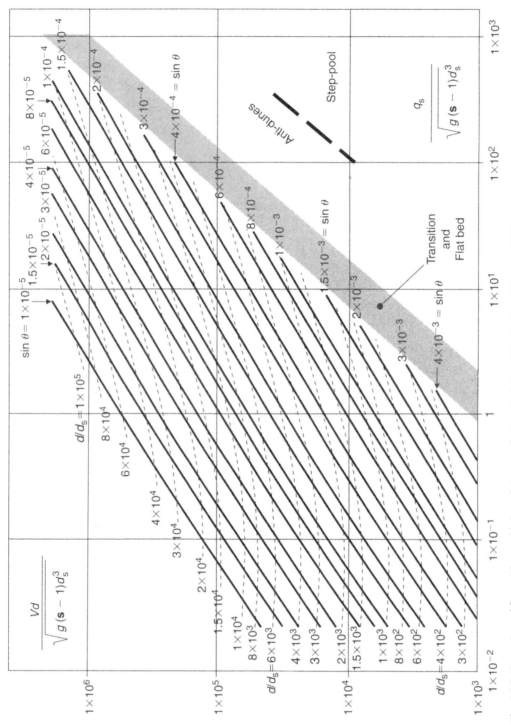

Fig. 16.7 Open channel flow with movable boundaries, pre-design calculations (Engelund and Hansen 1967).

3.1. Preliminary calculations assuming a flat bed, uniform equilibrium flow and $k_s = d_s$ yield some estimate of flow properties (V and d) and bed shear stress τ_o, and of the occurrence of sediment motion.
3.2. Pre-design calculations to assess type of bed form using the design chart of Engelund and Hansen (1967) (Fig. 16.7).
3.3. Flow calculation iterations until the type of bed form, bed resistance and flow conditions satisfy the continuity and momentum equations.
3.4. Sediment transport calculations include an estimate of the sediment transport capacity (equation (16.11)) and the application of the continuity equation for the sediment material to predict erosion or accretion.

16.6 Exercises

1. The dry density of a sand mixture is $1725\,kg/m^3$. Calculate the sand mixture porosity and the wet density of the mixture (*and* give units).
2. Considering a natural stream, the water discharge is $6.4\,m^2/s$ and the observed flow depth is 4.2 m. What is most likely type of bed form with a movable bed?
3. Considering a 1.2 mm sediment particle settling in water at 20°C, calculate the settling velocity.
4. Considering a stream with a flow depth of 3.2 m and a bed slope $\sin\theta = 0.0002$, the median grain size of the channel bed is 2.5 mm. Calculate the flow properties and predict the occurrence of sediment motion. (Assume that the equivalent roughness height of the channel bed equals the median grain size and that uniform equlibrium fow conditions are achieved.)
5. Considering a stream with a flow depth of 1.8 m and a bed slope $\sin\theta = 0.0015$, the median grain size of the channel bed is 0.85 mm. Predict the occurrence of bed-load motion and of suspension. (Assume that the equivalent roughness height of the channel bed equals the median grain size.)
6. A stream carries a discharge of $58\,m^3/s$. The channel is 33 m wide and the longitudinal bed slope is 9 m/km. The bed consists of a mixture of fine sands ($d_{50} = 1.1\,mm$). Assume that uniform equilibrium flow conditions are achieved. Calculate the flow properties, the occurrence of sediment motion and the sediment transport rate capacity.
7. A 20 m wide channel has a bed slope of 0.0011. The bed consists of a mixture of light particle ($\rho_s = 2350\,kg/m^3$) with a median particle size $d_{50} = 1.32\,mm$. The flow rate is $6.4\,m^3/s$. Calculate the bed-load transport rate at uniform equilibrium flow conditions. (Assume that the equivalent roughness height of the channel bed equals the median grain size.) Predict the suspended sediment transport rate.
8. Considering a 2000 m reach of an alluvial channel (55 m wide), the median grain size of the movable bed is 0.8 mm and the longitudinal bed slope is $\sin\theta = 0.00033$. The observed flow depth is 1.41 m and the mean sediment concentration of the inflow is 1.8%. Calculate the total sediment transport capacity and the rate of erosion (or accretion). (Assume uniform equilibrium flow conditions. Take into account the bed form and use the design chart of Engelund and Hansen (1967) to predict the type of bed form. Assume $k_s = 3d_{50}$.)

17

Interaction between flowing water and free surfaces: self-aeration and air entrainment

Summary

In this chapter, the mechanisms of free-surface aeration in turbulent flows are reviewed. Then dimensional analysis and similitude are developed including a discussion of scale effects affecting physical modelling. Measurements techniques are presented. Later simple applications are developed.

17.1 Introduction

Air–water flows have been studied recently compared to classical fluid mechanics. Although some researchers observed free-surface aeration and discussed possible effects (e.g. Leonardo da Vinci), the first successful experimental investigations were conducted during the mid-20th century: i.e. Ehrenberger (1926) in Austria and Straub and Anderson (1958) in North America. The latter data set is still widely used by engineers and researchers. Another milestone was the series of prototype experiments performed on the Aviemore Dam spillway in New Zealand under the supervision of I.R. Wood (Cain and Wood 1981). Laboratory and prototype investigations showed the complexity of the free-surface aeration process. Ian R. Wood further developed the basic principles of modern self-aerated flow calculations (e.g. Wood 1991). Recent developments were discussed in Falvey (1980), Wood (1991) and Chanson (1997a).

17.2 Free-surface aeration in turbulent flows: basic mechanisms

17.2.1 Presentation

Air entrainment, or free-surface aeration, is defined as the entrainment/entrapment of un-dissolved air bubbles and air pockets that are carried away within the flowing fluid (Fig. 17.1).

Fig. 17.1 Photographs of free-surface aeration in open channel flows. (a) Cascading waters at Craddle Mountain, Tasmania in July 2002 (courtesy of York-wee Tan and Jerry Lim). (b) Mount Crosby Weir overflow on 5 September 1999 – note free-surface aeration down the steep chute and at the plunge point at chute toe. (c) Jiroft Dam spillway operation (Iran) (courtesy of Amir Aghakoochak). Note the dark colour of water suggesting heavy sediment suspension.

The resulting *air–water mixture* consists of both air packets within water and water droplets surrounded by air. It also includes spray, foam and complex air–water structures. In turbulent flows, there are two basic types of air entrainment process. The entrainment of air packets can be localized or continuous along the air–water interface (Fig. 17.2). Examples of *local aeration* include air entrainment by plunging jet and at hydraulic jump. Air bubbles are entrained locally at the intersection of the impinging jet with the surrounding waters (Fig. 17.2(a)). The intersecting perimeter is a singularity in terms of both air entrainment and momentum exchange, and air is entrapped at the discontinuity between the high-velocity jet flow and the receiving pool of water. *Interfacial aeration* (or continuous aeration) is defined as the air entrainment process along an air–water interface, usually parallel to the flow direction: e.g. in chute flows (Figs 17.1 and 17.2(b)). An *intermediate case* is a high-velocity water jets discharging into air. The nozzle is a singularity, characterized by a high rate of aeration, followed by some interfacial aeration downstream at the jet free surfaces (Figs 17.1(c) and 17.2(c)).

17.2.2 Local/singular aeration mechanism: air entrapment at plunging jets

With local (singular) aeration, air entrainment results from some discontinuity at the impingement perimeter: e.g. plunging water jets, hydraulic jump flows (Fig. 17.3). One basic example is the vertical plunging jet (Figs 17.2(a) and 17.3). At plunge point, air may be entrapped when the impacting flow conditions exceed a critical threshold (McKeogh 1978, Ervine *et al.* 1980, Cummings and Chanson 1999). McKeogh (1978) showed first that the flow conditions at *inception of air entrainment* are functions of the jet turbulence level. For a given plunging jet configuration, the onset velocity increases with decreasing jet turbulence. For vertical water jets, the dimensionless onset velocity may be correlated by:

$$\frac{V_e \mu_w}{\sigma} = 0.0109(1 + 3.375 \exp(-80 Tu)) \tag{17.1}$$

where V_e is the onset velocity, μ_w is the liquid dynamic viscosity, σ is the surface tension and Tu is the ratio of the standard deviation of the jet velocity fluctuations about the mean to the jet impact velocity.

For jet impact velocities slightly larger than the onset velocity, air is entrained in the form of *individual bubbles and packets*. The entrained air may have the form of 'kidney-shaped' bubbles which may break-up into two 'daughter' bubbles, 'S-shape' packets, or elongated 'fingers' that may break-up to form several small bubbles by a tip-streaming mechanism, depending upon the initial size of the entrained air packet. The air entrainment rate is very small, hardly measurable with phase detection intrusive probes. At higher impact velocities, the amount of entrained air becomes significant and the air diffusion layer is clearly marked by the white plume generated by the entrained bubbles (Fig. 17.3(b) and (c)). Air entrainment is an unsteady rapidly varied process. An *air cavity is set into motion* between the impinging jet and the surrounding fluid and it is stretched by turbulent shear (Fig. 17.3(a)). The air cavity behaves as a ventilated air sheet and air pockets are entrained by discontinuous gusts at the lower tip of the elongated air cavity.

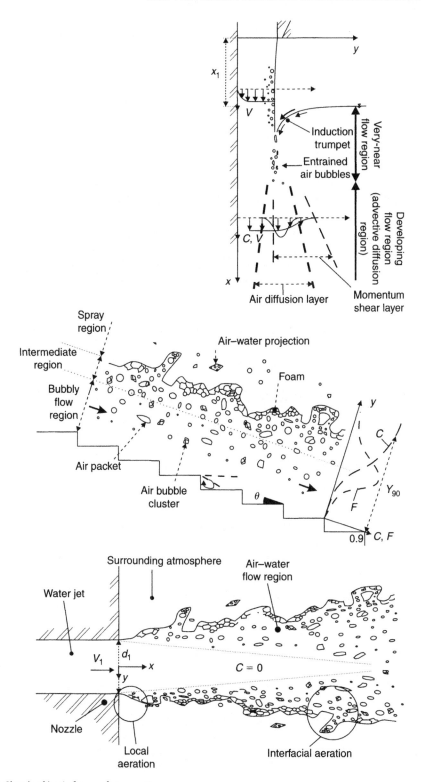

Fig. 17.2 Sketch of basic free-surface aeration processes.

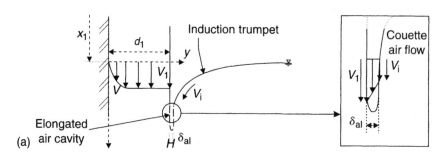

(b)

Fig. 17.3 Air entrainment at vertical plunging jet. (a) Detail of the air entrapment region and the very-near flow field. (b) Vertical supported (two-dimensional) plunging jet flow: supported jet thickness = 0.012 m, V_1 = 6.14 m/s, x_1 = 0.090 m, fresh water (high-shutter speed 1/1000 s). Note the jet support on the left and rising bubbles on the right – 'white waters' highlight the air–water shear layer and vortical structures.

Remarks

1. Equation (17.1) was obtained for air and water, and it was deduced from a number of large-size experimental data with both circular and two-dimensional jets (Cummings and Chanson 1999). Additional experimental studies confirmed the results including in sea water and salt water (e.g. Chanson *et al.* 2002a, Chanson and Manasseh 2003). For rough turbulent jets ($Tu > 2\%$), the onset velocity is about 1 m/s.

2. Initial aeration of the impinging jet free surface may further enhance the process. Van de Sande and Smith (1973) and Brattberg and Chanson (1998) discussed specifically this topic.

(c)

Fig. 17.3 (*Contd*) (c) Underwater photograph of the bubbly flow region below impingement of a vertical circular plunging jet – $d_1 = 0.012\,\mathrm{m}$, $V_1 = 2.5\,\mathrm{m/s}$, $x_1 = 0.05\,\mathrm{m}$, sea water (high-shutter speed 1/1000 s).

In the very-near flow field (i.e. $(x - x_1)/d_1 < 5$), the flow is dominated by air entrapment and the interactions between gas and liquid entrainment (Fig. 17.3(a)). Dominant flow features include an *induction trumpet* generated by the liquid entrainment and the *elongated air cavity* at jet impingement (thickness δ_{al}). There is a distinct discontinuity between the imping-ing jet flow and the induction trumpet as sketched in Fig. 17.3(a) which shows an instantaneous 'snapshot' of the entrapment region. Air entrainment takes place predominantly in the elon-gated cavity by a Couette flow motion (Fig. 17.3(a), right). The velocity discontinuity across the elongated air cavity is a function of the jet impact velocity V_1 and the liquid entrainment velocity in the induction trumpet V_i: $V_i \propto (V_1 - V_e)^{0.15}$ (Chanson 2002a). For two-dimensional plunging jets, the air entrainment rate q_{air} may be estimated as:

$$q_{air} = \int_{d_1}^{d_1+\delta_{al}} V_{air}\, \mathrm{d}y \approx \frac{V_1 + V_i}{2}\,\delta_{al} \tag{17.2}$$

Downstream of the entrapment region (i.e. $(x - x_1)/d_1 > 5$), the distributions of void fractions exhibit smooth, derivative profiles which follow closely simple analytical solutions of the advective diffusion equation for air bubbles (Chanson 1997). For two-dimensional vertical jets, it yields:

$$C = \frac{1}{2} \frac{q_{air}}{q_w} \frac{1}{\sqrt{4\pi D^{\#} \dfrac{x - x_1}{d_1}}}$$

$$\times \left(\exp\left[-\frac{1}{4D^{\#}} \frac{\left(\dfrac{y}{d_1} - 1\right)^2}{\dfrac{x - x_1}{d_1}} \right] + \exp\left[-\frac{1}{4D^{\#}} \frac{\left(\dfrac{y}{d_1} + 1\right)^2}{\dfrac{x - x_1}{d_1}} \right] \right) \quad (17.3)$$

where x is the streamwise direction, y is the distance normal to the jet centreline, x_1 is the free-jet length, V_1 is the jet velocity at impingement, q_{air} is the air flow rate per unit width, q_w is the water discharge per unit width, $D^{\#}$ is a dimensionless air bubble diffusivity and d_1 is the jet thickness at impact. (Full details of the integration are given in Appendix A, Section 17.6.)

With circular plunging jets, the analytical solution of the advective diffusion equation becomes:

$$C = \frac{Q_{air}}{Q_w} \frac{1}{4D^{\#} \dfrac{x - x_1}{r_1}} \exp\left[-\frac{1}{4D^{\#}} \frac{\left(\dfrac{r}{r_1}\right)^2 + 1}{\dfrac{x - x_1}{r_1}} \right] I_o\left(\frac{1}{2D^{\#}} \frac{\dfrac{r}{r_1}}{\dfrac{x - x_1}{r_1}} \right) \quad (17.4)$$

where I_o is the modified Bessel function of the first kind of order zero (Appendix A, Section 17.6).

Remarks

1. The void fraction or air concentration is defined as the volume of air per unit volume of air and water. In clear water, $C = 0$ and $C = 1$ in air.
2. Cummings and Chanson (1997a) and Brattberg and Chanson (1998) presented successful comparisons between equation (17.3) and experimental data. Chanson and Manasseh (2003) and Chanson et al. (2002a) compared successfully equation (17.4) with experimental data.
3. Estimates of dimensionless diffusivity $D^{\#}$ and air flow rate Q_{air}/Q_w are discussed in Appendix A (Section 17.6).

Discussion

Plunging jet entrainment is a highly efficient mechanism for producing large gas–liquid interfacial areas. Applications include minerals-processing flotation cells, waste-water treatment, oxygenation of mammalian-cell bio-reactors and riverine re-oxygenation weirs. Another form of plunging jet is the plunging breaking waves.

17.2.3 Interfacial aeration process: self-aeration down a steep chute

Examples of interfacial aeration include spillway chute flows and 'white waters' down a mountain stream (Figs 17.1(a), (b), 17.2(b) and 17.4). On smooth and stepped (skimming flow) chutes, the upstream flow is non-aerated but free-surface instabilities are observed. However, the location of inception of free-surface aeration is clearly defined (Figs 17.1(b) and 17.4). Downstream the flow becomes rapidly aerated. Self-aeration may induce significant flow bulking, air–water mass transfer drag reduction while it may prevent cavitation damage.

Photographs at Aviemore Dam spillway showed that air is entrained by the action of a multitude of irregular vortices acting next to the free surface (Cain 1978). Basically air bubble

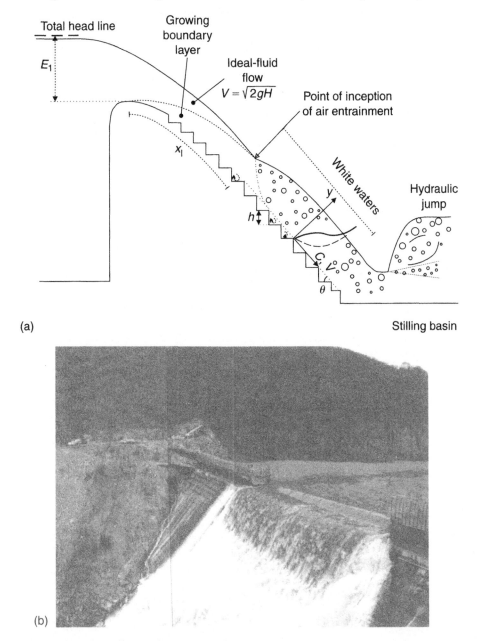

Fig. 17.4 Air entrainment in self-aerated open channel flows. (a) Definition sketch. (b) Skimming flow down Trigomil stepped spillway (Mexico) (courtesy of Drs Sanchez-Bribiesca and Gonzalez-Villareal) – $Q_w = 1017\,\text{m}^3/\text{s}$, chute width: $W = 75\,\text{m}$, step height: $h = 0.3\,\text{m}$, chute slope: $\theta = 51.34°$.

entrainment is caused by turbulence fluctuations acting next to the air–water free surface. Through the 'free surface', air is continuously trapped and released. Air bubbles may be entrained when the turbulent kinetic energy is large enough to overcome both surface tension and gravity effects. The turbulent velocity must be greater than the surface tension pressure and the bubble rise velocity component for the bubbles to be carried away:

$$v' > \text{maximum}\left(\sqrt{\frac{8\sigma}{\rho_w d_{ab}}}; \ u_r \cos\theta\right) \tag{17.5}$$

where v' is an instantaneous turbulent velocity normal to the flow direction, σ is the surface tension, ρ_w is the water density, d_{ab} is the diameter of the entrained bubble, u_r is the bubble rise velocity and θ is the channel slope. Equation (17.5) predicts the occurrence of air bubble entrainment for $v' > 0.1$–0.3 m/s. The condition is nearly always achieved in prototype chute flows because of the strong turbulence generated by boundary friction. Interfacial aeration involves both the entrainment of air bubbles and the formation of water droplets. The air–water mixture flow consists of water surrounding air bubbles ($C < 30\%$), air surrounding water droplets ($C > 70\%$) and an intermediate flow structure for $0.3 < C < 0.7$ (Fig. 17.2(b)).

Notes

1. Relevant monographs on self-aeration in high-velocity chute flows include Falvey (1980), Wood (1991), Chanson (1997a, 2001b) while specialized publications comprise Wood (1983), Chanson (1994) and Falvey (1990).
2. Measurements of bubble rise velocity are discussed in Appendix A (Section 3.5) in Chapter 3. Typical results are reported in Table 17.1. For millimetric bubbles, the rise velocity of air bubbles in still water is about 0.2–0.4 m/s for $1 < d_{ab} < 30$ mm.
3. Assuming a rise velocity of 0.25 m/s, equation (17.5) suggests that self-aeration occurs for turbulent velocities normal to the free surface >0.1–0.3 m/s, and bubbles in the range 8–40 mm are the most likely to be entrained. For steep slopes the action of the buoyancy force is reduced and larger bubbles are expected to be carried away. Equation (17.5) is an extension the work of Ervine and Falvey (1987) by Chanson (1993).
4. Rein (1998) and Chanson (1999b) discussed specifically the characteristics of the spray region (i.e. $C > 95\%$).

Table 17.1 Rise velocity of individual air bubble in fresh water and sea water

d_{ab} (mm)	u_r	
	Fresh water	Sea water
(1)	(2)	(3)
0.1	0.0054	0.0045
0.5	0.135	0.112
1	0.402	0.405
2	0.297	0.299
5	0.238	0.239
10	0.258	0.259
20	0.331	0.332
30	0.380	0.380
50	0.508	0.508

Note: Based upon measured fluid properties reported by Chanson et al. (2002).

5. Note that waves and wavelets propagate downstream along the free surface of super-critical flows. A phase detection probe, fixed in space, will record a fluctuating signal corresponding to both air–water structures and wave passages, adding complexity of the interpretation of the signal (Toombes 2002).

Downstream of the inception point of free-surface aeration, air and water are fully mixed, forming a homogeneous two-phase flow (Chanson 1995a, 1997a). The advective diffusion of air bubbles may be described by simple analytical models (Appendix B, Section 17.7). In smooth-chute flows and in skimming flows over stepped chutes, the air concentration profiles have a S-shape that may be modelled by:

$$C = 1 - \tanh^2 \left(K' - \frac{\dfrac{y}{Y_{90}}}{2D_o} + \frac{\left(\dfrac{y}{Y_{90}} - \dfrac{1}{3} \right)^3}{3D_o} \right) \quad \text{Skimming and smooth-chute flows} \quad (17.6)$$

DISCUSSION

On stepped chutes, low flows behave as a succession of free-falling nappes impacting on each step (i.e. nappe flow regime). At large discharges, the waters skim over the pseudo-bottom formed by the step edges (i.e. skimming flow regime). For intermediate flow rates, the flow is highly chaotic and aerated (i.e. transition flow regime). In transition flows down a stepped chute, the distributions of void fraction follow closely:

$$C = K' \left(1 - \exp \left(-\lambda \frac{y}{Y_{90}} \right) \right) \quad \text{Transition flows} \quad (17.7)$$

where K' and λ are dimensionless function of the mean air content only (Appendix B, Section 17.7).

Although void fraction distribution in transition and skimming flows exhibit different shapes, equations (17.6) and (17.7) derive from the same basic equation assuming different diffusivity profiles (Appendix B, Section 17.7).

where y is distance measured normal to the pseudo-invert, Y_{90} is the characteristic distance where $C = 90\%$, K' is an integration constant and D_o is a function of the mean void fraction only (Appendix B, Section 17.7).

17.2.4 Interfacial aeration process: self-aeration at water jet interfaces

Turbulent water jets discharging into the atmosphere are often characterized by a substantial amount of free-surface aeration. Applications include water jets at bottom outlets to dissipate energy, jet flows downstream of a spillway ski jump, mixing devices in chemical plants and spray devices, water jets for fire-fighting, jet cutting (e.g. coal mining) and with Pelton turbines.

A related case is the ventilated cavity flow, observed downstream of blunt bodies, on the extrados of foils and turbine blades and on spillways (i.e. aeration devices).

Downstream of the jet nozzle, interfacial aeration takes place along the jet interfaces. For a two-dimensional developing jet flows, the analytical solution of the diffusion equation for air bubble is:

$$C = \frac{1}{2}\left(1 - \mathrm{erf}\left(\frac{\dfrac{d_1}{2} - y}{2\sqrt{\dfrac{D_t}{V_1}x}}\right)\right) \tag{17.8}$$

where d_1 and V_1 are the jet thickness and velocity at the nozzle, x is the streamwise direction and y is the distance normal to the jet centreline (Fig. 17.2, Appendix C, Section 17.8). For circular jets, the theoretical solution is significantly more complicated and it is presented in Appendix C (Section 17.8).

Remarks

1. Equation (17.8) was first developed by Chanson (1989). It was successfully compared with experimental data by Chanson (1988) and Brattberg *et al.* (1998).
2. Estimates of turbulent diffusivity are discussed in Appendix C (Section 17.8).

Discussion

There is some similarity between air entrainment in high-velocity water jets and in supercritical open channel flows. (The channel invert is somehow analogue to the jet centreline.) Both flow configurations are high-speed turbulent flows with interfacial aeration rather than local aeration. In water jets and open channel flows, the air bubbles are gradually diffused and dispersed within the mean flow, and the distributions of air concentrations have a similar S-shape although the characteristics of the shear flow are significantly different.

17.3 Dimensional analysis and similitude

17.3.1 Introduction

Analytical and numerical studies of air–water flows are particularly complex because of the large number of relevant equations. Experimental investigations are also difficult but recent advances in instrumentation brought new measuring systems enabling successful experiments (paragraph 17.4). Traditionally model studies are performed with geometrically similar models and the geometric scaling ratio L_r is defined as the ratio of prototype to model dimensions (Fig. 17.5). Laboratory studies of air–water flows require however the selection of an adequate similitude.

17.3.2 Applications

In a study of air–water flows, the relevant parameters needed for any dimensional analysis include the fluid properties and physical constants, the channel geometry and inflow conditions,

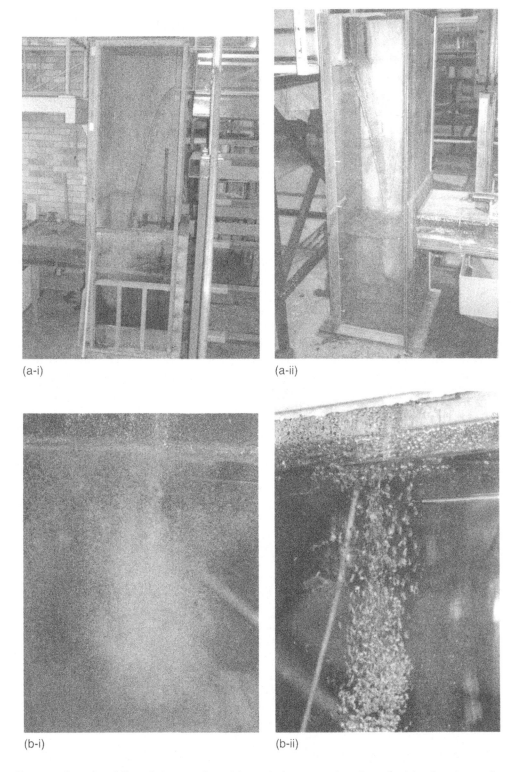

Fig. 17.5 Physical modelling of air–water flows. (a) Dropshaft operation (Froude similitude). Left: prototype dropshaft (3 m high, $We \sim 200$). Right: model dropshaft (1 m high, $We \sim 22$). (b) Air entrainment at vertical circular plunging jets (Froude similitude) – the probe on the right-hand side gives the scale for the entrained air bubbles. Left $d_1 = 12.5$ mm, $Fr_1 = 8.5$, $We_1 \sim 1500$. Right: $d_1 = 6.83$ mm, $Fr_1 = 8.5$, $We_1 \sim 370$.

the air–water flow properties including the entrained air bubble characteristics, and, possibly, the geometry of the air supply system in closed-conduit systems.

Air entrainment at vertical plunging jets

Considering the simple steady, vertical, circular plunging jet, a simplified dimensional analysis yields a relationship between the air–water flow properties beneath the free surface, the fluid properties and physical constants, flow geometry, and impingement flow properties:

$$
C, \frac{Fd_1}{V_1}, \frac{V}{\sqrt{gd_1}}, \frac{u'}{V_1}, \frac{d_{ab}}{d_1}, \dots
$$

$$
= F_1 \left(\frac{x - x_1}{d_1}; \frac{r}{d_1}; \frac{x_1}{d_1}; \frac{V_1}{\sqrt{gd_1}}; \frac{\rho_w V_1^2 d_1}{\sigma}; \frac{u_1'}{V_1}; \frac{g\mu_w^4}{\rho_w \sigma^3}; \text{salinity}; \dots \right) \qquad (17.9)
$$

where C is the void fraction, F is the bubble count rate, V is the velocity, g is the gravity acceleration, d_1 is the jet diameter at impact, u' is a characteristic turbulent velocity, V_1 is the jet impact velocity, d_{ab} is a characteristic size of entrained bubble, x is the coordinate in the flow direction measured from the nozzle, x_1 is the free-jet length, r is the radial coordinate, ρ_w and μ_w are the water density and dynamic viscosity respectively, σ is the surface tension between air and water, u_1' is a characteristic turbulent velocity at impingement (Fig. 17.2). In equation (17.9) right-hand side, the fourth and fifth terms are the inflow Froude and Weber numbers respectively while the seventh term is the Morton number. In addition, biochemical properties of the water solution may be considered.

Remarks
1. Any combination of these numbers is also dimensionless and may be used to replace one of the combinations. One parameter among the Froude, Reynolds and Weber numbers can be replaced by the Morton number $Mo = g\mu_w^4/\rho_w\sigma^3$, also called the liquid parameter, since:

$$
Mo = \frac{We^3}{Fr^2 Re^4}
$$

The Morton number is a function only of fluid properties and gravity constant. For the same fluids (air and water) in both model and prototype, Mo is a constant.
2. The bubble count rate F is defined as the number of bubbles detected by the probe sensor per second. For a given void fraction and velocity, the bubble count rate is proportional to the air–water interface area.

Air entrainment in steep chute flows

Considering a supercritical flow at uniform equilibrium flows (i.e. normal flow conditions) down a prismatic rectangular channel, a complete dimensional analysis yields:

$$
C, \frac{V}{\sqrt{gd}}, \frac{u'}{V}, \frac{d_{ab}}{d}, \dots = F_2 \left(\frac{y}{d}; \frac{q_w}{\sqrt{gd^3}}; \rho_w\left(\frac{q_w}{\mu_w}\right); \frac{g\mu_w^4}{\rho_w\sigma^3}; \frac{W}{d}; \theta; \frac{k_s}{d}; \dots \right) \qquad (17.10a)
$$

where d is the flow depth at uniform equilibrium, q_w is the water discharge per unit width, W is the channel width, θ is the invert slope, k_s is the equivalent roughness height, g is the gravity acceleration. For air–water flows, the equivalent clear-water depth is defined as:

$$d = \int_{y=0}^{y=Y_{90}} (1 - C)\, dy$$

where y is the distance normal to the invert, C is the local void fraction and Y_{90} is the depth where $C = 0.9$.

In equation (17.10a) right-hand side, the second, third and fourth dimensionless terms are Froude, Reynolds and Morton numbers respectively, and the last three terms characterize the chute geometry and the skin friction effects on the invert and sidewalls.

Further simplifications may be derived by considering the depth-averaged air–water flow properties. For a smooth-chute flow at uniform equilibrium, it yields:

$$F_4\left(\frac{U_w}{\sqrt{gd}};\ \rho_w \frac{U_w d}{\mu_w};\ \frac{g\mu_w^4}{\rho_w \sigma^3};\ C_{mean};\ \frac{W}{d};\ \theta;\ \frac{k_s}{d} \right) = 0 \qquad (17.10\text{b})$$

where U_w and d are respectively the mean flow velocity ($U_w = q_w/d$) and equivalent clear-water flow depth at uniform equilibrium flow conditions and C_{mean} is the depth-averaged void fraction defined as:

$$C_{mean} = \frac{1}{Y_{90}} \int_{y=0}^{y=Y_{90}} C\, dy$$

DISCUSSION

Considering normal flow conditions down a rectangular stepped channel (horizontal steps), a dominant flow feature is the momentum exchange between the free stream and the cavity flow within the steps (Chanson *et al.* 2002b). Dimensional analysis must include the step cavity characteristics and it yields:

$$C, \frac{V}{\sqrt{gd}}, \frac{u'}{V}, \frac{d_{ab}}{d}, \ldots = F_3\left(\frac{x}{d};\ \frac{y}{d};\ \frac{q_w}{\sqrt{gd^3}};\ \rho_w\left(\frac{q_w}{\mu_w}\right);\ \frac{g\mu_w^4}{\rho_w\sigma^3};\ \frac{d}{h};\ \frac{W}{h};\ \theta;\ \frac{k_s'}{h} \right)$$

where d is the flow depth at uniform equilibrium, x is the streamwise direction measured from the upstream step edge, h is the step height and k_s' the skin roughness height. In the above equation, the last four terms characterize the step cavity shape and the skin friction effects on the cavity walls.

By considering the depth-averaged air–water flow properties, the above equation becomes:

$$F_5\left(\frac{U_w}{\sqrt{gd}};\ \rho_w \frac{U_w d}{\mu_w};\ \frac{g\mu_w^4}{\rho_w\sigma^3};\ C_{mean};\ \frac{d}{h};\ \frac{W}{h};\ \theta;\ \frac{k_s'}{h} \right) = 0$$

where U_w is the mean flow velocity ($U_w = q_w/d$) and C_{mean} is the depth-averaged void fraction.

17.3.3 Dynamic similarity and scale effects

Despite their simplistic assumptions, equations (17.9) and (17.10a) demonstrate a large number of relevant dimensionless parameters and dynamic similarity of air bubble entrainment at plunging jets and in steep chutes might be impossible with geometrically similar models. In free-surface flows and wave motion, most laboratory studies are based upon a Froude similitude (e.g. Henderson 1966, Hughes 1993, Chanson 1999a) while the entrapment of air bubbles and the mechanisms of air bubble break-up and coalescence are dominated by surface tension effects implying the need for a Weber similitude. For geometrically similar models, it is impossible to satisfy simultaneously Froude and Weber similarities. In small size models, the air entrainment process may be affected by significant scale effects (Fig. 17.5). Wood (1991) and Chanson (1997a) presented comprehensive reviews. Kobus (1984) illustrated some applications.

Table 17.2 Summary of well-documented scale effects affecting air–water flow studies

Study (1)	Similitude (2)	Definition of scale effects (3)	Limiting conditions (4)	Experimental flow conditions (5)
Plunging jet flows Chanson et al. (2002a), Fig. 17.5(b)	Froude	Void fraction and bubble count rate distributions, bubble sizes	$We_1 > 1 \times 10^3$ $V_1/u_r > 10$	Vertical circular jets: $d_1 = 0.05$, 0.0125, 0.0068 m, $7 < Fr_1 < 10$ $L_r = 1, 2, 3.66$
Stepped chutes BaCaRa (1991)	Froude[a]	Flow resistance and energy dissipation	$L_r < 25$	Model studies: $\theta = 53.1°$, $h = 0.06, 0.028, 0.024, 0.014$ m $L_r = 10, 21.3, 25, 42.7$
Boes (2000)	Froude[a]	Void fraction and velocity distributions	$Re > 1 \times 10^5$	Model studies: $\theta = 30°$ and $50°$, $W = 0.5$ m, $h = 0.023–0.093$ m $L_r = 6.6, 13, 26$ (30°)/6.5, 20 (50°)
Chanson et al. (2002b)	Froude[a]	Flow resistance	$Re > 1 \times 10^5$ $h > 0.02$ m	Prototype and model studies $\theta = 5–50°$, $W = 0.2–15$ m, $h = 0.005–0.3$ m, $3 \times 10^4 < Re < 2 \times 10^8$, $10 < We < 6.5 \times 10^6$
Gonzalez and Chanson (2004)	Froude[a]	Void fraction, bubble count rate, velocity and turbulence level distributions, bubble sizes and clustering	$L_r < 2$	$\theta = 16°$, $W = 1$ m, $h = 0.10, 0.05$ m, $1.2 \times 10^5 < Re < 1.3 \times 10^6$ $L_r = 1, 2$
Other studies Spillway aeration device Pinto and Neidert (1982)	Froude[a]	Air demand of aerator	$L_r < 30$	Model studies: $\theta = 15°$, $W = 0.15$ m, $5 < Fr < 17.4$ $L_r = 8, 15, 30, 50$
Dropshaft Chanson (2002b), Fig. 17.5(a)	Froude	Bubble penetration depth and neutrally buoyant particle recirculation times	$L_r < 3.1$	Shaft dimensions: 0.76×0.76 m^2, 0.24×0.24 m^2, $\Delta z_0 = 1.7$ and 0.55 m, $2 \times 10^3 < Re < 6 \times 10^5$ $L_r = 1, 3.1$

Notes: h: step height; L_r: geometric scaling ratio; $Re = Vd/\nu_w$; W: channel width; $We = \rho_w V^2 d/\sigma$; Δz_0: drop in invert elevation; θ: chute slope; [a]: two-dimensional models.

A few studies investigated systematically air–water flows with geometrically similar models under controlled flow conditions (Table 17.2). These were based upon a Froude similitude with undistorted models and sometimes two-dimensional models. Results are summarized in Table 17.2, Column (4) indicating conditions to avoid scale effects. The outcomes demonstrated that scale effects may be significant. At the limit no scale effect is observed at full scale only ($L_r = 1$) using the same fluids in model and prototype. Basically dynamic similarity of air entrainment is impossible with geometrically similar models because of too many relevant parameters (e.g. equations (17.9) and (17.10a)).

Despite a limited number of systematic studies, results listed in Table 17.2 highlight the limitations of dynamic similarity and physical modelling of air–water flows. They show further that the selection of the criteria to assess scale affects is further critical.

Remarks
1. Scale effects are discrepancies between model and prototype resulting when one or more dimensionless parameters have different values in the model and prototype.
2. For a wide channel (e.g. spillway chute), the problem becomes a two-dimensional study. If the sidewall effects are assumed small, it is often convenient to use a two-dimensional model.
3. The above dimensional considerations were developed for simple flow conditions and geometries. In real hydraulic structures (e.g. dropshaft, Fig. 17.5(a)), the number of relevant parameters increases with the complexity of the system.

An example: scale effects on steep stepped chutes
For stepped chutes (Fig. 17.4(b)), several studies demonstrated that a Froude similitude with geometric similarity and same fluids in model and prototype does not describe the complexity of skimming flows. BaCaRa (1991) described a systematic laboratory investigation of the M'Bali Dam spillway with model scales of $L_r = 10, 21.3, 25$ and 42.7. For the smallest models ($L_r = 25$ and 42.7), the flow resistance was improperly reproduced. Chanson *et al.* (2002b) re-analysed more than 38 model studies and four prototype investigations with channel slopes ranging from $5.7°$ up to $55°$. They concluded that physical modelling of flow resistance may be conducted based upon a Froude similitude if laboratory flow conditions satisfy $h > 0.020\,\text{m}$ and $Re > 1 \times 10^5$. They added that true similarity of air entrainment was achieved only for model scales $L_r < 10$.

However detailed studies of local air–water flow properties yielded more stringent conditions (Table 17.2). One study showed that turbulence levels, entrained bubble sizes and interfacial areas were not properly scaled by a Froude similitude with $L_r = 2$.

Discussion: fresh water versus salt water
Comparative studies of bubble entrainment in fresh water and sea water are scarce (e.g. Scott 1975, Kolaini 1998, Slauenwhite and Johnson 1999). Some studies considered the size of bubbles produced by a frit, showing that bubble coalescence was drastically reduced in salt water compared to freshwater experiments ($L_r = 1$).

An experimental study in the developing flow region of plunging jets was conducted systematically with fresh water, sea water and salty fresh water (Chanson *et al.* 2002b). The results indicated lesser air entrainment in sea water than in fresh water, all inflow parameters being identical ($L_r = 1$). It was hypothesized that surfactants, biological and chemical elements harden the induction trumpet and diminish air entrapment at impingement in sea water. The results implied further that classical dimensional analysis is incomplete unless physical, chemical and biological properties other than density, viscosity and surface tension are taken into account.

Notes

1. Spillway aeration devices are designed to introduce air into high-velocity flows. Such aerators include basically a deflector and air is supplied beneath the deflected waters (Vischer *et al.* 1982, Chanson 1989, 1997a, Falvey 1990). Downstream of the aerator, the entrained air can reduce or prevent cavitation erosion.
2. Cavitation is the formation of vapour bubbles within a homogeneous liquid caused by excessive stress (Franc *et al.* 1995). Cavitation may occur in low-pressure regions where the liquid has been accelerated (e.g. marine propellers, baffle blocks, spillway chutes). Cavitation modifies the hydraulic characteristics of a system, and it is characterized by damaging erosion, additional noise, vibrations and energy dissipation.
3. A dropshaft is a vertical shaft connecting two channels at different invert elevations (Fig. 17.5(a)). The design is commonly used in sewers and stormwater systems. The Romans built a number of dropshafts along some aqueducts: e.g. Yzeron, Cherchell, Montjeu aqueducts.
4. Sea water is a complex mixture of 96.5% water, 2.5% salts and smaller amounts of other substances. The most abundant salts are sodium chloride ($NaCl$), sulphate (SO_4), Magnesium (Mg), Calcium (Ca) and Potassium (Riley and Skirrow 1965, Open University Course Team 1995).

17.4 Basic metrology in air–water flow studies

17.4.1 Introduction

In hydraulic engineering, most air–water flows are characterized by large amounts of entrained air. Void fractions are commonly >5–10%, and flows are of high-velocity with ratios of flow velocity to bubble rise velocity >10 or even 20. Classical measurement devices (e.g. Pitot tube, ADV, PIV, LDV) are affected by entrained bubbles and might lead to inaccurate readings. When the void fraction C exceeds 5–10%, and is <90–95%, the most robust instrumentation is the intrusive phase detection probes: optical fibre probe and conductivity/resistivity probe. Although the first designs were resistivity probes, both optical fibre and resistivity probe systems are commonly used today. The intrusive probe is designed to pierce bubbles and droplets (Fig. 17.6(a)). For example, the probe design shown in Fig. 17.6(a) has a small frontal area of the first tip to facilitate interface piercing.

The principle behind the optical probe is the change in optical index between the two phases. The principle behind the conductivity, or electrical probe, is the difference in electrical resistivity between air and water. The resistance of air is one thousand times larger than that of water and a needle resistivity probe gives accurate information on the local void fluctuations. A typical probe signal output is shown in Fig. 17.6(b). Although the signal is theoretically rectangular, the probe response is not exactly square because of the finite size of the tip, the wetting/drying time of the interface covering the tip and the response time of the probe and electronics.

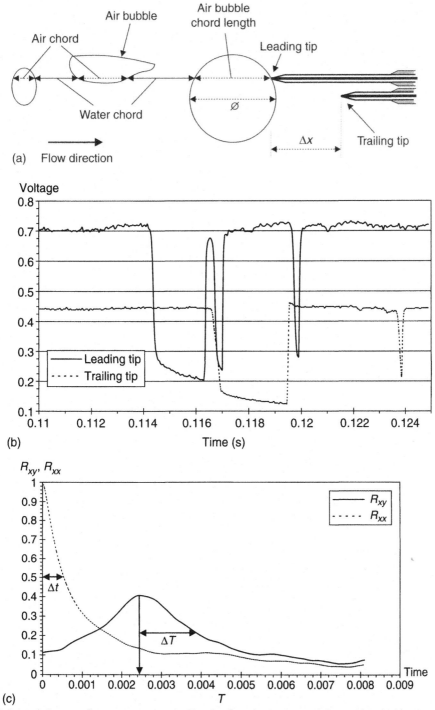

Fig. 17.6 Local air–water flow measurements in skimming flow down a stepped chute with a double-tip conductivity probe – $\theta = 16°$, $h = 0.10$ m, $q_w = 0.188$ m^2/s, $d_c/h = 1.5$, $Re = 7.5 \times 10^5$, step edge 8, scan rate: 40 kHz per sensor for 20 s, $\varnothing = 0.025$ mm, $\Delta x = 8$ mm. (a) Sketch of bubble impact on phase-detection probe tips (dual-tip probe design). (b) Voltage outputs from a double-tip conductivity probe ($y = 39$ mm, $C = 0.09$, $V = 3.05$ m/s, $F = 121$ Hz). (c) Normalized auto-correlation and cross-correlation functions ($y = 39$ mm, $C = 0.09$, $V = 3.05$ m/s, $F = 121$ Hz).

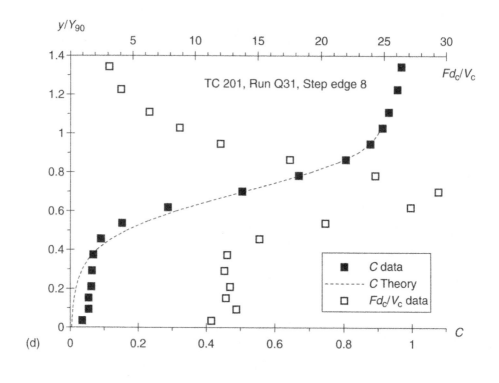

(d)

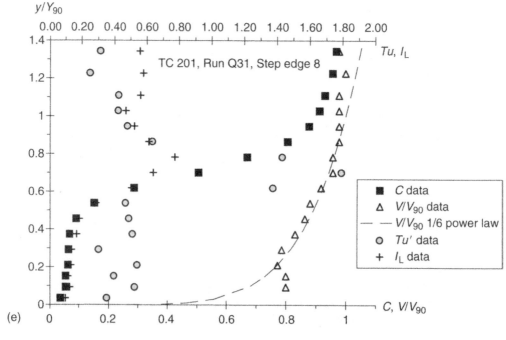

(e)

Fig. 17.6 (*Contd*) (d) Dimensionless distributions of void fraction C and bubble count rate Fd_c/V_c – comparison with equation (17.6). (e) Dimensionless distributions of air–water velocity V/V_{90}, turbulence intensity Tu and integral length scale I_L – comparison with a 1/6 power law.

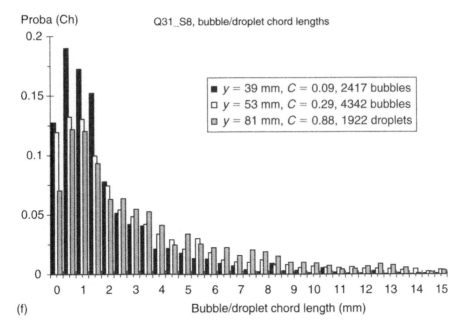

Fig. 17.6 (*Contd*) (f) Probability distribution functions of bubble and droplet chord sizes in 0.5 mm intervals. Local air–water flow properties: $y = 39$ mm, $C = 0.09$, $V = 3.0$ m/s, $F = 121$ Hz / $y = 53$ mm, $C = 0.29$, $V = 3.2$ m/s, $F = 217$ Hz / $y = 81$ mm, $C = 0.88$, $V = 3.5$ m/s, $F = 96$ Hz.

Notes
1. Basic references on phase-detection intrusive probes include Jones and Delhaye (1976), Bachalo (1994) and Chanson (1997a, 2002c).
2. Since the needle probe was developed by Prof. S.G. Bankoff (Neal and Bankoff 1963), key references on conductivity/resistivity probes regroup Herringe (1973), Serizawa *et al.* (1975) and Chanson (1995a, 1997a). For optical fibre probes, specialized references are Cartellier (1992) and Chang *et al.* (2003).
3. The probe design shown in Fig. 17.6(a) has a small frontal area of the first tip well suited to pierce small bubbles. The design further minimizes wake disturbance from the leading tip onto the displaced second tip (offset $< 0.2\Delta x$) (Chanson 1995a).

Remarks
While intrusive probe measurements give local flow properties including void fraction and bubble count rate, an acoustic technique may provide useful information on bubble size distribution, onset of bubble entrainment and entrainment regime. Bubbles generate sounds upon formation and deformation (Minnaert 1933, Leighton 1994) that are responsible for most of the noise created by an impinging water (e.g. plunging jet). Most underwater acoustic sensors are made from robust piezoelectric crystals and a key advantage is their robustness for use in the field and in hostile environments. Chanson and Manasseh (2003) presented comparative results in plunging jet flows using intrusive probes and hydrophones, and they outlined some limitations of the acoustic measurement technique in turbulent shear flows.

17.4.2 Signal processing and data analysis

The basic probe outputs are the void fraction, bubble count rate and bubble chord time distributions with both single-tip and double-tip probe designs. The void fraction C is the proportion of time that the probe tip is in the air. The bubble count rate F is the number of bubbles impacting the probe tip per second. The bubble chord times provide information on the air–water flow structure. For one-dimensional flows, chord sizes distributions may be further derived. A dual-tip probe design (Fig. 17.6(a)) provides additionally the air–water velocity, specific interface area, chord length size distributions and turbulence level. This technique assumes that the probe tips are aligned along a streamline and that bubbles and droplets are little affected by the leading tip.

With a dual-tip probe, the velocity measurement is based upon the successive detection of air–water interfaces by two sensors. In turbulent air–water flows, the successive detection of all bubbles by each tip is highly improbable and it is common to use a cross-correlation technique (e.g. Crowe *et al.* 1998). The time-averaged air–water velocity equals:

$$V = \frac{\Delta x}{T} \tag{17.11}$$

where Δx is the distance between tips and T is the time for which the cross-correlation function R_{xy} is maximum (Fig. 17.6(c)). The shape of the cross-correlation function provides further information on the velocity fluctuations (Chanson and Toombes 2002). The turbulent intensity may be derived from the broadening of the cross-correlation function compared to the auto-correlation function:

$$Tu = \frac{u'}{V} = 0.851 \left(\frac{\sqrt{\Delta T^2 - \Delta t^2}}{T} \right) \tag{17.12}$$

where ΔT as a time scale satisfying: $R_{xy}(T + \Delta T) = 0.5 \, R_{xy}(T)$, R_{xy} is the normalized cross-correlation function, and Δt is the characteristic time for which the normalized auto-correlation function R_{xx} equals 0.5 (Fig. 17.6(c)). The auto-correlation function R_{xx} provides some information on the air–water flow structure. A dimensionless integral length scale is:

$$I_{\mathrm{L}} = 0.851 \frac{\Delta t}{T} \tag{17.13}$$

Figure 17.6(d) and (e) presents some example of void fraction, bubble count rate, velocity and turbulence intensity distributions measured in a skimming flow down a stepped cascade. All data presented in Fig. 17.6 were recorded at the same cross-section. (Details are given in Chanson and Toombes 2002.)

A time-series analysis gives information on the frequency distribution of the signal which is related to the air and water (or water and air) length scale distribution of the flow. Chord sizes may be calculated from the raw probe signal outputs. The results provide a complete characterization of the streamwise distribution of air and water chords, including the existence of bubble/droplet clusters. Figure 17.6(f) presents probability distribution functions of bubble and droplet chord sizes in 0.5 mm intervals. Bubble chord sizes are indicated in white and black, while droplet chord sizes in grey. Data for chord sizes exceeding 15 mm were not shown for clarity.

The measurement of air–water interface area is a function of void fraction, velocity and bubble sizes. The specific air–water interface area a is defined as the air–water interface area per unit volume of air and water. For any bubble shape, bubble size distribution and chord length distribution, it may be derived from continuity:

$$a = \frac{4F}{V} \qquad (17.14)$$

where equation (17.14) is valid in bubbly flows ($C < 0.3$). In high air content regions, the flow structure is more complex and the specific interface area a becomes simply proportional to the number of air–water interfaces per unit length of flow ($a \propto 2F/V$).

With relative ease, intrusive phase-detection probes may provide detailed information on bubble count rate, specific interface area and bubble chord sizes. Such information is essential to gain a better understanding of air–water mass transfer in hydraulic engineering applications. It further assists comprehension of the interactions between turbulence and entrained air.

Remarks

1. A dual-tip probe provides only streamwise measurements. It does not give information on transverse flow properties.
2. The signal output analysis is basically identical for conductivity and optical fibre probes.
3. In turbulent shear flows, bubble size distributions are possibly best fitted by a log-normal distribution, although both Gamma and Weibull distributions provided also good fit. Basic experimental results include Cummings and Chanson (1997b) and Chanson *et al.* (2002a) in the developing flow region of vertical plunging jets, Chanson (1997b) in smooth-chute flows, Chanson and Toombes (2002) in skimming flow on stepped chute and Brattberg *et al.* (1998) in high-velocity jets.
4. The streamwise distribution of bubbles provides information on their spatial arrangement and the existence of bubble cluster. In a cluster, the bubbles are close together and the packet is surrounded by a sizeable volume of water. The existence of bubble clusters may be related to break-up, coalescence, bubble wake interference and to other processes. As the bubble response time is significantly smaller than the characteristic time of the flow, bubble trapping in large-scale turbulent structures may also be a clustering mechanism in bubbly flows. Preliminary studies include Chanson and Toombes (2002).

Discussion: Velocity measurements in air–water flows

Some studies suggested that interfacial velocities may be measured with a single-tip phase-detection probe based upon the voltage rise time associated with a water–air transition. This technique is, however, restricted to specific applications and probe designs. It was believed that the drying process on the probe sensor is strongly affected by the presence of water impurities and by sensor shape irregularities, yielding a wide scatter

Table 17.3 Comparison of air–water velocity measurement techniques

Feature	Double-tip probe design		Single-tip probe: Single event analysis	Remarks
	Single event analysis	Cross-correlation analysis		
(1)	(2)	(3)	(4)	(5)
Probe type	Double-tip resistivity/optical fibre probe	Double-tip resistivity/optical fibre probe	Single-tip optical fibre probe	
Scan rate	10 kHz	10 kHz	10 MHz	Typical values
Velocity measurement method	Individual bubble event analysis	Cross-correlation	Time-series analysis of individual bubble event	
Post-processing calculations	Complicated	Simple	Very complicated	
Remarks	Void fraction < 20%	Void fractions between 0 and 1	Chang et al. (2003)	

of the calibration data (e.g. Sene 1984, Cummings 1996). Recently, however, Chang et al. (2003) proposed a new processing technique deriving the air–water velocity from a time-series analysis of single-tip optical probe signals.

With phase-detection intrusive probes, velocity measurements are commonly conducted with double-tip probe systems. The signals may be analysed using two methods: the analysis of individual bubbles impacting successively both sensors, or a cross-correlation analysis. Table 17.3 summarizes the comparative advantages of each technique.

17.4.3 Unsteady flow measurements

Air–water flow measurements in unsteady flows are difficult, although prototype observations of sudden spillway releases and flash floods highlighted strong aeration of the leading edge of the wave associated with chaotic flow motion and energy dissipation. Figure 17.7(a) shows a laboratory experiment of dam break wave propagation down a stepped waterway. Note the 'white waters' at the surge leading edge.

In unsteady air–water flows, the measurement processing technique must be adapted (e.g. Stutz and Reboud 2000, Chanson 2003b). Basically local, instantaneous void fractions and bubble count rates must be calculated over a short time interval $\tau = \Delta X/V$ where V is the velocity and ΔX is a control volume streamwise length which must be selected to contain a minimum of 5–20 bubbles. For the experiment shown in Fig. 17.7(a), the control volume was set as: $\Delta X \geq 70$ mm and experimental results are shown in Fig. 17.7(b).

The void fraction becomes the proportion of time the probe sensor is in air during the time interval τ while the bubble count rate equals $F = N_{ab}/\tau$ where N_{ab} is the number of bubbles detected during the time interval τ. Bubble and water chords may be similarly recorded based upon the time spent by the bubble/droplet on the probe tip.

Velocity measurements may be performed using a dual-tip probe, but the processing technique must be based upon individual bubble/droplet events impacting successively the two

(a)

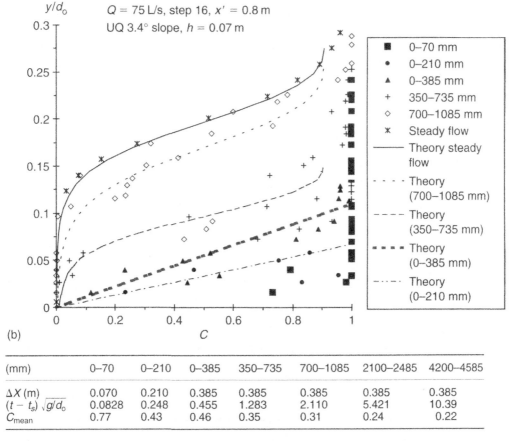

(b)

(mm)	0–70	0–210	0–385	350–735	700–1085	2100–2485	4200–4585
ΔX (m)	0.070	0.210	0.385	0.385	0.385	0.385	0.385
$(t - t_s) \sqrt{g/d_o}$	0.0828	0.248	0.455	1.283	2.110	5.421	10.39
C_{mean}	0.77	0.43	0.46	0.35	0.31	0.24	0.22

Fig. 17.7 (*Contd*)

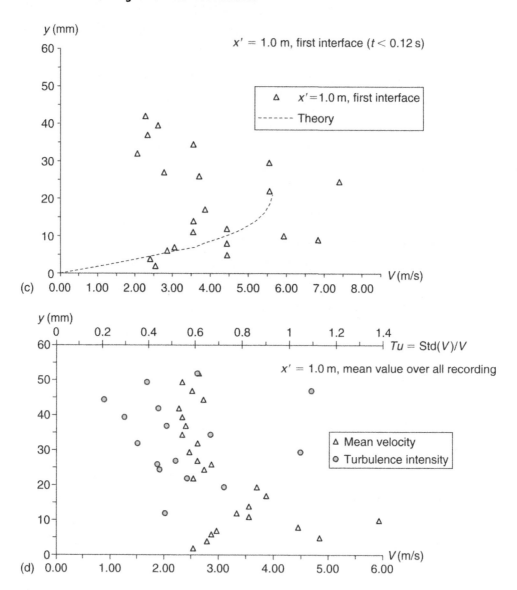

Fig. 17.7 (*Contd*) Unsteady air–water flow measurements in the leading edge of dam break wave. (a) Advancing flood waves down stepped chute – looking upstream at an advancing wave on step 16 with an array of conductivity probes in foreground – $Q(t = 0+) = 0.055\,\text{m}^3/\text{s}$, $d_o = 0.241\,\text{m}$, 3.4° slope, $h = 0.07\,\text{m}$ ($W = 0.5\,\text{m}$). (b) Dimensionless void fraction distributions in dam break wave front – Step 16, $Q(t = 0+) = 0.075\,\text{m}^3/\text{s}$, $x' = 0.8\,\text{m}$ – comparison with steady flow data and with equations (17.6) and (17.7) – refer table in Fig. b. (c) Interfacial velocity distribution of the first air-to-water interface ($t < 0.12\,\text{s}$) – comparison with the first Stokes problem solution – Step 16, $Q(t = 0+) = 0.065\,\text{m}^3/\text{s}$, $x' = 1\,\text{m}$. (d) Median interfacial velocity and average turbulence intensity (over about 5 s) in the dam break wave front – Step 16, $Q(t = 0+) = 0.065\,\text{m}^3/\text{s}$, $x' = 1\,\text{m}$.

probe sensors. The velocity is deduced from the time lag for air–water interface detections between leading and trailing tips respectively. For each meaningful event, the interfacial velocity is calculated as: $V = \Delta x/\delta t$ where Δx is the distance between probe sensors and δt is the interface travelling time between probe sensors.

An example: Air entrainment in dam break wave front

Typical unsteady void fraction distributions are presented in Fig. 17.7(b) for a dam break wave (Fig. 17.7(a)). In Fig. 17.7(b), the legend indicates the location and size of the control volume behind the leading edge of wave front: e.g. 350–735 mm means a 385 mm long control volume located between 350 and 735 mm behind the leading edge. Void fraction data are plotted as functions of the dimensionless distance y/d_o, where y is the distance normal to the invert and d_o is a measure of the surging flow rate $Q(t = 0+)$:

$$d_o = \frac{9}{4} \sqrt[3]{\frac{Q(t = 0+)^2}{gW^2}}$$

where W is the channel width. In Fig. 17.7(b), the data are compared with the corresponding steady flow data while the depth-averaged void fraction C_{mean} is given in the figure caption. In steady flow, the mean air content was $C_{mean} = 0.24$.

For the same type of experiment, typical air–water velocity measurements are shown in Fig. 17.7(c) and (d). In Fig. 17.7(c), each data point represents the velocity of the first air-to-water interface at each location y. Figure 17.7(d) presents the mean velocity and the ratio of interfacial velocity standard deviation to mean velocity for the first 5 s of the flow. (For large interface counts, the ratio is the turbulence intensity Tu.) At the leading edge of dam break wave, instantaneous velocity measurements suggest a boundary layer region with a potential region above (Fig. 17.7(c)). Boundary layer velocity data were compared successfully with an analytical solution of the Navier–Stokes equations (first Stokes problem):

$$\frac{V}{U} = \text{erf}\left(\frac{y}{2\sqrt{\nu_T t}}\right)$$

where U is the free-stream velocity, ν_T is the kinematic viscosity, t is the time from the first water detection by a reference probe and y is the distance normal to the invert. Erf is the Gaussian error function defined as:

$$\text{erf}(u) = \frac{2}{\sqrt{\pi}}\left(\int_0^u \exp(-z^2)\, dz\right)$$

Figure 17.7(d) highlight high levels of turbulence in the surging flow that are consistent with steady air–water flow measurements in stepped chutes (Chanson and Toombes 2002).

17.5 Applications

17.5.1 Application to plunging jet flows

General considerations

At a plunging jet, air entrainment takes place when the jet impact velocity V_1 exceeds a critical threshold, called onset or inception velocity. For rough turbulent jet flows, the inception velocity V_e is about 1 m/s for a vertical jet discharging into fresh water and sea water (paragraph 17.2.2). V_e tends to decrease slightly with decreasing angle θ between the jet flow direction and the pool free surface, but it should tend to the onset velocity of air entrainment at hydraulic jumps ($V_e \sim 1$ m/s) for $\theta = 0$.

For plunging jet velocities greater than the onset velocity, air entrainment takes place. The air entrainment rate is often expressed as:

$$\frac{Q_{air}}{Q_w} = F\left(\frac{V_1 - V_e}{\sqrt{gd_1}}\right) \qquad (17.15)$$

where Q_{air} is the entrained air discharge, Q_w is the plunging jet flow rate, and d_1 and V_1 are respectively the jet thickness/diameter and velocity at impact. Some results are detailed in Appendix A (Section 17.6). It is important, however, to note that most results were obtained in deep receiving pool with no or slow transverse flow motion in the receiving pool. Some studies suggested a decrease in air entrainment rate with decreasing pool depth.

Beneath the entrapment point, the bubbly flow consists of a downward plume surrounded by a swarm of rising bubbles. In the downward flow region, the 'diffusion cone' consists a developing flow region, a redistribution zone and a fully developed flow region (Fig. 17.8). In the developing air–water flow region, the air content is zero on the jet centreline and the velocity is the ideal-fluid flow velocity. The air diffusion layer and the momentum shear layer

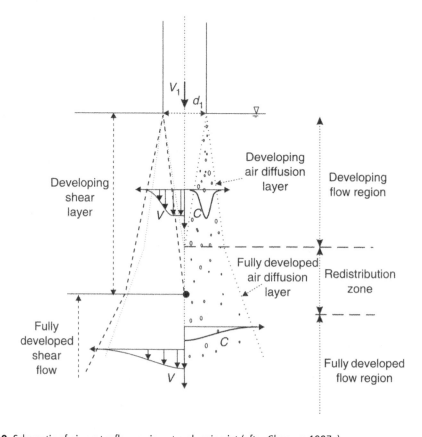

Fig. 17.8 Schematic of air–water flow region at a plunging jet (after Chanson 1997a).

are developing, and there is some momentum transfer from the jet core to the surrounding liquid. The entrained bubbles are advected in regions of high shear stresses and they are broken up into bubbles of smaller sizes. Downstream of the location where the developing air diffusion layers meet, the air content is rapidly redistributed from zero air concentration to maximum void fraction on the centreline. Experimental evidences showed that the developing shear layers and air diffusion layers do not intersect of the same location as sketched in Fig. 17.8. Further downstream, both void fraction and velocity distributions are fully developed. The air content and velocity are maximum on the jet centreline. The jet velocity decays with streamwise distance as fluid is entrained and momentum is exchanged with the pool.

In the surrounding swarm of rising bubbles, the flow motion is driven by buoyancy, although it is strongly affected by large-scale vortical structures. Several researchers observed that fine bubbles could be trapped in large eddies for a substantial time, while one study noted submerged air bubbles more than 5 min after stopping the plunging jet (Chanson *et al.* 2002a).

Remarks

1. Basic reviews on air entrainment at plunging jets include Bin (1993) and Chanson (1997a).
2. A theoretical value of the maximum bubble penetration depth D_p can be deduced from the continuity and momentum equations for diffusing jets:

$$\frac{D_p}{d_1} = 0.0240 \left(\frac{V_1}{u_r} \right)^2 \frac{(\sin\theta)^3}{(\tan\theta_3)^2} \left(1 + \sqrt{1 - 20.81 \left(\frac{u_r}{V_1} \right)^2 \frac{\tan\theta_3}{(\sin\theta)^2}} \right)^2 \quad \text{Two-dimensional jet}$$

$$\frac{D_p}{d_1} = 0.040 \left(\frac{V_1}{u_r} \right) \left(\frac{\sin\theta}{\tan\theta_3} \right)^2 \left(1 - 12.5 \left(\frac{u_r \tan\theta_3}{V_1 \sin\theta} \right) + \sqrt{1 - 25 \left(\frac{u_r \tan\theta_3}{V_1 \sin\theta} \right)} \right) \quad \text{Circular jet}$$

where D_p is the penetration depth measured normal to the pool free surface, d_1 and V_1 are the jet thickness and velocity at impact, u_r is the bubble rise velocity, θ is the jet impact angle with the receiving pool of water and θ_3 is the outer spread angle in the fully developed flow region. For circular jets, $\theta_3 \approx 14°$ on models and prototypes.

The above equations are based upon the method of Ervine and Falvey (1987) extended by Chanson and Cummings (1992).

Discussion

A related application is the analysis of experimental observations performed with phase-detection intrusive probes. Two examples are discussed below.

Case no. 1: Air–water flow measurements in circular plunging jet flow

Experiments were conducted with a circular jet discharging vertically into a seawater pool. The nozzle (12.5 mm diameter) was located 0.05 m above the pool free surface and the flow rate was 0.394 L/s. Void fraction measurements beneath the pool free surface are reported below. Estimate the dimensionless air flow rate and bubble diffusivity.

$x - x_1 = 10$ mm				$x - x_1 = 25$ mm			
y (cm)	C (%)	y (cm)	C (%)	y (cm)	C (%)	y (cm)	C (%)
34.62	1.55	36.6	11.785	34.5	1.48	35.46	0.48
34.65	1.61	36.63	9.17	34.53	1.2	36.16	0.77
34.68	2.47	36.66	6.255	34.56	1.47	36.19	1.83
34.71	2.23	36.69	3.775	34.59	2.02	36.22	3.78
34.74	2.58	36.72	4.34	34.62	2.325	36.25	1.705
34.77	3.98	36.75	2.74	34.65	2.655	36.28	2.395
34.8	5.84	36.78	1.49	34.68	2.625	36.31	2.605
34.83	7.16	36.81	1.345	34.71	3.145	36.34	3.445
34.86	8.47			34.74	3.32	36.37	4.01
34.89	9.72			34.77	4.425	36.4	5.395
34.92	11.82			34.8	6.87	36.43	7
34.95	18.00			34.83	6.605	36.46	9.835
34.98	19.22			34.86	6.62	36.49	11.235
35.01	21.36			34.89	6.99	36.52	14.04
35.04	26.23			34.92	9.58	36.55	16.635
35.07	23.17			34.95	10.685	36.58	16.86
35.1	25.36			34.98	11.515	36.61	16.82
35.13	19.77			35.01	12.805	36.64	15.81
35.16	12.33			35.04	13	36.67	14.46
35.19	4.27			35.07	12.25	36.7	10.725
35.22	1.61			35.1	10.84	36.73	13.01
35.25	0.87			35.13	10.295	36.76	8.71
36.3	1.59			35.16	8.615	36.79	8.51
36.33	3.59			35.19	7.66	36.82	5.965
36.36	6.38			35.22	6.025	36.85	2.865
36.39	13.43			35.25	4.23	36.88	2.865
36.42	18.84			35.28	2.715	36.91	2.04
36.45	21.20			35.31	2.175	36.94	1.45
36.48	23.63			35.34	1.98	36.97	1.145
36.51	20.56			35.37	1.45	37	0.645
36.54	17.15			35.4	0.925	37.03	0.51
36.57	12.67			35.43	0.77		

Note: Data by Chanson *et al.* (2002a), Run Sea_6.

First the inflow conditions must be calculated. The impingement velocity V_1 is calculated using the Bernoulli principle:

$$V_1 = \sqrt{V_n^2 + 2gx_1} \qquad \text{Bernoulli principle}$$

where x_1 is the free-jet length ($x_1 = 0.05$ m) and V_n is the nozzle velocity. The jet diameter at impact is deduced from the continuity equation: $Q_w = V_1(\pi/4)d_1^2$. It yields: $V_1 = 3.36$ m/s and $d_1 = 0.0122$ m.

Then the void fraction distributions may be compared with the analytical solution of the advective diffusion equation for bubbles (equation (17.4)). In practice, however, the data are best fitted by:

$$C = \frac{Q_{air}}{Q_w} \left(\frac{1}{4D^\# \dfrac{x - x_1}{r_o}} \right) \exp\left[-\frac{1}{4D^\#} \frac{\left(\dfrac{r}{r_o}\right)^2 + 1}{\dfrac{x - x_1}{r_o}} \right] I_o \left(\frac{1}{2D^\#} \frac{\dfrac{r}{r_o}}{\dfrac{x - x_1}{r_o}} \right) \qquad (17.16)$$

where r_o is an arbitrary location slightly greater than the jet radius at impingement (i.e. $r_o \geqslant d_1/2$).

Equation (17.16) is compared with the data in Fig. 17.9. The results illustrate nicely the advective diffusion of bubbles. The values of r_o, $D^{\#}$ and Q_{air}/Q_w were determined from the best fit of the data and are given below:

	$r_o/(d_1/2)$	$D^{\#}$	Q_{air}/Q_w
$x - x_1 = 10\,\text{mm}$	1.18	6.5×10^{-3}	0.193
$x - x_1 = 25\,\text{mm}$	1.26	6.3×10^{-3}	0.170

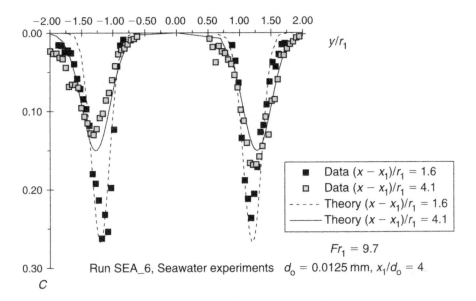

Fig. 17.9 Dimensionless void fraction distributions at a vertical plunging jet in sea water (*Data*: Chanson *et al.* 2002a) – comparison with equation (17.16).

Case no. 2: Air–water flow measurements in a horizontal hydraulic jump flow
Air–water flow measurements were conducted in a hydraulic jump in a horizontal rectangular channel ($W = 0.25\,\text{m}$) (Fig. 17.10). The inflow depth and velocity were 0.014 m and 3.47 m/s

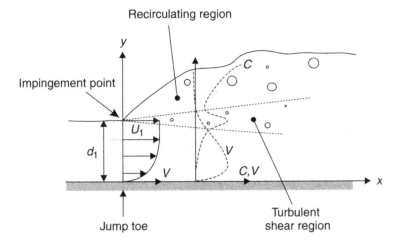

Fig. 17.10 Sketch of hydraulic jump flow.

respectively. Void fractions distributions at two locations x measured from the jump toe are reported below. Estimate the dimensionless air flow rate and bubble diffusivity.

$x = 0.20\,m$				$x = 0.4\,m$			
y (m)	C	y (m)	C	y (m)	C	y (m)	C
0.0035	0.047	0.0435	0.22925	0.0035	0.015315	0.1035	0.10762
0.0055	0.054635	0.0455	0.25071	0.0085	0.043355	0.1085	0.124535
0.0075	0.06776	0.0477	0.26922	0.0135	0.038625	0.114	0.12261
0.0095	0.084465	0.0495	0.19388	0.0185	0.06594	0.1185	0.163925
0.0115	0.11108	0.0515	0.173465	0.0235	0.069495	0.1235	0.329935
0.0135	0.12007	0.0535	0.214735	0.0285	0.075655	0.1285	0.40778
0.0155	0.144925	0.0555	0.21593	0.0335	0.094265	0.1335	0.447725
0.0175	0.164925	0.0575	0.19198	0.0385	0.093995	0.1385	0.66902
0.0195	0.16661	0.0595	0.28343	0.0435	0.128375	0.1435	0.808595
0.0215	0.17608	0.0615	0.240255	0.0485	0.12797	0.1485	0.78993
0.0235	0.200355	0.0635	0.30089	0.0535	0.131055	0.1535	0.9174
0.0255	0.218165	0.0685	0.34248	0.0585	0.122345		
0.0275	0.250595	0.0735	0.42479	0.0635	0.124945		
0.0295	0.244205	0.0785	0.54227	0.0685	0.124915		
0.0315	0.250625	0.0835	0.643515	0.0735	0.11539		
0.0335	0.26841	0.0885	0.7204	0.0785	0.14902		
0.0355	0.25474	0.0935	0.852945	0.0835	0.122625		
0.0375	0.270525	0.0985	0.8683	0.0885	0.11488		
0.0395	0.26519	0.1035	0.896872	0.0935	0.08806		
0.0415	0.2556			0.0985	0.11999		

Note: Data from Chanson and Brattberg (2000), Run T8_5.

A hydraulic jump is the limiting case of a two-dimensional supported jet. In the turbulent shear region (Fig. 17.10), the void fraction distributions may be compared with the analytical solution of the advective diffusion equation for bubbles, assuming that the channel invert acts as a symmetry line. Experimental observations showed however that void fraction data are best fitted by:

$$C = \frac{q_{air}}{q_w} \frac{1}{\sqrt{4\pi D^{\#} \frac{x}{d_o}}} \left(\exp\left(-\frac{1}{4D^{\#}} \frac{\left(\frac{y}{d_o} - 1\right)^2}{\frac{x}{d_o}} \right) + \exp\left(-\frac{1}{4D^{\#}} \frac{\left(\frac{y}{d_o} + 1\right)^2}{\frac{x}{d_o}} \right) \right) \quad (17.17)$$

where d_o is an arbitrary location slightly greater than the water depth at impingement (i.e. $d_o \geqslant d_1$), x and y are the longitudinal and vertical distances measured from the jump toe and bed respectively (Fig. 17.10).

Equation (17.17) is compared with the data in Fig. 17.11. In addition measured distributions of bubble count rate and velocity are reported on the same graph for completeness. The values of d_o, $D^{\#}$ and Q_{air}/Q_w were determined from the best fit of the data and are given below:

	d_o/d_1	$D^{\#}$	q_{air}/q_w
$x = 200\,mm$	2.54	0.04	0.73
$x = 400\,mm$	4.54	0.09	0.79

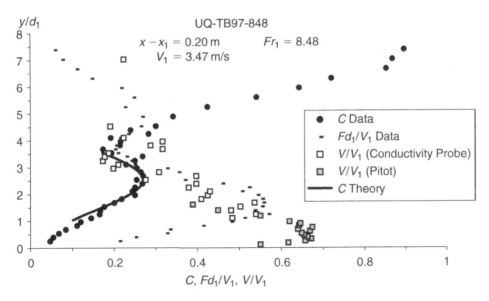

Fig. 17.11 Dimensionless distributions of void fraction, bubble count rate and velocity in a hydraulic jump at $x = 0.2\,m$ (data: Chanson and Brattberg 2000) – comparison with equation (17.17).

DISCUSSION

The above example corresponds to a hydraulic jump flow with partially developed inflow conditions. Resch and Leutheusser (1972) first showed the different air–water flow patterns between partially developed and fully developed hydraulic jumps. Recent investigation (Chanson and Qiao 1994, Chanson and Brattberg 2000) studied particularly the air–water flow properties of partially developed hydraulic jumps.

17.5.2 Application to steep chute flows

On an uncontrolled chute spillway, the flow is accelerated by the gravity force component in the flow direction (Fig. 17.4). For an ideal-fluid flow, the velocity can be deduced from the Bernoulli equation:

$$V_{\max} = \sqrt{2g\,(H_1 - z_o - d\cos\theta)} \qquad \text{Ideal-fluid flow} \qquad (17.18)$$

where H_1 is the upstream total head, z_o is the bed elevation, θ is the channel slope and d is the local flow depth. In practice, friction losses occur and the flow velocity on the chute is less than the ideal-fluid velocity, called the maximum flow velocity. At the upstream end of the chute, a bottom boundary layer is generated by bottom friction and develops in the flow direction (Fig. 17.4). When the outer edge of the boundary layer reaches the free surface, the flow becomes fully developed.

In the boundary layer, model and prototype data indicate that the velocity distribution follows closely a power law:

$$\frac{V}{V_{\max}} = \left(\frac{y}{\delta}\right)^{1/N_{bl}} \qquad 0 < y/\delta < 1 \qquad (17.19)$$

where δ is the boundary layer thickness defined as $\delta = y(V = 0.99 \times V_{\text{max}})$ and y is the distance normal to the channel bed. The velocity distribution exponent equals typically $N_{\text{bl}} = 6$ for smooth concrete chutes. For smooth inverts, the boundary growth may be estimated as:

$$\frac{\delta}{x} = 0.0212 \, (\sin\theta)^{0.11} \left(\frac{x}{k_s}\right)^{-0.10} \qquad \text{Smooth concrete chute } (\theta > 30°) \qquad (17.20)$$

where x is the distance from the crest measured along the chute invert, θ is the chute slope and k_s is the equivalent roughness height. Equation (17.20) is a semi-empirical formula which fits well model and prototype data (Wood et al. 1983, Chanson 1997). For stepped chutes with skimming flow, the turbulence generated by the steps enhances the boundary layer growth. The following formula can be used in first approximation:

$$\frac{\delta}{x} = 0.06106 \, (\sin\theta)^{0.133} \left(\frac{x}{h\cos\theta}\right)^{-0.17} \qquad \text{Stepped chute (skimming flow)} \qquad (17.21)$$

where h is the step height. Equation (17.21) was checked with model and prototype data (Chanson 2001b). It applies only to skimming flow on steep chutes (i.e. $\theta > 30°$).

Free-surface aeration takes place downstream of the intersection of the outer edge of the developing boundary layer with the free surface. Air entrainment is clearly identified by the 'white water' appearance of the free-surface flow (e.g. Fig. 17.4). For smooth inverts, the flow properties at the inception of free-surface aeration may be estimated as:

$$\frac{x_I}{k_s} = 13.6 \, (\sin\theta)^{0.0796} \, (F_*)^{0.713} \qquad \text{Smooth chute} \qquad (17.22)$$

$$\frac{d_I}{k_s} = \frac{0.223}{(\sin\theta)^{0.04}} \, (F_*)^{0.643} \qquad \text{Smooth chute} \qquad (17.23)$$

where x_I and d_I are the distance and flow depth respectively at the intersection of the boundary layer outer edge with the free surface, $F_* = q_w/\sqrt{g \sin\theta k_s^3}$, q_w is the discharge per unit width, g is the gravity constant, k_s is the roughness height and θ is the channel slope. For stepped channel the characteristics of the inception of free-surface aeration are best estimated as:

$$\frac{x_I}{h\cos\alpha} = 9.719 \, (\sin\theta)^{0.0796} \, (F_*)^{0.713} \qquad \text{Stepped chute} \qquad (17.24)$$

$$\frac{d_I}{h\cos\alpha} = \frac{0.4034}{(\sin\theta)^{0.04}} \, (F_*)^{0.592} \qquad \text{Stepped chute} \qquad (17.25)$$

where F_* is now the Froude number defined in terms of the step roughness height: $F_* = q_w/\sqrt{g \sin\theta (h\cos\theta)^3}$ and $h\cos\theta$ is the step roughness height.

In the fully developed flow region, the flow is gradually varied until it reaches equilibrium (i.e. normal flow conditions). Normal flow conditions may be calculated using the momentum principle (Chapter 2). Downstream of the inception point, both the acceleration and boundary layer development affect the flow properties and complete calculations of the flow properties can be tedious on a steep channel. In practice, however, the combination of the flow calculations in developing flow and in uniform equilibrium flow give a general trend which may be used for a preliminary design (Fig. 17.12). Figure 17.12 provides some information on the mean flow velocity V at the end of the chute as a function of the theoretical velocity V_{max} (equation (17.18)), the upstream total head above spillway toe H_1, the critical depth d_c, the invert slope θ and the Darcy friction factor f. In Fig. 17.12, the general trend is shown for smooth and stepped spillways (concrete chutes), with slopes ranging from 45° to 55° (i.e. 1V:1H to 1V:0.7H). Experimental results obtained on smooth-invert prototype spillways and stepped chutes are also shown.

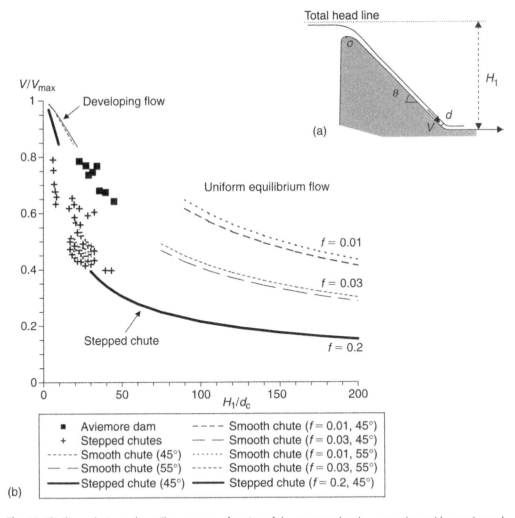

Fig. 17.12 Flow velocity at the spillway toe as a function of the upstream head – comparison with experimental data (smooth chute: Aviemore Dam spillway (Cain 1978); stepped chutes: Chamani and Rajaratnam (1999), Yasuda and Ohtsu (1999), Matos (2000), Chanson and Toombes (2002)). (a) Definition sketch. (b) Results.

Notes

1. Air entrainment in open channels is also called free-surface aeration, self-aeration, insufflation or white waters.
2. It must be stressed that equations (17.22)–(17.25) are rough correlations often presented on a log–log graph. Their accuracy is no better than ±30%.
3. For smooth concrete chutes, the Darcy friction factor is typically: $f \sim 0.01$–0.03. Air–water flow measurements at Aviemore Dam spillway yielded: $f \sim 0.022$.
4. In skimming flow on stepped chutes, the flow resistance is predominantly form drag and it is consistently larger than on smooth-invert channels. A detailed re-analysis of prototype and model observations gave $f \sim 0.2$ on stepped chutes (Chanson *et al.* 2002b).

In the gradually varied flow region downstream of the inception point, experimental data show a gradual increase in mean air content along smooth and stepped chutes (Wood 1985, Chanson 1993, 2001b). Assuming a slow variation of the rate of air entrainment, a gradual change in velocity with distance and a hydrostatic pressure distribution, the continuity equation for the air phase yields in a prismatic channel:

$$\frac{\mathrm{d}}{\mathrm{d}x'} C_{\mathrm{mean}} = \frac{u_r \, d_* \cos\theta}{q_w}(C_e - C_{\mathrm{mean}})(1 - C_{\mathrm{mean}})^2 \tag{17.26}$$

where d_* is the flow depth at the reference location ($x = x_*$), u_r is a bubble rise velocity and $x' = (x - x_*)/d_*$ and x is the curvilinear coordinate along the invert (Wood 1985, Chanson 1993). The limit of equation (17.26) is $C_{\mathrm{means}} = C_e$ in uniform equilibrium flows (see below). For a channel of constant width and channel slope, an analytical solution of equation (17.26) is:

$$\frac{1}{(1 - C_e)^2} \ln\left(\frac{1 - C_{\mathrm{mean}}}{C_e - C_{\mathrm{mean}}}\right) - \frac{1}{(1 - C_e)(1 - C_{\mathrm{mean}})} = k_o x' + K_o \tag{17.27}$$

where K_o and k_o are:

$$k_o = \frac{u_r d_* \cos\theta}{q_w}$$

$$K_o = \frac{1}{1 - C_e}\left(\frac{1}{1 - C_e} \ln\left(\frac{1 - C_*}{C_e - C_*}\right) - \frac{1}{1 - C_*}\right)$$

and C_* is the mean air concentration at the reference location ($x = x_*$).

Equation (17.26) allows the calculations of the average air concentration C_{mean} as a function of the distance along the chute independently of the velocity, roughness and flow depth. If the reference location is the inception point of free-surface aeration, $x_* = x_I$ and $d_* = d_I$. Calculations may be performed assuming $u_r = 0.4\,\mathrm{m/s}$ and $C_* = 0$ for smooth chutes,[1] and $u_r = 0.4\,\mathrm{m/s}$ and $C_* = 0.20$ for stepped chutes. On stepped chutes, $C_* = 0.20$ is used to account for the sudden flow aeration in the rapidly varied flow immediately downstream of the inception point (e.g. Chanson 2001b, pp. 152 and 172–174).

---

[1] Based upon Aviemore Dam spillway modelling (Wood 1985, Chanson 1993).

Remark: Uniform equilibrium air content on steep chutes

Far downstream, at uniform equilibrium, the depth-averaged void fraction C_{mean} tends to a constant C_e function of the channel slope θ only on smooth and stepped inverts. Uniform equilibrium air content data are shown in Table 17.4. The results are valid for both smooth and stepped chutes (Wood 1983, Chanson 1997a, 2001). For slopes $< 50°$, the equilibrium mean air concentration may be approximated by: $C_e = 0.9 \sin \theta$.

Table 17.4 Depth-averaged void fraction in uniform equilibrium self-aerated flows down steep chutes

Slope θ (degrees) (1)	$C_e{}^a$ (2)	Y_{90}/d^a (3)
0.0	0.0	1.0
7.5	0.16	1.19
15.0	0.24	1.318
22.5	0.31	1.45
30.0	0.41	1.70
37.5	0.57	2.32
45.0	0.62	2.65
60.0	0.69	3.12
75.0	0.72	3.58

Notes: [a]: Data from Straub and Anderson (1958); *d*: equivalent clear-water depth (paragraph 3.2); Y_{90}: characteristic depth where $C = 0.90$.

Notes

1. The above development (equation (17.26)) was derived by Wood (1985) for smooth-invert chutes and extended by Chanson (1993). It was successfully compared with prototype and model observations.
2. The result was applied successfully to stepped chutes by Chanson (2001).
3. Note that an alternative method relates the uniform equilibrium mean air content C_e to the friction slope S_f instead of the bed slope $S_o = \sin \theta$. The technique is more appropriate for flat chutes downstream of high-head gates, but it introduces some coupling between the air entrainment calculations (equation (17.26)) and the energy equation (i.e. backwater equation).

Drag reduction in self-aerated chute flows

Air–water flow measurements show smooth, continuous distributions of air concentrations and air–water velocity in smooth and stepped chutes. Although the presence of air bubbles does not affect the velocity profile, experimental observations demonstrate some drag reduction which increases with the mean air concentration C_{mean} (Fig. 17.13) (Wood 1983, Chanson 1994, 2004a, b).

On smooth-invert chutes, the presence of air reduces the shear stress between flow layers. An estimate of the drag reduction is:

$$\frac{f_e}{f} = 0.5\left(1 + \tanh\left(0.628 \frac{0.514 - C_{mean}}{C_{mean}(1 - C_{mean})}\right)\right) \tag{17.28}$$

where tanh is the hyperbolic tangent function, f_e is the air–water flow friction factor and f is the non-aerated flow friction factor. Equation (17.28) characterizes the reduction in skin friction

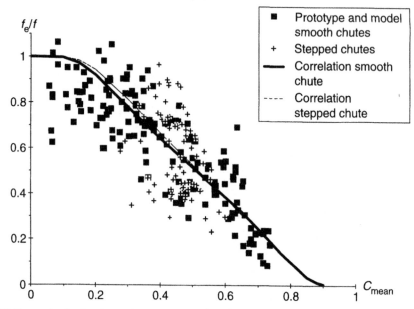

Fig. 17.13 Drag reduction by air entrainment on smooth and stepped chutes. Comparison between smoothchute data (after Chanson 1994), stepped chute data (after Chanson 2004a, b), and equations (17.28) and (17.29).

associated with air entrainment causing a thickening of the 'viscous' sublayer. It does satisfy basic boundary conditions: i.e. $f_e/f = 1$ in clear-water flow ($C_{mean} = 0$) and f_e/f negligible in air flow ($C_{mean} = 1$).

In skimming flows, separation occurs at each step edge and a shear layer develops with cavity recirculation beneath. It is believed that drag reduction results from interactions between entrained bubbles and developing mixing layer. The reduction in flow resistance may be correlated by:

$$\frac{f_e}{f} = 0.5\left(1 + \tanh\left(0.71\frac{0.52 - C_{mean}}{C_{mean}\,(1 - C_{mean})}\right)\right) \qquad (17.29)$$

Equation (17.29) is shown in Fig. 17.13 assuming $f = 0.24$. The trend is very close to equation (17.28) (Fig. 17.13) although the drag reduction mechanism is entirely different.

Remarks
1. In Fig. 17.12, calculations of the flow velocity must be based upon the air–water flow friction factor f_e.
2. All data shown in Fig. 17.12 derive from detailed air–water flow measurements showing consistently some drag reduction caused by free-surface aeration.

17.6 Appendix A – Air bubble diffusion in plunging jet flows (after Chanson 1997a)

Turbulent water jets intersecting the free surface of a pool of water are characterized by a substantial amount of air entrainment. In the bubbly flow region, the air bubble diffusion is a form of advective diffusion (Chapter 6). For a small control volume, the continuity equation for air in the air–water flow is:

$$\text{div}(C \times \vec{V}) = \text{div}(D_t \times \overrightarrow{\text{grad }} C) \tag{17A.1}$$

where C is the void fraction, $\vec{V}$ is the velocity vector, D_t is the air bubble turbulent diffusivity. Equation (17A.1) implies a constant air density (i.e. neglecting compressibility effects), it neglects buoyancy effects and is valid for a steady flow situation.

Two-dimensional plunging jets

Considering a two-dimensional free-falling jet, air bubbles are supplied by point sources located at $(x = x_1, y = +d_1/2)$ and $(x = x_1, y = -d_1/2)$ in the two-dimensional plane, where d_1 is the jet thickness at impact. Assuming an uniform velocity distribution, for a diffusion coefficient independent of the transverse location and for small control volume (dx, dy) limited between two streamlines, the continuity equation (17A.1) becomes a simple diffusion equation:

$$\frac{V_1}{D_t} \frac{\partial C}{\partial x} = \frac{\partial^2 C}{\partial y^2} \tag{17A.2}$$

where x is the streamwise direction, y is the distance normal to the jet centreline or jet support and V_1 is the jet velocity at impingement. The boundary conditions of the two-dimensional free-jet flow are: $C(x < x_1, y) = 0$ and two point sources of equal strength $0.5q_{air}$ located at $(x_1, +d_1/2)$ and $(x_1, -d_1/2)$.

The problem can be solved by superposing the contribution of each point source. The solution of the diffusion equation is:

$$C = \frac{1}{2} \frac{q_{air}}{q_w} \frac{1}{\sqrt{4\pi \, D^{\#} \dfrac{x - x_1}{d_1}}} \left(\exp\left(-\frac{1}{4D^{\#}} \frac{\left(\dfrac{y}{d_1} - 1 \right)^2}{\dfrac{x - x_1}{d_1}} \right) + \exp\left(-\frac{1}{4D^{\#}} \frac{\left(\dfrac{y}{d_1} + 1 \right)^2}{\dfrac{x - x_1}{d_1}} \right) \right)$$

Two-dimensional free-falling plunging jet   (17A.3)

where q_{air} is the volume air flow rate per unit width, q_w is the water discharge per unit width and $D^{\#}$ is a dimensionless diffusivity: $D^{\#} = D_t/(V_1 d_1)$.

Considering a two-dimensional supported jet, the air bubbles are supplied by a point source located at $(x = x_1, y = +d_1)$ in the two-dimensional plane and the strength of the source is q_{air}. The diffusion equation can be solved by applying the method of images and assuming an infinitesimally long support. It yields:

$$C = \frac{q_{air}}{q_w} \frac{1}{\sqrt{4\pi D^{\#} \dfrac{x - x_1}{d_1}}} \left(\exp\left(-\frac{1}{4D^{\#}} \frac{\left(\dfrac{y}{d_1} - 1 \right)^2}{\dfrac{x - x_1}{d_1}} \right) + \exp\left(-\frac{1}{4D^{\#}} \frac{\left(\dfrac{y}{d_1} + 1 \right)^2}{\dfrac{x - x_1}{d_1}} \right) \right)$$

Two-dimensional supported plunging jet   (17A.4)

Note that d_1 is the thickness of the supported jet at impact.

Note
The hydraulic jump is the limiting case of a supported plunging jet when a high veloc-ity horizontal flow impinges into a low-velocity flow and a roller forms. Equations (17A.3) and (17A.4) are valid in hydraulic jump flows with partially developed inflow conditions (Chanson and Brattberg 2000).

Circular plunging jets
Considering a circular plunging jet, assuming an uniform velocity distribution, for a constant dif-fusivity (in the radial direction) independent of the longitudinal location and for a small control volume delimited by streamlines, equation (17A.1) becomes an advective diffusion equation:

$$\frac{V_1}{D_t}\frac{\partial C}{\partial x} = \frac{1}{r}\frac{\partial}{\partial r}\left(r\frac{\partial C}{\partial r}\right) \tag{17A.5}$$

where x is the longitudinal direction, r is the radial direction and the diffusivity term D_t aver-ages the effects of the turbulent diffusion and of the longitudinal velocity gradient. The boundary conditions of the axi-symmetric problem are: $C(x < x_1, r) = 0$ and a circular source of total strength Q_{air} at $(x - x_1 = 0, r = r_1)$.

The problem can be solved analytically by applying a superposition method. The general solution of the air bubble diffusion equation is solved by superposing all the infinitesinal point sources:

$$C = \frac{Q_{air}}{Q_w}\frac{1}{4D^\#\frac{x-x_1}{r_1}}\exp\left(-\frac{1}{4D^\#}\frac{\left(\frac{r}{r_1}\right)^2+1}{\frac{x-x_1}{r_1}}\right)I_0\left(\frac{1}{2D^\#}\frac{\frac{r}{r_1}}{\frac{x-x_1}{r_1}}\right)$$

Circular plunging jet   (17A.6)

where Q_w is the water discharge and I_0 is the modified Bessel function of the first kind of order zero (Table 17A.1), $D^\# = D_t/V_1 r_1$ and r_1 is the jet radius at impingement ($r_1 = d_1/2$).

Discussion
It is interesting to note that equations (17A.4)–(17A.6) are valid both in the developing bub-bly region and in the fully aerated flow region. In other words, they are valid both close to and away from the jet impact. Further equation (17A.6) is a three-dimensional solution of the advective diffusion equation.

Applications
The re-analysis of experimental data suggests that the dimensionless air flow rate may be estimated for two-dimensional vertical jets as:

$$\frac{Q_{air}}{Q_w} = 2.9 \times 10^{-3}\left(\frac{x_1}{d_1}-0.52\right)\left(\frac{V_1-V_e}{\sqrt{gd_1}}\right)^{1.8}\qquad \frac{V_1-V_e}{\sqrt{gd_1}} < 7.5 \tag{17A.7a}$$

$$\frac{Q_{air}}{Q_w} = 5.75\left(\frac{x_1}{d_1}-0.52\right)\left(\frac{V_1-V_e}{\sqrt{gd_1}}+6.6\right)\qquad 7.5 < \frac{V_1-V_e}{\sqrt{gd_1}} \tag{17A.7b}$$

Table 17A.1   Values of the modified Bessel function of the first kind of order zero

u (1)	I_o (2)	u (1)	I_o (2)	u (1)	I_o (2)
0	1	2.0	2.280	4.0	11.30
0.1	1.003	2.1	2.446	4.1	12.32
0.2	1.010	2.2	2.629	4.2	13.44
0.3	1.023	2.3	2.830	4.3	14.67
0.4	1.040	2.4	3.049	4.4	16.01
0.5	1.063	2.5	3.290	4.5	17.48
0.6	1.092	2.6	3.553	4.6	19.09
0.7	1.126	2.7	3.842	4.7	20.86
0.8	1.167	2.8	4.157	4.8	22.79
0.8	1.213	2.9	4.503	4.9	24.91
1	1.266	3.0	4.881	5.0	27.24
1.1	1.326	3.1	5.294	5.1	29.79
1.2	1.394	3.2	5.747	5.2	32.58
1.3	1.469	3.3	6.243	5.3	35.65
1.4	1.553	3.4	6.785	5.4	39.01
1.5	1.647	3.5	7.378	5.5	42.69
1.6	1.750	3.6	8.028	5.6	46.74
1.7	1.864	3.7	8.739	5.7	51.17
1.8	1.990	3.8	9.517	5.8	56.04
1.9	2.128	3.9	10.37	5.9	61.38

where V_1 is the jet impact velocity, V_e is the inception velocity (paragraph 2.2), d_1 is the free-jet thickness at impact and x_1 is the free-jet length. The dimensionless air bubble diffusivity may be estimated as:

$$\frac{D_t}{V_1 d_1} = 0.363 \left(5.3 \times 10^{-5} \left(\frac{x_1}{d_1} \right) - 4.27 \times 10^{-5} \right) \left(\frac{V_1 d_1}{\nu_w} \right)^{0.462} \qquad 2 \times 10^4 < \frac{V_1 d_1}{\nu_w} < 2 \times 10^5$$

(17A.8)

where ν_w is the water kinematic viscosity.

For circular vertical jets, the air flow rate may be estimated as:

$$\frac{Q_{air}}{Q_w} = \frac{7.1 \times 10^{-4}}{1 + 156 \exp\left(-2.38 \frac{x_1}{d_1} \right)} \left(\frac{V_1 - V_e}{\sqrt{g d_1}} \right)^{2.45} \qquad 1.8 < \frac{V_1 - V_e}{\sqrt{g d_1}} < 9 \qquad (17A.9)$$

where d_1 is the jet radius at impact. The dimensionless air bubble diffusivity may be estimated as:

$$\frac{D_t}{V_1 d_1} = 0.5 \left(\left(3.45 \times 10^{-3} + 8.9 \times 10^{-4} \frac{x_1}{d_1} \right) \right.$$
$$\left. - 2.19 \times 10^{-8} \frac{V_1 d_1}{\nu_w} \right) \qquad 2 \times 10^4 < \frac{V_1 d_1}{\nu_w} < 2 \times 10^5 \qquad (17A.10)$$

Equations (17A.7) and (17A.9) are compared with experimental data obtained in the developing flow region of large size plunging jets in Fig. 17A.1, while equations (17A.8) and (17A.10) are compared with experimental data in Fig. 17A.2.

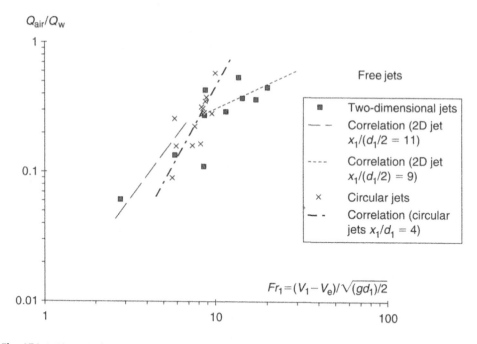

Fig. 17A.1 Dimensionless air entrainment rate at vertical plunging jets – comparison between experimental data and equations (17A.7) and (17A.9).

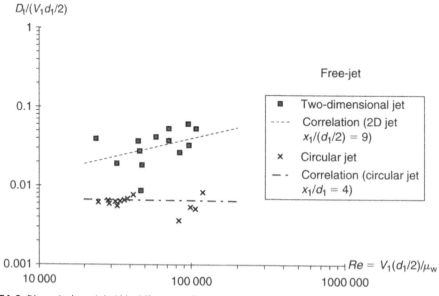

Fig. 17A.2 Dimensionless air bubble diffusion coefficient – comparison between experimental data and equations (17A.8) and (17A.10).

Notes

1. Equation (17A.7) was first proposed by Brattberg and Chanson (1998) who measured both void fraction and velocity distributions below impingement. The entrained air flow rate was estimated as:

$$q_{air} = \int_{-\infty}^{+\infty} C V \, dy$$

where C and V were measured void fraction and air–water velocity respectively. The results were obtained for $4 < x_1/d_1 < 16$.

2. Equation (17A.8) was deduced from the experiments of Chanson (1995a), Cummings (1996) and Brattberg and Chanson (1998) for $0.2 < x_1/d_1 < 9$.

3. Equations (17A.9) and (17A.10) compared favourably with experimental data obtained in fresh water, salt water and sea water for $0.2 < x_1/d_1 < 9$ (Chanson et al. 2002a, Chanson and Manasseh 2003).

4. It must be stressed that equations (17A.7)–(17A.10) are crude correlations. In practice, $D^\#$ and Q_{air}/Q_w must be deduced from measured distributions of void fraction distributions (paragraph 17.5.1.2).

17.7 Appendix B – Air bubble diffusion in self-aerated supercritical flows

In supercritical open channel flows, free-surface aeration is often observed. The phenomenon, called 'white waters', occurs when turbulence acting next to the free surface is large enough to overcome both surface tension for the entrainment of air bubbles and buoyancy to carry downwards the bubbles. Assuming a homogeneous air–water mixture for $C < 90\%$, the advective diffusion of air bubbles may be analytically predicted. At uniform equilibrium, the air concentration distribution is a constant with respect to the distance x in the flow direction. The continuity equation for air in the air–water flow yields:

$$\frac{\partial}{\partial y}\left(D_t \frac{\partial C}{\partial y} \right) = \cos\theta \, \frac{\partial}{\partial y}(u_r C) \tag{17B.1}$$

where C is the void fraction, D_t is the air bubble turbulent diffusivity, u_r is the bubble rise velocity, θ is the channel slope and y is measured perpendicular to the mean flow direction. The bubble rise velocity in a fluid of density $\rho_w(1 - C)$ equals:

$$u_r^2 = [(u_r)_{Hyd}]^2 (1 - C) \tag{17B.2}$$

where $(u_r)_{Hyd}$ is the rise velocity in hydrostatic pressure gradient (Chanson 1995a, 1997a). A first integration of the continuity equation for air in the equilibrium flow region leads to:

$$\frac{\partial C}{\partial y'} = \frac{1}{D'} C \sqrt{1 - C} \tag{17B.3}$$

where $y' = y/Y_{90}$, $D' = D_t/((u_r)_{Hyd} \cos\theta \, Y_{90})$ is a dimensionless turbulent diffusivity and Y_{90} is the location where $C = 0.90$. D' is the ratio of the air bubble diffusion coefficient to the rise velocity component normal to the flow direction times the characteristic transverse dimension of the shear flow.

Assuming a homogeneous turbulence across the flow (i.e. D' constant), the integration of equation (17B.3) yields:

$$C = 1 - \tanh^2\left(K' - \frac{y'}{2D'}\right) \qquad (17B.4)$$

where tanh is the hyperbolic tangent function and K' a dimensionless integration constant. A relationship between D' and K' is deduced for $C = 0.9$ for $y' = 1$:

$$K' = K^* + \frac{1}{2D'} \qquad (17B.5)$$

where $K^* = \tanh^{-1}(\sqrt{0.1}) = 0.327\,450\,15\ldots$ The diffusivity and the mean air content C_{mean} defined in terms of Y_{90} are related by:

$$C_{mean} = 2D'\left(\tanh\left(K^* + \frac{1}{2D'}\right) - \tanh(K^*)\right) \qquad (17B.6)$$

Notes

1. The dimensionless bubble diffusion coefficient D' and integration constant K' are functions of the depth-averaged air content C_{mean} only. They may be estimated as:

$$D' = \frac{0.848\,C_{mean} - 0.00302}{1 + 1.1375\,C_{mean} - 2.2925\,C_{mean}^2} \qquad C_{mean} < 0.7$$

$$K' = 0.32745015 + \frac{1}{2D'}$$

2. The depth-averaged air concentration is commonly defined in term of the characteristic air–water depth Y_{90}:

$$C_{mean} = \frac{1}{Y_{90}} \int_0^{Y_{90}} C \, dy$$

Advanced void fraction distribution models may be developed assuming a non-constant bubble diffusivity. Assuming that the diffusivity distribution satisfies:

$$D' = \frac{C\sqrt{1 - C}}{\lambda(K' - C)}$$

the integration of equation (17B.1) yields:

$$C = K'\left(1 - \exp\left(-\lambda \frac{y}{Y_{90}}\right)\right) \qquad (17B.7)$$

where y is the distance measured normal to the pseudo-invert, Y_{90} is the characteristic distance where $C = 90\%$, K' and λ are dimensionless function of the mean air content only:

$$K' = \frac{0.9}{1 - \exp(-\lambda)}$$

$$C_{\text{mean}} = K' - \frac{0.9}{\lambda}$$

Note that the depth-averaged air content satisfies $C_{\text{mean}} > 0.45$. Equation (17B.7) applies to highly aerated (or fragmented) flows, like transition flows on stepped chutes.

In skimming flows and smooth-chute flows, the air concentration profiles have a S-shape that correspond to:

$$D' = \frac{D_0}{1 - 2\left(\dfrac{y}{Y_{90}} - \dfrac{1}{3}\right)^2}$$

for which the integration of the air bubble diffusion equation yields:

$$C = 1 - \tanh^2\left(K' - \frac{\dfrac{y}{Y_{90}}}{2D_0} + \frac{\left(\dfrac{y}{Y_{90}} - \dfrac{1}{3}\right)^3}{3D_0} \right) \tag{17B.8}$$

where K' is an integration constant and D_0 is a function of the mean void fraction only:

$$K' = K^* + \frac{1}{2D_0} - \frac{8}{81D_0} \quad \text{with } K^* = \tanh^{-1}\left(\sqrt{0.1}\right) = 0.32745015 \ldots$$

$$C_{\text{mean}} = 0.7622\,(1.0434 - \exp(-3.614D_0))$$

Discussion

The theoretical models were compared with model and prototype experimental data on smooth and stepped chutes. In each case, the dimensionless turbulent diffusivity was deduced from the mean air content. Results are presented in Fig. 17B.1. Figure 17B.1(a) shows the dimensionless air bubble diffusivity D_t/V_*Y_{90} as a function of the shear Reynolds number V_*Y_{90}/ν_w. Chanson (1995a, 1997a) discussed the results in more length, including some analogies with sediment-laden open channel flows.

Using a mixing length model, an estimate of the depth-averaged momentum exchange coefficient across the air–water flow is:

$$\nu_T = \frac{K}{6}V_*Y_{90}$$

where K is the von Karman constant ($K = 0.4$).

The ratio of the air bubble diffusion coefficient over the momentum transfer coefficient (ν_T) becomes:

$$\frac{D_t}{\nu_T} = \frac{6}{K}\frac{D_t}{V_*Y_{90}} \tag{17B.9}$$

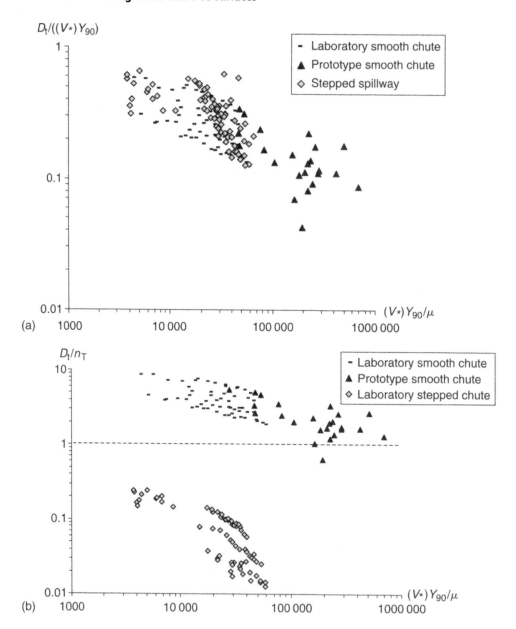

Fig. 17B.1 Dimensionless air bubble diffusion coefficient in self-aerated open channel flows. (a) Dimensionless air bubble diffusivity D_t/V_*Y_{90} as a function of the shear Reynolds number V_*Y_{90}/ν_w. (b) Ratio of air bubble diffusivity to momentum exchange coefficient D_t/ν_T as a function of the shear Reynolds number V_*Y_{90}/ν_w.

Figure 17B.1(b) presents experimental results. In smooth chutes, the results show that self-aerated flows tend to exhibit large values of diffusion coefficients implying usually $D_t/\nu_T > 1$ while $D_t/\nu_T < 1$ is usually observed in skimming flow down stepped chutes. The ratio D_t/ν_T describes the combined effects of (1) the difference in the diffusion of a discrete particle (e.g. air bubble, sediment) and the diffusion of a small coherent fluid structure and (2) the influence of the particles on the turbulence field (e.g. turbulence damping or drag reduction). In Fig. 17B.1(b), the reader shall note that, on large smooth-chute prototypes, the ratio of the turbulent diffusivity over the eddy viscosity is less than unity while it is larger than one on models. In skimming flow down stepped chutes, a comparable trend is seen with the ratio D_t/ν_T decreasing with increasing shear Reynolds number. Such a result suggests that scale-model studies of self-aerated flows might not describe accurately the air bubble diffusion process in self-aerated open channel flows.

Notes

1. In Fig. 17B.1, laboratory data include the re-analysis of the data of Straub and Anderson (1958) and Aivazyan (1986) on smooth chutes and a re-analysis of the data of Boes (2000), Chamani and Rajaratnam (1999), Chanson and Toombes (2001), Gonzalez and Chanson (2004) and Yasuda and Ohtsu (1999) on stepped chutes. Prototype data include the re-analysed data of Aivazyan (1986) and Cain (1978).
2. The above analysis is an extension of the work of Chanson (1995a, 1997a). Note however that there was a typographic error in this development and that the equation (17B.9) is correct.

17.8 Appendix C – Air bubble diffusion in high-velocity water jets

Free-surface aeration is observed along the air–water interfaces of turbulent water jets discharging into the atmosphere. Within the air–water flow, simple analytical solutions of air bubble diffusion may be developed for both two-dimensional free-shear layers and circular jets.

Two-dimensional free-shear layers

For a two-dimensional free-shear layer, an analytical solution of air bubble diffusion may be developed in a simple manner. Consider the free-shear layer of a two-dimensional developing jet, the continuity equation becomes:

$$V_x \frac{\partial C}{\partial x} + C \frac{\partial V_x}{\partial x} = \frac{\partial}{\partial y}\left(D_t \frac{\partial C}{\partial y}\right) \tag{17C.1}$$

where x is the streamwise direction ($x = 0$ at nozzle) and y is the distance normal to the the jet centreline, V_x is the velocity component in the x-direction and D_t is the turbulent diffusivity of air bubbles.

The analytical solution of equation (17C.1) is:

$$C = \frac{1}{2}\left(1 - \mathrm{erf}\left(\frac{\frac{d_1}{2} - y}{2\sqrt{\frac{D_t}{V_1}x}}\right)\right) \tag{17C.2}$$

where V_1 is the jet velocity at nozzle, d_1 is the jet nozzle thickness and the diffusivity D_t averages the effect of the turbulence on the transverse dispersion and of the longitudinal velocity

Table 17C.1 Approximate solution of the roots of the equation $J_0(u) = 0$ and $J_1(u) = 0$

Root (1)	$J_0(u_i) = 0$ (2)	$J_1(u_i) = 0$ (3)
u_1	2.4048	3.8317
u_2	5.5201	7.0156
u_3	8.6537	10.1735
u_4	11.7915	13.3237
u_5	14.0309	16.4706
u_6	18.0711	19.6159
u_7	18.0711 + 3.14159	19.6159 + 3.14159
u_{n+1}	$u_n + \pi$	$u_n + \pi$

Reference: Spigel (1974).

gradient. D_t is further assumed independent of the transverse direction y. The Gaussian error function is defined as:

$$\text{erf}(u) = \frac{2}{\sqrt{\pi}} \int_0^u \exp(-t^2)\, dt$$

Circular jets

For a circular water jet discharging into the atmosphere, the continuity equation for air becomes:

$$\frac{V_x}{D_t} \frac{\partial C}{\partial x} + \frac{C}{D_t} \frac{\partial V_x}{\partial x} = \frac{1}{r} \frac{\partial}{\partial r}\left(r \frac{\partial C}{\partial r} \right) \tag{17C.3}$$

where x is the longitudinal direction, r is the radial direction, V_x is the velocity component in the x-direction and D_t is the turbulent diffusivity.

If the diffusivity term D_t averages the effects of the turbulent diffusion and the longitudinal velocity gradient, the solution of equation (17C.3) is a series of Bessel functions:

$$C = 0.9 - \frac{1.8}{r_{90}} \sum_{n=1}^{+\infty} \frac{J_0(r\alpha_n)}{\alpha_n J_1(r_{90}\alpha_n)} \exp\left(-\frac{D_t}{V_1} \alpha_n^2 x \right) \tag{17C.4}$$

where r_{90} is the radial distance where $C = 0.9$, J_0 is the Bessel function of the first kind of order zero, α_n is the positive root of: $J_0(r_{90}\alpha_n) = 0$ (Table 17C.1) and J_1 is the Bessel function of the first kind of order one.

Remarks

1. The Bessel function of the first kind of order zero is defined as:

$$J_0(u) = 1 - \frac{u^2}{2^2} + \frac{u^4}{2^2 \times 4^2} - \frac{u^6}{2^2 \times 4^2 \times 6^2} + \cdots$$

2. The Bessel function of the first kind of order one is:

$$J_1(u) = \frac{u}{2} - \frac{u^3}{2^2 \times 4} + \frac{u^5}{2^2 \times 4^2 \times 6} - \frac{u^7}{2^2 \times 4^2 \times 6^2 \times 8} + \cdots$$

3. Chanson (1997a) integrated numerically equation (17C.4) for several values of the dimensionless diffusivity D'' defined as:

$$D'' = \frac{D_t x}{V_o r_{90}^2}$$

4. Equation (17C.4) is valid close to and away from the deflector edge. It is a three-dimensional solution of the diffusion equation. It is valid also when the clear-water core of the jet disappears and the jet becomes fully aerated.

Discussion

Theoretical results were compared with experimental data. The dimensionless air bubble diffusivity may be estimated as:

$$\frac{D_t}{V_1 d_1} = 3.6 \times 10^{-4} \left(\frac{V_1}{\sqrt{gd_1}} \right)^{0.93} \qquad \text{Two-dimensional} \qquad (17C.5)$$

$$\frac{D_t}{V_1 d_1} = 5 \times 10^{-8} \left(\frac{V_1}{\sqrt{gd_1}} \right)^{2.2} \qquad \text{Circular jets} \qquad (17C.6)$$

where d_1 is the jet thickness and diameter at nozzle for two-dimensional and circular jets respectively. Experimental results are compared with equations (17C.5) and (17C.6) in Fig. 17C.1.

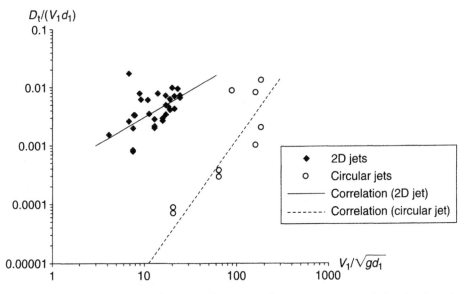

Fig. 17C.1 Dimensionless air bubble diffusion coefficient in high-velocity water jets discharging into air – comparison with equation (17C.5) and (17C.6).

Notes

1. Equation (17C.5) was deduced from experiments conducted with $6 < Fr_1 < 24$ and $4 < V_1 < 12$ m/s (Brattberg et al. 1998).
2. Equation (17C.6) derived from the re-analysis of experiments with $20 < Fr_1 < 162$ and $11 < V_1 < 60$ m/s.
3. It must be stressed that equations (17C.5) and (17C.6) are crude correlations. In practice, D_t must be deduced from measured distributions of void fraction distributions.

17.9 Exercises

1. Considering a chute spillway ending with a flip bucket, describe the basic types of air entrainment upstream of the flip bucket, downstream of the ski jump, and in the downstream plunge pool.
2. A vertical circular jet discharge into a plunge pool. The nozzle is located 0.1 m above the pool free surface and its diameter is 30 mm. Calculate the onset of air entrainment in terms of impact velocity V_1 and nozzle flow rate Q_w. *Assume the fluid to be tap water at 20°C.*
3. A two-dimensional water jet plunges into a deep pool. The jet thickness and velocity at impact are respectively 28 mm and 6.5 m/s. Calculate and plot the void fraction distribution at 100 and 200 mm beneath the free surface. *Assume the fluid to be tap water at 20°C. Estimate the air entrainment rate and bubble diffusivity using Appendix A (Section 17.6).*
4. At a vertical circular plunging jet, the water discharge is 0.011 m³/s and the nozzle velocity is 8 m/s. For a free-jet length of 150 mm, the measured air entrainment rate at the plunge point is 0.0032 m³/s. Calculate and plot the void fraction distribution at 150 and 250 mm beneath the free surface. *Assume the fluid to be tap water at 20°C. Estimate the bubble diffusivity using Appendix A (Section 17.6).*
5. Predict the onset conditions for air entrainment down a 60° chute for 1, 10 and 50 mm bubbles. *Assume the fluid to be tap water at 20°C.*
6. Calculate the rise velocity of a 5 mm bubble in sea water at 20°C.
7. In a smooth spillway chute, the depth-averaged void fraction and characteristic depth Y_{90} are 0.31 and 0.95 m respectively. Calculate and plot the void fraction distribution.
8. An overflow spillway is to be designed with an uncontrolled crest followed by a stepped chute and a hydraulic jump dissipator. The maximum spillway capacity will be 4300 m³/s, and the flow velocity and depth at the end of the chute are expected to be 14 m/s and 2.9 m.
   (a) Calculate the prototype Froude, Reynolds and Weber number.
   A 17:1 scale model of the spillway is to be built. (a) Based upon a Froude similitude, calculate the corresponding model Reynolds and Weber numbers.
9. What are the basic differences between optical fibre and resistivity probes?
10. The data processing of a dual-tip probe gives the following auto-correlation and cross-correlation functions. For a 8.1 mm spacing between probe sensors, calculate the local velocity, dimensionless integral length scale and turbulence intensity.

t (s)	R_{xx}	R_{xy}	t (s)	R_{xx}	R_{xy}	t (s)	R_{xx}	R_{xy}
0	1	0.087637	0.00295	0.121275	0.326785	0.00595	0.050955	0.114978
0.00005	0.96586	0.088616	0.003	0.118072	0.327271	0.006	0.049636	0.112499
0.0001	0.904278	0.090061	0.00305	0.115306	0.327358	0.00605	0.048382	0.109345
0.00015	0.837674	0.091651	0.0031	0.112391	0.327258	0.0061	0.047107	0.105075
0.0002	0.772436	0.093276	0.00315	0.109255	0.326918	0.00615	0.045534	0.099868
0.00025	0.711249	0.095248	0.0032	0.106087	0.325491	0.0062	0.043833	0.094609
0.0003	0.655413	0.09751	0.00325	0.103387	0.322909	0.00625	0.04223	0.089896
0.00035	0.605642	0.099922	0.0033	0.100965	0.319289	0.0063	0.040903	0.085858
0.0004	0.562258	0.102742	0.00335	0.098579	0.315016	0.00635	0.039846	0.083035
0.00045	0.524616	0.105734	0.0034	0.096554	0.30943	0.0064	0.039394	0.081668
0.0005	0.491912	0.108786	0.00345	0.094845	0.303674	0.00645	0.039291	0.081287
0.00055	0.463086	0.111869	0.0035	0.093291	0.297704	0.0065	0.039019	0.081311
0.0006	0.437589	0.115012	0.00355	0.091654	0.291496	0.00655	0.038766	0.081208
0.00065	0.415059	0.118489	0.0036	0.08964	0.285305	0.0066	0.038915	0.081469
0.0007	0.395063	0.121975	0.00365	0.087022	0.278854	0.00665	0.039606	0.082232
0.00075	0.377694	0.125662	0.0037	0.084282	0.272668	0.0067	0.040581	0.083108
0.0008	0.362256	0.130004	0.00375	0.081846	0.266608	0.00675	0.041433	0.084016
0.00085	0.348675	0.13437	0.0038	0.07964	0.260237	0.0068	0.042165	0.08514
0.0009	0.336638	0.13915	0.00385	0.077468	0.253814	0.00685	0.042818	0.086322
0.00095	0.32585	0.144207	0.0039	0.075573	0.247429	0.0069	0.043392	0.087231
0.001	0.316006	0.149515	0.00395	0.073941	0.24149	0.00695	0.044006	0.08755
0.00105	0.307062	0.15509	0.004	0.072307	0.235819	0.007	0.044416	0.087621
0.0011	0.298849	0.160946	0.00405	0.070747	0.230018	0.00705	0.044313	0.087475
0.00115	0.291762	0.166954	0.0041	0.069671	0.22392	0.0071	0.043951	0.087057
0.0012	0.286022	0.172454	0.00415	0.068971	0.217947	0.00715	0.043471	0.086331
0.00125	0.281381	0.177021	0.0042	0.068495	0.212139	0.0072	0.042921	0.08559
0.0013	0.276504	0.180688	0.00425	0.068287	0.206875	0.00725	0.042795	0.084427
0.00135	0.271244	0.183773	0.0043	0.06799	0.201912	0.0073	0.043508	0.082352
0.0014	0.266031	0.185948	0.00435	0.067501	0.197466	0.00735	0.04455	0.079925
0.00145	0.261056	0.18734	0.00445	0.066085	0.189998	0.0074	0.045083	0.077406
0.0015	0.256012	0.187326	0.0045	0.065064	0.185716	0.00745	0.045225	0.074522
0.00155	0.250623	0.186435	0.00455	0.063727	0.181157	0.0075		0.071277
0.0016	0.245439	0.18541	0.0046	0.062505	0.176889	0.00755		0.068265
0.00165	0.240399	0.184553	0.00465	0.061469	0.172818	0.0076		0.065275
0.0017	0.235888	0.184371	0.0047	0.060153	0.168729	0.00765		0.06255
0.00175	0.231475	0.185341	0.00475	0.058641	0.164603	0.0077		0.060497
0.0018	0.227046	0.187463	0.0048	0.057051	0.160242	0.00775		0.059174
0.00185	0.22271	0.190021	0.00485	0.055339	0.156255	0.0078		0.058194
0.0019	0.218401	0.193484	0.0049	0.053853	0.153033	0.00785		0.057373
0.00195	0.214449	0.198104	0.00495	0.052914	0.150285	0.0079		0.056551
0.002	0.210557	0.203444	0.005	0.052618	0.148278	0.00795		0.055692
0.00205	0.206111	0.209427	0.00505	0.052643	0.146752	0.008		0.05452
0.0021	0.201267	0.216444	0.0051	0.052923	0.145461	0.00805		0.053224
0.00215	0.195988	0.224443	0.00515	0.053662	0.144303	0.0081		0.051831
0.0022	0.190544	0.233122	0.0052	0.055273	0.143139	0.00815		0.050639
0.00225	0.185114	0.241678	0.00525	0.057347	0.142009	0.0082		0.049857
0.0023	0.179563	0.250238	0.0053	0.059371	0.140995	0.00825		0.049428
0.00235	0.174289	0.258437	0.00535	0.061346	0.139422	0.0083		0.049424
0.0024	0.169248	0.266899	0.0054	0.063119	0.1377	0.00835		0.049984
0.00245	0.164254	0.274792	0.00545	0.064326	0.135406	0.0084		0.050505
0.0025	0.15909	0.282163	0.0055	0.064833	0.13316	0.00845		0.050937
0.00255	0.153975	0.288934	0.00555	0.064109	0.13072	0.0085		0.051781
0.0026	0.148747	0.295428	0.0056	0.062619	0.128092	0.00855		0.053256
0.00265	0.143654	0.302344	0.00565	0.060692	0.125513	0.0086		0.054781
0.0027	0.138966	0.309573	0.0057	0.058455	0.123491	0.00865		0.056288
0.00275	0.13483	0.315676	0.00575	0.056274	0.122002	0.0087		0.057542
0.0028	0.131087	0.320177	0.0058	0.054468	0.120626	0.00875		0.058128
0.00285	0.127764	0.323319	0.00585	0.053067	0.119141	0.0088		0.058157
0.0029	0.124473	0.325432	0.0059	0.052011	0.117174			

Answer: $Tu = 0.46$

11. An overflow spillway is to be designed with an uncontrolled broad crest followed by a stepped chute and a hydraulic jump dissipator. The width of the crest, chute and dissipation basin will be 100 m. The crest level will be at 316.1 m R.L. and the design head above crest level will be 3.1 m. The chute slope will be set at 51° and the step height will be 0.5 m. The elevation of the chute toe will be set at 278.3 m R.L. The stepped chute will be followed (without transition section) by a horizontal stilling basin.
   (a) Calculate the maximum discharge capacity of the spillway.
   (b) Compute the location of and flow properties at the inception point of free-surface aeration.
   (c) Calculate the flow velocity and depth-averaged void fraction at the toe of the chute.
   (d) Calculate the residual power at the end of the chute (give the SI unit). Comment.
   Notes: In calculating the crest discharge capacity, assume that the discharge capacity of the broad crest is 2% smaller than that of an ideal broad crest (for the same upstream head above crest). In computing the velocity at the spillway toe, allow for energy losses by using results presented in the book. The residual power equals $\rho g Q_w H_{res}$ where Q_w is the total water discharge and H_{res} is the residual total head at chute toe taking the chute toe elevation as datum. Assume the Darcy friction factor of non-aerated stepped chute to be 0.24.

Appendix A:
Constants and fluid properties

A.1 Acceleration of gravity

The gravity varies with the local geology and topography. Measured values of g are reported below:

Location (1)	g (m/s^2) (2)	Location (1)	g (m/s^2) (2)	Location (1)	g (m/s^2) (2)
Addis Ababa, Ethiopia	9.7743	Helsinki, Finland	9.81090	Quito, Ecuador	9.7726
Algiers, Algeria	9.79896	Kuala Lumpur, Malaysia	9.78034	Sapporo, Japan	9.80476
Anchorage, USA	9.81925	La Paz, Bolivia	9.7745	Reykjavik, Iceland	9.82265
Ankara, Turkey	9.79925	Lisbon, Portugal	9.8007	Taipei, Taiwan	9.7895
Aswan, Egypt	9.78854	Manila, Philippines	9.78382	Teheran, Iran	9.7939
Bangkok, Thailand	9.7830	Mexico City, Mexico	9.77927	Thule, Greenland	9.82914
Bogota, Colombia	9.7739	Nairobi, Kenya	9.77526	Tokyo, Japan	9.79787
Brisbane, Australia	9.794	New Delhi, India	9.79122	Vancouver, Canada	9.80921
Buenos Aires, Argentina	9.7949	Paris, France	9.80926	Ushuaia, Argentina	9.81465
Christchurch, New Zealand	9.8050	Perth, Australia	9.794		
Denver, USA	9.79598	Port-Moresby, P.N.G.	9.782		
Edmonton, Canada	9.81145	Pretoria, South Africa	9.78615		
Guatemala, Guatemala	9.77967	Québec, Canada	9.80726		

Reference: Morelli (1971).

A.2 Properties of water

Temperature (°C) (1)	Density ρ_w (kg/m^3) (2)	Dynamic viscosity μ_w (Pa s) $\times 10^{-3}$ (3)	Surface tension σ (N/m) (4)	Vapour pressure P_v (Pa) $\times 10^3$ (5)	Bulk modulus of elasticity E_b (Pa) $\times 10^9$
0	999.9	1.792	0.0762	0.6	2.04
5	1000.0	1.519	0.0754	0.9	2.06
10	999.7	1.308	0.0748	1.2	2.11
15	999.1	1.140	0.0741	1.7	2.14
20	998.2	1.005	0.0736	2.5	2.20
25	997.1	0.894	0.0726	3.2	2.22
30	995.7	0.801	0.0718	4.3	2.23
35	994.1	0.723	0.0710	5.7	2.24
40	992.2	0.656	0.0701	7.5	2.27

Reference: Streeter and Wylie (1981).

A.3 Gas properties

Basic equations

The *state equation* of perfect gas is:

$$P = \rho R T \tag{A.1}$$

where P is the absolute pressure (in Pascal), ρ is the gas density (in kg/m³), T is the absolute temperature (in Kelvin) and R is the gas constant (in J/kg K) (see table below).

For a perfect gas, the *specific heat* at constant pressure C_p and the specific heat at constant volume C_v are related to the gas constant as:

$$C_p = \frac{\gamma}{\gamma - 1} R \tag{A.2a}$$

$$C_p = C_v R \tag{A.2b}$$

where γ is the specific heat ratio (i.e. $\gamma = C_p/C_v$).

During an *isentropic transformation* of perfect gas, the following relationships hold:

$$\frac{P}{\rho^{\gamma}} = \text{constant} \tag{A.3a}$$

$$T P^{(1-\gamma)/\gamma} = \text{constant} \tag{A.3b}$$

Physical properties

Gas	Formula	Gas constant R (J/kg K)	Specific heat (J/kg K)		Specific heat ratio γ
			C_p	C_v	
(1)	(2)	(3)	(4)	(5)	(6)
Perfect gas					
Mono-atomic gas	(e.g. He)		$\frac{5}{2}R$	$\frac{3}{2}R$	$\frac{5}{3}$
Di-atomic gas	(e.g. O₂)		$\frac{7}{2}R$	$\frac{5}{2}R$	$\frac{7}{5}$
Poly-atomic gas	(e.g. CH₄)		4R	3R	$\frac{4}{3}$
Real gas[a]					
Air		287	1.004	0.716	1.40
Helium	He	2077.4	5.233	3.153	1.67
Nitrogen	N₂	297	1.038	0.741	1.40
Oxygen	O₂	260	0.917	0.657	1.40
Water vapour	H₂O	462	1.863	1.403	1.33

[a] At low pressures and at 299.83 K.
Reference: Streeter and Wylie (1981).

Atmospheric parameters

The standard atmosphere or normal pressure at sea level equals:

$$P_{std} = 1\,atm = 360\,mmHg = 101\,325\,Pa \tag{A.4}$$

where Hg is the chemical symbol of mercury. Unit conversion tables are provided in Appendix B.

The atmospheric pressure varies with the elevation above the sea level (i.e. altitude). For dry air, the atmospheric pressure at the altitude z equals

$$P_{atm} = P_{std}\,\exp\left(-\int_0^z \frac{0.0034841g}{T}dz\right) \tag{A.5}$$

where T is the absolute temperature in Kelvin and equation (A.5) is expressed in SI units. In the troposphere (i.e. $z < 10\,000$ m), the air temperature T decreases with altitude, on the average, at a rate of 6.5×10^{-3} K/m (i.e. 6.5 K/km). Table A.1 presents the distributions of average air temperatures (Miller 1971) and corresponding atmospheric pressures with the altitude (equation (A.5)).

Table A.1   Distributions of air temperature and air pressure as functions of the altitude (for dry air and standard acceleration of gravity)

Altitude z (m) (1)	Mean air temperature (K) (2)	Atmospheric pressure (equation (A.5)) (Pa) (3)	Atmospheric pressure (equation (A.5)) (atm) (4)
0	288.2	1.013×10^5	1.000
500	285.0	9.546×10^4	0.942
1000	281.7	8.987×10^4	0.887
1500	278.4	8.456×10^4	0.834
2000	275.2	7.949×10^4	0.785
2500	272.0	7.468×10^4	0.737
3000	268.7	7.011×10^4	0.692
3500	265.5	6.576×10^4	0.649
4000	262.2	6.164×10^4	0.608
4500	259.0	5.773×10^4	0.570
5000	255.7	5.402×10^4	0.533
5500	252.5	5.051×10^4	0.498
6000	249.2	4.718×10^4	0.466
6500	246.0	4.404×10^4	0.435
7000	242.8	4.106×10^4	0.405
7500	239.5	3.825×10^4	0.378
8000	236.3	3.560×10^4	0.351
8500	233.0	3.310×10^4	0.327
9000	229.8	3.075×10^4	0.303
9500	226.5	2.853×10^4	0.282
10 000	223.3	2.644×10^4	0.261

Viscosity of air

Viscosity and density of air at 1.0 atm:

Temperature (K) (1)	μ_{air} (Pa s) $\times 10^{-6}$ (3)	ρ_{air} (kg/m^3) (3)
300	18.4	1.177
400	22.7	0.883
500	26.7	0.705
600	29.9	0.588

The viscosity of air at standard atmosphere is commonly fitted by the Sutherland formula (Sutherland 1883):

$$\mu_{air} = 17.16 \times 10^{-6} \left(\frac{T}{273.1} \right)^{3/2} \frac{383.7}{T + 110.6} \tag{A.6}$$

A simpler correlation is:

$$\frac{\mu_{air}(T)}{\mu_{air}(T_o)} = \left(\frac{T}{T_o} \right)^{0.76} \tag{A.7}$$

where μ_{air} is in Pa s, and the temperature T and reference temperature T_o are expressed in Kelvin.

Appendix B:
Unit conversions

B.1   Introduction

The systems of units derived from the metric system have gradually been replaced a single system, called the Système International d'Unités (SI unit system, or International System of Units). The basic SI units are the metre, kilogramme, second, Ampere, Kelvin, mole and candela. Supplementary units are the radian and the steradian. All other SI units derive from the basic units. Conversion tables are provided in this appendix. Basic references in unit conversions include Degremont (1979) and ISO (1979).

Unit symbols are written in small letters (i.e. m for metre, kg for kilogramme) but a capital is used for the first letter when the name of the unit derives from a surname (e.g. Pa after Blaise Pascal, N after Isaac Newton). Multiples and submultiples of SI units are formed by adding one prefix to the name of the unit: e.g. km for kilometre, cm for centimetre, dam for decametre, μm for micrometre (or micron).

Multiple/submultiple factor	Prefix	Symbol
1×10^9	giga	G
1×10^6	mega	M
1×10^3	kilo	k
1×10^2	hecto	d
1×10^1	deca	da
1×10^{-1}	deci	d
1×10^{-2}	centi	c
1×10^{-3}	milli	m
1×10^{-6}	micro	μ
1×10^{-9}	nano	n

B.2 Units and conversion factors

Quantity (1)	Unit (symbol) (2)	Conversion (3)	Comments (4)
Length	1 inch (in)	$= 25.4 \times 10^{-3}$ m	Exactly
	1 foot (ft)	$= 0.3048$ m	Exactly
	1 yard (yd)	$= 0.9144$ m	Exactly
	1 mil	$= 25.4 \times 10^{-6}$ m	1/1000 inch
	1 mile	$= 1.609\,344$ m	Exactly
Area	1 square inch (in^2)	$= 6.4516 \times 10^{-4}$ m^2	Exactly
	1 square foot (ft^2)	$= 0.092\,903\,06$ m^2	Exactly
Volume	1 litre (L)	$= 1.0 \times 10^{-3}$ m^3	Exactly. Previous symbol: l
	1 cubic inch (in^3)	$= 16.387\,064 \times 10^{-6}$ m^3	Exactly
	1 cubic foot (ft^3)	$= 28.316\,8 \times 10^{-3}$ m^3	Exactly
	1 gallon UK (gal UK)	$= 4.546\,09 \times 10^{-3}$ m^3	
	1 gallon US (gal US)	$= 3.785\,41 \times 10^{-3}$ m^3	
	1 barrel US	$= 158.987 \times 10^{-3}$ m^3	For petroleum, etc.
Velocity	1 foot per second (ft/s)	$= 0.3048$ m/s	Exactly
	1 mile per hour (mph)	$= 0.447\,04$ m/s	Exactly
Acceleration	1 foot per second squared (ft/s^2)	$= 0.3048$ m/s^2	Exactly
Mass	1 pound (lb or lbm)	$= 0.453\,592\,37$ kg	Exactly
	1 ton UK	$= 1016.05$ kg	
	1 ton US	$= 907.185$ kg	
Density	1 pound per cubic foot (lb/ft^3)	$= 16.0185$ kg/m^3	
Force	1 kilogram-force (kgf)	$= 9.806\,65$ N (exactly)	Exactly
	1 pound force (lbf)	$= 4.448\,221\,615\,2605$ N	
Moment of force	1 foot pound force (ft lbf)	$= 1.355\,82$ N m	
Pressure	1 Pascal (Pa)	$= 1$ N/m^2	
	1 standard atmosphere (atm)	$= 101\,325$ Pa	
		$= 760$ mm of mercury at normal pressure (i.e. mmHg)	Exactly
	1 bar	$= 10^5$ Pa	Exactly
	1 Torr	$= 133.322$ Pa	
	1 conventional metre of water (m of H$_2$O)	$= 9.806\,65 \times 10^3$ Pa	Exactly
	1 conventional meter of mercury (m of Hg)	$= 1.333\,224 \times 10^5$ Pa	
	1 Pound per Square Inch (PSI)	$= 6.894\,7572 \times 10^3$ Pa	
Temperature	T (Celsius)	$= T$ (Kelvin) $- 273.16$	0 Celsius is 0.01 K below the temperature of the triple point of water.
	T (Fahrenheit)	$= T$ (Celcius) $\frac{9}{5} + 32$	
	T (Rankine)	$= \frac{9}{5}$ T (Kelvin)	
Dynamic viscosity	1 Pa s	$= 0.006\,720$ lbm/ft/s	
	1 Pa s	$= 10$ Poises	Exactly
	1 N s/m^2	$= 1$ Pa s	Exactly
	1 Poise (P)	$= 0.1$ Pa s	Exactly
	1 milliPoise (mP)	$= 1.0 \times 10^{-4}$ Pa s	Exactly
Kinematic viscosity	1 square foot per second (ft^2/s)	$= 0.092\,903\,0$ m^2/s	
	1 m^2/s	$= 10.7639$ ft^2/s	
	1 m^2/s	$= 10^4$ Stokes	
Surface tension	1 dyne/cm	$= 0.99987 \times 10^{-3}$ N/m	
	1 dyne/cm	$= 5.709 \times 10^{-6}$ pound/inch	

(Contd)

Quantity (1)	Unit (symbol) (2)	Conversion (3)	Comments (4)
Work energy	1 Joule (J)	$= 1 \, \text{N m}$	
	1 Joule (J)	$= 1 \, \text{W s}$	
	1 Watt hour (W h)	$= 3.600 \times 10^3 \, \text{J}$	Exactly
	1 electronvolt (eV)	$= 1.602 \, 19 \times 10^{-19} \, \text{J}$	
	1 Erg	$= 10^{-7} \, \text{J}$	Exactly
	1 foot pound force (ft lbf)	$= 1.355 \, 82 \, \text{J}$	
Power	1 Watt (W)	$= 1 \, \text{J/s}$	
	1 foot pound force per second (ft lbf/s)	$= 1.355 \, 82 \, \text{W}$	
	1 horsepower (hp)	$= 745.700 \, \text{W}$	

References

Aivazyan, O.M. (1986) Stabilized aeration on chutes. *Gid. Stroit.*, No. 12, 33–40 (in Russian) (Translated in *Hydrotechnical Construction*, 1987, Plenum Publication, 713–722).

Ancey, C. (2000) *Debris Flows and Related Phenomena*. Lecture notes, Gran Combin Summer School, Aosta, Italy, June (Also in *Geomorphological Fluid Mechanics*, N.J. Balmforth and A. Provenzale (Eds), Springer, Berlin, 2001, pp. 528–547).

Anderson, A.G. (1942) Distribution of suspended sediment in a natural stream. *Trans. Amer. Geophys. Union*, **23**(Part 2), 678–683.

Apelt, C.J. (1983) Hydraulics of minimum energy culverts and bridge waterways. *Aust. Civil Eng. Trans., I.E.Aust.*, **CE25**(2), 89–95.

APHA, AWWA, and WPCF (1985) *Standard Methods for the Examination of Water and Wastewater*. APHA Publication, 6th edn.

APHA, AWWA, and WPCF (1989) *Standard Methods for the Examination of Water and Wastewater*. APHA Publication, 7th edn.

Alembert, Jean le Rond d' (1752) *Essai d'une Nouvelle Théorie de la Résistance des Fluides*. (*Essay on a New Theory on the Resistance of Fluids*.) (David: Paris, France).

BaCaRa (1991) *Etude de la Dissipation d'Energie sur les Evacuateurs à Marches*. (*Study of the Energy Dissipation on Stepped Spillways*.) Rapport d'Essais, Projet National BaCaRa, CEMA-GREF-SCP, Aix-en-Provence, France, October, 111 pages (in French).

Bachalo, W.D. (1994) Experimental methods in multiphase flows. *Intl. Jl. Multiphase Flow*, **20** (Suppl.), 261–295.

Baker, R. (1994) *Brushes Clough Wedge Block Spillway*. Progress Report No. 3. SCEL Project Report No. SJ542-4, University of Salford, UK, November, 47 pages.

Bakhmeteff, B.A. (1912). *O Neravnomernom Dwijenii Jidkosti v Otkrytom Rusle*. (*Varied Flow in Open Channel*.) St Petersburg, Russia (in Russian).

Bakhmeteff, B.A., and Matzke, A.E. (1936) The hydraulic jump in terms of dynamic similarity. *Trans., ASCE*, **101**, 630–647. Discussion: **101**, 648–680.

Barnes, D., Liss, P.J., Gould, B.W., and Vallentine, H.R. (1981) *Water and Wastewater Engineering Systems*. (Pitman: London, UK).

Barnett, A.G. (1983) Exact and aproximate solutions of the advective–diffusion equation. *Proc. 20th IAHR Cong.*, Moscow, **3**, 180–190.

Barnett, A.G. (2002) Implicit high-resolution methods for modelling one-dimensional open channel flow. Discussion. *Jl. Hyd. Res., IAHR*, **40**(4), 540–541.

Barré de Saint-Venant, A.J.C. (1871a) Théorie du mouvement non permanent des eaux, avec application aux crues de rivières et à l'introduction des marées dans leur lit. *Comptes Rendus des Séances de l'Académie des Sciences* (Paris, France), Séance 17 July 1871, **73**, 147–154 (in French).

Barré de Saint-Venant, A.J.C. (1871b) Théorie du mouvement non permanent des eaux, avec application aux crues de rivières et à l'introduction des marées dans leur lit. *Comptes Rendus des séances de l'Académie des Sciences* (Paris, France), **73**(4), 237–240 (in French).

Bartsch-Winkler, S., Emmanuel, R.P., and Winkler, G.R. (1985) Reconnaissance hydrology and suspended sediment analysis, Turnagain Arm Estuary, Upper Cook Inlet. *US Geolog. Surv. Circ.*, **967**, 48–52.

Bazin, H. (1865a) Recherches expérimentales sur l'ecoulement de l'eau dans les canaux découverts. (Experimental research on water flow in open channels.) *Mémoires Présentés par Divers Savants à l'Académie des Sciences* (Paris, France), **19**, 1–494 (in French).

Bazin, H. (1865b) Recherches expérimentales sur la propagation des ondes. (Experimental research on wave propagation.) *Mémoires Présentés par Divers Savants à l'Académie des Sciences* (Paris, France), **19**, 495–644 (in French).

Beer, T., and Young, P.C. (1983) Longitudinal dispersion in natural streams. *Jl. Environ. Eng.*, *ASCE*, **109**(5), 1049–1067.

Bélanger, J.B. (1828) *Essai sur la Solution Numérique de Quelques Problèmes Relatifs au Mouvement Permanent des Eaux Courantes. (Essay on the Numerical Solution of Some Problems Relative to Steady Flow of Water.)* (Carilian-Goeury: Paris, France) (in French).

Benet, F., and Cunge, J.A. (1971) Analysis of experiments on secondary undulations caused by surge waves in trapezoidal channels. *Jl. Hyd. Res.*, *IAHR*, **9**(1), 11–33.

Bhargava, D.S., and Ojha, C.S.P. (1990) Genesis of free hydraulic jumps for better mixing. *Wat. Res.*, **24**(8), 1003–1010.

Bin, A.K. (1993) Gas entrainment by plunging liquid jets. *Chem. Eng. Sci.*, **48**(21), 3585–3630.

Blanckaert, K., and Graf, W.H. (2001) Mean flow and turbulence in open-channel bend. *Jl. Hyd. Eng.*, *ASCE*, **127**(10) 835–847.

Blasius, H. (1913) Das ähnlichkeitsgesetz bei reibungsvorgängen in flüssigkeiten. *Forschg. Arb. Ing.-Wes.*, **134** (Berlin, Germany) (in German).

Boes, R.M. (2000) *Zweiphasenstroömung und Energieumsetzung an Grosskaskaden. (Two-Phase Flow and Energy Dissipation on Cascades.)* Ph.D. Thesis, *VAW-ETH*, Zürich, Switzerland (in German) (Also Mitt. der Versuchsanstalt fur Wasserbau, Hydrologie und Glaziologie, ETH-Zurich, Switzerland, No. 166).

Bornschein, A., and Pohl, R. (2003) Dam break during the flood in Saxon/Germay in August 2002. *Proc. 30th IAHR Biennial Cong.*, Thessaloniki, Greece, J. Ganoulis, and P. Prinos (Eds), **C2**, 229–236.

Bossut, Abbé C. (1772) *Traité Elémentaire d'Hydrodynamique. (Elementary Treaty on Hydrodynamics.)* 1st edn (Paris, France) (in French) (2nd edn: 1786, Paris, France; 3rd edn: 1796, Paris, France).

Boussinesq, J.V. (1877) Essai sur la théorie des eaux courantes. (Essay on the theory of water flow.) *Mémoires Présentés par Divers Savants à l'Académie des Sciences* (Paris, France), **23**, Series 3, No. 1, Suppl. 24, 1–680 (in French).

Boussinesq, J.V. (1896) Théorie de l'ecoulement tourbillonnant et tumultueux des liquides dans les lits rectilignes à grande section (tuyaux de conduite et canaux découverts) quand cet ecoulement s'est régularisé en un régime uniforme, c'est-à-dire, moyennement pareil à travers toutes les sections Normales du lit. (Theory of turbulent and tumultuous flow of liquids in prismatic channels of large cross-sections (pipes and open channels) when the flow is uniform, i.e. constant in average at each cross-section along the flow direction.) *Comptes Rendus des Séances de l'Académie des Sciences* (Paris, France), **122**, 1290–1295 (in French).

Bowie, G., Mills, W.B., Fordella, D.B., Campbell, C.L., Pagenkopf, J.R., Rupp, G.L., Johnson, K.M., Chan, P.W.H., Cherini, S., and Chamberlin, C.E. (1985) *Rates, Constants, and Kinetics Formulations in Surface Water Quality Modeling.* Tetra Tech. Report, No. EPA/600/3-85-040, 2nd edn, 475 pages.

Boxall, J.B., Guymer, I., and Marion, A. (2003) Transverse mixing in sinuous natural open channel flows. *Jl. Hyd. Res.*, *IAHR*, **41**(3), 153–165.

Boys, P.F.D. du (1879) Etude du Régime et de l'Action exercée par les Eaux sur un Lit á Fond de Graviers indéfiniment affouillable. (Study of flow regime and force exerted on a gravel bed of infinite depth.) *Ann. Ponts et Chaussées*, Paris, France, série 5, **19**, 141–195 (in French).

Brattberg, T., and Chanson, H. (1998) Air entrapment and air bubble dispersion at two-dimensional plunging water jets. *Chem. Eng. Sci.*, **53**(24, December), 4113–4127. Errata: 1999, **54**(12), 1925.

Brattberg, T., Chanson, H., and Toombes, L. (1998) Experimental investigations of free-surface aeration in the developing flow of two-dimensional water jets. *Jl. Fluids Eng., Trans. ASME*, **120**(4), 738–744.

Bresse, J.A. (1860) *Cours de Mécanique Appliquée Professé à l'Ecole des Ponts et Chaussées.* (*Course in Applied Mechanics Lectured at the Pont-et-Chaussées Engineering School.*) (Mallet-Bachelier: Paris, France) (in French).

Brodie, I. (1988) *Breaching of Coastal Lagoons. A Model Study.* Masters Thesis, M.Sc., University of New South Wales, Australia.

Brown, G.O. (2002) Henry Darcy and the making of a law. *Wat. Res. Res.*, **38**(7), Paper 11, 11-1–11-12.

Brown, P.P., and Lawler, D.F. (2003) Sphere drag and settling velocity revisited. *Jl. Environ. Eng., ASCE*, **129**(3), 222–231.

Buat, P.L.G. du (1779) *Principes d'Hydraulique*, vérifiés par un grand nombre d'expériences faites par ordre du gouvernement. (*Hydraulic Principles*, verified by a large number of experiments.) 1st edn (Imprimerie de Monsieur: Paris, France) (in French) (2nd edn: 1786, Paris, France, 2 volumes; 3rd edn: 1816, Paris, France, 3 volumes).

Bushnell, D.M., and Hefner, J.N. (1990) Viscous drag reduction in boundary layers. AIAA Publication, *Progress in Astronautics and Aeronautics*, **123**, Washington DC, USA, 517 pages.

Cain, P. (1978) *Measurements within Self-Aerated Flow on a Large Spillway.* Ph.D. Thesis, Ref. 78-18, Department of Civil Engineering, University of Canterbury, Christchurch, New Zealand.

Cain, P., and Wood, I.R. (1981). Measurements of self-aerated flow on a spillway. *Jl. Hyd. Div., ASCE*, **107**(HY11), 1425–1444.

Capart, H., and Young, D.L. (1998) Formation of a jump by the dam-break wave over a granular bed. *Jl. Fluid Mech.*, **372**, 165–187.

Carslaw, H.S., and Jaeger, J.C. (1959) *Conduction of Heat in Solids.* (Oxford University Press: London, UK) 2nd edn, 510 pages.

Cartellier, A. (1992) Simultaneous void fraction measurement, bubble velocity, and size estimate using a single optical probe in gas–liquid tow-phase flows. *Rev. Sci. Instrum.*, **63**(11), 5442–5453.

Chamani, M.R., and Rajaratnam, N. (1999) Characteristics of skimming flow over stepped spillways. *Jl. Hyd. Eng., ASCE*, **125**(4), 361–368. Discussion: **126**(11), 860–872. Closure: **126**(11), 872–873.

Chang, K.A., Lim, H.J., and Su, C.B. (2003) Fiber optic reflectometer for velocity and fraction ratio measurements in multiphase flows. *Rev. Scientif. Inst.*, **74**(7), 3559–3565. Discussion: **75** (in print).

Chanson, H. (1988) *A Study of Air Entrainment and Aeration Devices on a Spillway Model.* Ph.D. Thesis, Ref. 88-8, Department of Civil Engineering, University of Canterbury, New Zealand.

Chanson, H.(1989) Study of air entrainment and aeration devices. *Jl. Hyd. Res., IAHR*, **27**(3), 301–319.

Chanson, H. (1993) Self-aerated flows on chutes and spillways. *Jl. Hyd. Eng., ASCE*, **119**(2), 220–243. Discussion: **120**(6), 778–782.

Chanson, H. (1994) Drag reduction in open channel flow by aeration and suspended load. *Jl. Hyd. Res., IAHR*, **32**(1), 87–101.

Chanson, H. (1995a) *Air Bubble Entrainment in Free-Surface Turbulent Flows. Experimental Investigations.* Report CH46/95, Department of Civil Engineering, University of Queensland, Australia, June, 368 pages.

Chanson, H. (1995b) *Flow Characteristics of Undular Hydraulic Jumps. Comparison with Near-Critical Flows.* Report CH45/95, Department of Civil Engineering, University of Queensland, Australia, June, 202 pages.

Chanson, H. (1997a) *Air Bubble Entrainment in Free-Surface Turbulent Shear Flows.* (Academic Press: London, UK), 401 pages. {http://www.uq.edu.au/~e2hchans/reprints/book2.htm}.

Chanson, H. (1997b) Air bubble entrainment in open channels. Flow structure and bubble size distributions. *Intl. Jl. Multiphase Flow,* **23**(1), 193–203.

Chanson, H. (1999a) *The Hydraulics of Open Channel Flows: An Introduction.* (Edward Arnold: London, UK) 1st edn, 512 pages (ISBN 0 340 74067 1). {http://www.uq.edu.au/~e2hchans/reprints/errata.htm}.

Chanson, H. (1999b) Turbulent open-channel flows: drop-generation and self-aeration. Discussion. *Jl. Hyd. Eng., ASCE,* **125**(6), 668–670.

Chanson, H. (2000) Boundary shear stress measurements in undular flows: application to standing wave bed forms. *Wat. Res. Res.,* **36**(10), 3063–3076.

Chanson, H. (2001a) Flow field in a tidal bore: a physical model. *Proc. 29th IAHR Cong.,* Beijing, China, Theme E, Tsinghua University Press, Beijing, G. Li (Ed.), pp. 365–373. (CD-ROM, Tsinghua University Press).

Chanson, H. (2001b) *The Hydraulics of Stepped Chutes and Spillways.* (Balkema: Lisse, The Netherlands), 418 pages. {http://www.uq.edu.au/~e2hchans/reprints/book4.htm}

Chanson, H. (2002a) Very strong free-surface aeration in turbulent flows: entrainment mechanisms and air–water flow structure at the pseudo free-surface. In *Interaction of Strong Turbulence with Free Surfaces,* World Scientific, Advances in Coastal and Ocean Engineering Series, **8**, Singapore, M. Brocchini, and D.H. Peregrine (Eds), **8**, 65–98 (ISBN 981-02-4952-7).

Chanson, H., (2002b) *An Experimental Study of Roman Dropshaft Operation: Hydraulics, Two-Phase Flow, Acoustics.* Report CH50/02, Department of Civil Engineering, University of Queensland, Brisbane, Australia, 99 pages. {http://www.uq.edu.au/~e2hchans/reprints/ch5002.zip}

Chanson, H. (2002c) Air–water flow measurements with intrusive phase-detection probes. Can we improve their interpretation? *Jl. Hyd. Eng., ASCE,* **128**(3), 252–255.

Chanson, H. (2003a) Mixing and dispersion in tidal bores. A review. *Proc. Intl. Conf. on Estuaries and Coasts* (ICEC-2003), Hangzhou, China, November.

Chanson, H. (2003b) *Sudden Flood Release Down a Stepped Cascade. Unsteady Air–Water Flow Measurements. Applications to Wave Run-Up, Flash Flood and Dam Break Wave.* Report CH51/03, Department of Civil Engineering, University of Queensland, Brisbane, Australia, 142 pages. {http://www.uq.edu.au/~e2hchans/reprints/ch5103.zip}

Chanson, H. (2003c) Two-phase flow characteristics of an unsteady dam break wave flow. *Proc. 30th IAHR Biennial Cong.,* Thessaloniki, Greece, J. Ganoulis, and P. Prinos (Eds), **C2**, 237–244.

Chanson, H. (2003d) History of minimum energy loss weirs and culverts. 1960–2002. *Proc. 30th IAHR Biennial Cong.,* Thessaloniki, Greece, J. Ganoulis, and P. Prinos (Eds), **E**, 379–387.

Chanson, H. (2004a) Drag reduction in skimming flow on stepped spillways by aeration. *Jl. Hyd. Research, IAHR,* **42** (in print).

Chanson, H. (2004b) *The Hydraulics of Open Channel Flows: An Introduction.* (Butterworth-Heinemann: Oxford, UK) 2nd edn (ISBN 0 7506 5978 5).

Chanson, H. (2004c) Enhancing students' motivation in the undergraduate teaching of hydraulic engineering: the role of field works. *Jl. Prof. Issues Eng. Educ. Pract., ASCE,* **130** (in print).

Chanson, H. (2004d) Free-Surface Aeration in Dam Break Waves: An Experimental Study. *Proc. Intl. Conf. on Hydraulics of Dams and River Structures,* Tehran, Iran, Balkema Publisher.

Chanson, H., and Brattberg, T. (2000) Experimental study of the air-–water shear flow in a hydraulic jump. *Intl. Jl. Multiphase Flow,* **26**(4), 583–607.

Chanson, H., and Cummings, P.D. (1992) *Aeration of the Ocean due to Plunging Breaking Waves.* Research Report No. CE142, Department of Civil Engineering, University of Queensland, Australia, November, 42 pages.

Chanson, H., and James, P. (1998) Teaching case studies in reservoir siltation and catchment erosion. *Intl. Jl. Eng. Educ.,* **14**(4), 265–275. {http://www.uq.edu.au/~e2hchans/res_silt.html}

Chanson, H., and Manasseh, R. (2003) Air entrainment processes in a circular plunging jet. Void fraction and acoustic measurements. *Jl. Fluids Eng., Trans. ASME*, **125**(5, September), 910–921.

Chanson, H., and Qiao, G.L. (1994a) *Air Bubble Entrainment and Gas Transfer at Hydraulic Jumps*. Research Report No. CE149, Department of Civil Engineering, University of Queensland, Australia, August, 68 pages.

Chanson, H., and Qiao, G.L. (1994b) Drag reduction in hydraulics flows. *Proc. Intl. Conf. on Hydraulics in Civil Eng., I.E.Aust.*, Brisbane, Australia, 15–17 February, pp. 123–128.

Chanson, H., and Toombes, L. (2000) Stream reaeration in nonuniform flow: macroroughness enhancement. Discussion. *Jl. Hyd. Eng., ASCE*, **126**(3), 222–224.

Chanson, H., and Toombes, L. (2001) *Experimental Investigations of Air Entrainment in Transition and Skimming Flows Down a Stepped Chute. Application to Embankment Overflow Stepped Spillways*. Research Report No. CE158, Department of Civil Engineering, University of Queensland, Brisbane, Australia, July, 74 pages.

Chanson, H., and Toombes, L. (2002) Air–water flows down stepped chutes: turbulence and flow structure observations. *Intl. Jl. Multiphase Flow*, **27**(11), 1737–1761.

Chanson, H., Aoki, S., and Maruyama, M. (2000) *Experimental Investigations of Wave Runup Downstream of Nappe Impact. Applications to Flood Wave Resulting from Dam Overtopping and Tsunami Wave Runup*. Coastal/Ocean Engineering Report, No. COE00-2, Department of Architecture and Civil Engineering, Toyohashi University of Technology, Japan, 38 pages.

Chanson, H., Aoki, S., and Hoque, A. (2002a) *Similitude of Air Bubble Entrainment and Dispersion in Vertical Circular Plunging Jet Flows. An Experimental Study with Freshwater, Salty Freshwater and Seawater*. Coastal/Ocean Engineering Report, No. COE02-1, Department of Architecture and Civil Engineering, Toyohashi University of Technology, Japan, 94 pages. {http://www.uq.edu.au/~e2hchans/reprints/tut02.zip}

Chanson, H., Yasuda, Y., and Ohtsu, I. (2002b) Flow resistance in skimming flows and its modelling. *Can. Jl. Civ. Eng.*, **29**(6), 809–819.

Chanson, H., Aoki, S., and Hoque, A. (2002c) Scaling bubble entrainment and dispersion in vertical circular plunging jet flows: freshwater versus seawater. *Proc. 5th Intl. Conf. on Hydrodynamics ICHD 2002, Tainan*, Taiwan, H.H. Hwung, J.F. Lee, and K.S. Hwang (Eds), pp. 431–436.

Chanson, H., Brown, R., Ferris, J., and Warburton, K. (2003) *A Hydraulic, Environmental and Ecological Assessment of a Sub-tropical Stream in Eastern Australia: Eprapah Creek, Victoria Point QLD on 4 April 2003*. Report No. CH52/03, Department of Civil Engineering, University of Queensland, Brisbane, Australia, June, 189 pages (ISBN 1864997044).

Chen, C.L. (1990) Unified theory on power laws for flow resistance. *Jl. Hyd. Eng., ASCE*, **117**(3), 371–389.

Chen, Jiyu, Liu, Cangzi, Zhang, Chongle, and Walker, H.J. (1990) Geomorphological development and sedimentation in qiantang estuary and Hangzhou Bay. *Jl. Coast. Res.*, **6**(3), 559–572.

Cheng, N.S. (1997) Simplified settling velocity formula for sediment particle. *Jl. Hyd. Eng., ASCE*, **123**(2), 149–152.

Chyan, Shuzhong, and Zhou, Chaosheng (1993) *The Qiantang Tidal Bore*. Hydropower Publication, Beijing, China, The World's Spectacular Sceneries, 152 pages (in Chinese).

Coleman, N.L. (1970) Flume studies of the sediment transfer coefficient. *Wat. Res. Res.*, **6**(3), 801–809.

Coleman, S.E., Andrews, D.P., Webby, M.G. (2002) Overtopping breaching of noncohesive homogeneous embankments. *Jl. Hyd. Eng., ASCE*, **128**(9), 829–838.

Coles, D. (1956) The law of wake in the turbulent boundary layer. *Jl. Fluid Mech.*, **1**, 191–226.

Comolet, R. (1979) Vitesse d'ascension d'une bulle de gaz isolée dans un liquide peu visqueux. (The terminal velocity of a gas bubble in a liquid of very low viscosity.) *Jl. Mécaniq. Appliq.*, **3**(2), 145–171 (in French).

Coriolis, G.G. (1836) Sur l'établissement de la formule qui donne la figure des remous et sur la correction qu'on doit introduire pour tenir compte des différences de vitesses dans les divers points d'une même section d'un courant. (On the establishment of the formula giving the backwater curves and on the correction to be introduced to take into account the velocity differences at various points in a cross-section of a stream.) *Ann. Ponts et Chaussées*, 1st Semester, Series 1, **11**, 314–335 (in French).

Couette, M. (1890) Etude sur les frottements des liquides. (Study on the frictions of liquids.) *Ann. Chim. Phys.* (Paris, France), **21**, 433–510 (in French).

Courant, R., Isaacson, E., and Rees, M. (1952) On the solution of non-linear hyperbolic differential equations by finite differences. *Comm. Pure Appl. Math.*, **5**, 243–255.

Cousteau, J.Y., and Richards, M. (1984) *Jacques Cousteau's Amazon Journey.* (The Cousteau Society: Paris, France) (Also RD Press, Australia, 1985).

Crank, J. (1956) *The Mathematics of Diffusion.* (Oxford University Press: London, UK).

Creager, W.P. (1917) *Engineering of Masonry Dams.* (John Wiley & Sons: New York, USA).

Crowe, C., Sommerfield, M., and Tsuji, Y. (1998) *Multiphase Flows with Droplets and Particles.* (CRC Press: Boca Raton, USA), 471 pages.

Csanady, G.T. (1973) Turbulent diffusion in the environment. *Geophy. Astrophy. Monograp.*, **3**, D. Reidel Publication, Dordrecht, The Netehrlands, 248 pages.

Cummings, P.D. (1996) *Aeration due to Breaking Waves.* Ph.D. Thesis, Department. of Civil Engineering, University of Queensland, Australia.

Cummings, P.D., and Chanson, H. (1997a) Air entrainment in the developing flow region of plunging jets. Part 1: theoretical development. *Jl. Fluids Eng., Trans. ASME*, **119**(3), 597–602.

Cummings, P.D., and Chanson, H. (1997b) Air entrainment in the developing flow region of plunging jets. Part 2: experimental. *Jl. Fluids Eng., Trans. ASME*, **119**(3), 603–608. {Databank: http://www.uq.edu.au/~e2hchans/data/jfe97.html}

Cummings, P.D., and Chanson, H. (1999) An experimental study of individual air bubble entrainment at a planar plunging jet. *Chem. Eng. Res. Design, Trans. IChemE, Part A*, **77**(A2), 159–164.

Cunge, J.A. (1969) On the subject of a flood propagation computation method (muskingum method). *Jl. Hyd. Res., IAHR*, **7**(2), 205–230.

Cunge, J.A. (1975a) Numerical methods of solution of the unsteady flow equations. In *Unsteady Flow in Open Channels*. WRP Publication, Fort Collins, USA, K. Mahmood, and V. Yevdjevich (Eds), **2**, 539–586.

Cunge, J.A. (1975b) Two-dimensional modelling of flood plains. In *Unsteady Flow in Open Channels*. WRP Publication, Fort Collins, USA, K. Mahmood, and V. Yevdjevich (Eds), **2**, 705–762.

Cunge, J.A. (2003) Undular bores and secondary waves – experiments and hybrid finite-volume modelling. *Jl. Hyd. Res., IAHR*, **41**(5), 557–558.

Cunge, J.A., and Berthier, P. (1962) *Digital Computers and Hydroelectric Design Problems.* Water Power.

Cunge, J.A., and Wegner, M. (1964) Intégration numérique des equations d'ecoulement de barré de Saint Venant par une schéma implicite de différences finies. *Jl. Houille Blanche*, **1**, 33–39.

Cunge, J.A., Holly Jr, F.M., and Verwey, A. (1980) *Practical Aspects of Computational River Hydraulics.* (Pitman: Boston, USA), 420 pages.

Dai, Z., and Zhou, C. (1987). The qiantang bore. *Intl. Jl. Sedim. Res.*, **1**(November), 21–26.

Danilevslkii, V.V. (1940) History of Hydroengineering in Russia before the Nineteenth Century. *Gosudarstvennoe Energeticheskoe Izdatel'stvo*, Leningrad, USSR (in Russian) (English translation: *Israel Program for Scientific Translation*, IPST No. 1896, Jerusalem, Israel, 1968, 190 pages).

Darcy, H.P.G. (1856) *Les Fontaines Publiques de la Ville de Dijon.* (*The Public Fountains of the City of Dijon.*) Victor Dalmont, Paris, France, 647 pages (in French).

Darcy, H.P.G. (1858) Recherches expérimentales relatives aux mouvements de l'eau dans les tuyaux. (Experimental research on the motion of water in pipes.) *Mémoires Présentés à l'Académie des Sciences de l'Institut de France*, **14**, 141 (in French).

Darcy, H.P.G., and Bazin, H. (1865) *Recherches Hydrauliques. (Hydraulic Research.)* Imprimerie Impériales, Paris, France, Parties 1ère et 2ème (in French).

Deng, Z.Q., Singh, V.P., and Bengtsson, L. (2001) Longitudinal dispersion coefficient in straight rivers. *Jl. Hyd. Eng., ASCE*, **127**(11), 919–927.

Deng, Z.Q., Bengtsson, L., Singh, V.P., and Adrian, D.D. (2002) Longitudinal dispersion coefficient in single-channel streams. *Jl. Hyd. Eng., ASCE*, **128**(10), 901–916.

Donnelly, C., and Chanson, H. (2002) Environmental impact of a tidal bore on tropical rivers. *Proc. 5th Intl. River Management Symp.*, Brisbane, Australia, September, 3–6, 9 pages (CD-ROM).

Dorava, J.M., Montgomery, D.R., Palcsak, B.B., and Fitzpatrick, F.A. (2001) Geomorphic processes and riverine habitat. AGU, *Water Science and Application* **4**, Washington DC, USA, 253 pages.

Dressler, R.F. (1952) Hydraulic resistance effect upon the dam-break functions. *Jl. Res., Natl. Bureau Stand.*, **49**(3), 217–225.

Dressler, R. (1954) Comparison of theories and experiments for the hydraulic dam-break wave. *Proc. Intl. Assoc. of Scientific Hydrology Assemblée Générale*, Rome, Italy, **3**(38), 319–328.

Dupuit, A.J.E. (1848) *Etudes Théoriques et Pratiques sur le Mouvement des Eaux Courantes. (Theoretical and Practical Studies on Flow of Water.)* Dunod, Paris, France (in French).

Ehrenberger, R. (1926) Wasserbewegung in steilen rinnen (susstennen) mit besonderer beruck-sichtigung der selbstbelüftung. (Flow of water in steep chutes with special reference to self-aeration.) *Zeitschrift des Österreichischer Ingenieur und Architektverein*, No. 15/16 and 17/18 (in German) (Translated by Wilsey, E.F., U.S. Bureau of Reclamation).

Elder, J.W. (1959) The dispersion of marked fluid in turbulent shear flow. *Jl. Fluid Mech.*, **5**(4), 544–560.

Elfrink, B., and Baldock, T. (2002) Hydrodynamics and sediment transport in the swash zone: a review and perspectives. *Coast. Eng.*, **45**(3–4), 149–167.

Engelund, F., and Hansen, E. (1967) *A Monograph on Sediment Transport in Alluvial Streams.* (Teknisk Forlag: Copenhagen, Denmark).

Engelund, F., and Hansen, E. (1972) *A Monograph on Sediment Transport in Alluvial Streams.* (Teknisk Forlag: Copenhagen, Denmark), 3rd edn, 62 pages.

Ervine, D.A., and Falvey, H.T. (1987) Behaviour of turbulent water jets in the atmosphere and in plunge pools. *Proc. Instn. Civ. Engrs.*, London, Part 2, March 1987, **83**, 295–314. Discussion: Part 2, March–June 1988, **85**, 359–363.

Ervine, D.A., McKeogh, E.J., and Elsawy, E.M. (1980) Effect of turbulence intensity on the rate of air entrainment by plunging water jets. *Proc. Instn. Civ. Engrs*, Part 2, June, pp. 425–445.

Escande, L., Nougaro, J., Castex, L., and Barthet, H. (1961) Influence de quelques paramètres sur une onde de crue subite à l'aval d'un barrage. (The influence of certain parameters on a sudden flood wave downstream from a dam.) *Jl. Houille Blanche*, **5**, 565–575 (in French).

Fair, G.M., Geyer, J.C., and Okun, D.A. (1968) *Water and Wastewater Engineering.* (John Wiley: New York), 2 volumes.

Falvey, H.T. (1980) *Air–Water Flow in Hydraulic Structures.* USBR Engineering Monograph, No. 41 (Denver: Colorado, USA).

Falvey, H.T. (1990) *Cavitation in Chutes and Spillways.* USBR Engineering Monograph, No. 42 (Denver: Colorado, USA), 160 pages.

Faure, J., and Nahas, N. (1961) Etude numérique et expérimentale d'intumescences à forte courbure du front. (A numerical and experimental study of steep-fronted solitary waves.) *Jl. Houille Blanche*, **5**, 576–586. Discussion: **5**, 587 (in French).

Faure, J., and Nahas, N. (1965) Comparaison entre observations réelles, calcul, etudes sur modèles distordu ou non, de la propagation d'une onde de submersion. (Comparison between field observations, calculations, distorted and undistorted model studies of a dam break wave.) *Proc. 11th IAHR Biennial Cong.* (Leningrad, Russia), **III**, Paper 3.5, 1–7 (in French).

Fawer, C. (1937) *Etude de Quelques Ecoulements Permanents à Filets Courbes.* (*Study of some Steady Flows with Curved Streamlines.*) Thesis, Lausanne, Switzerland, Imprimerie La Concorde, 127 pages (in French).

Ferrell, R.T., and Himmelblau, D.M. (1967) Diffusion coefficients of nitrogen and oxygen in water. *Jl. Chem. Eng. Data*, **12**(1), 111–115.

Fick, A.E. (1855) On liquid diffusion. *Philos. Mag.*, **4**(10), 30–39.

Fischer, H.B., List, E.J., Koh, R.C.Y., Imberger, J., and Brooks, N.H. (1979) *Mixing in Inland and Coastal Waters.* (Academic Press: New York, USA).

Fourier, J.B.J. (1822) *Théorie Analytique de la Chaleur.* (*Analytical Theory of Heat.*) Didot, Paris, France (in French).

Franc, J.P., Avellan, F., Belahadji, B., Billard, J.Y., Briancon-Marjollet, L., Frechou, D., Fruman, D.H., Karimi, A., Kueny, J.L., and Michel, J.M. (1995) *La Cavitation. Mécanismes Physiques et Aspects Industriels.* (*The Cavitation. Physical Mechanisms and Industrial Aspects.*) Presses Universitaires de Grenoble, Collection Grenoble Sciences, France, 581 pages (in French).

Frazao, S.S., and Zech, Y. (2002) Undular bores and secondary waves – experiments and hybrid finite-volume modelling. *Jl. Hyd. Res., IAHR*, **40**(1), 33–43.

Galay, V. (1987) *Erosion and Sedimentation in the Nepal Himalaya. An Assessment of River Processes.* CIDA, ICIMOD, IDRC and Kefford Press, Singapore (Also Report No. 4/3/010587/1/1, Sequence 259, Government of Nepal).

Garbrecht, G. (1987a) *Hydraulics and Hydraulic Research: A Historical Review.* (Balkema Publishers: Rotterdam, The Netherlands).

Gauckler, P.G. (1867) Etudes Théoriques et Pratiques sur l'Ecoulement et le Mouvement des Eaux. (Theoretical and Practical Studies of the Flow and Motion of Waters.) *Comptes Rendues de l'Académie des Sciences*, Paris, France, Tome 64, pp. 818–822 (in French).

Goertler, H. (1942) Berechnung von aufgaben der freien turbulenz auf grund eines neuen näherungsansatzes. *Z.A.M.M.*, **22**, 244–254 (in German).

Gonzalez, C.A., and Chanson, H. (2004) Interactions between cavity flow and main stream skimming flows: an experimental study. *Can. Jl. Civ. Eng.*, **31** (in print).

Gordon, A.D. (1981) The behaviour of lagoon inlets. *Proc. 5th Aust. Conf. Coastal and Ocean Eng.*, Perth WA, pp. 62–63.

Gordon, A.D. (1990) Coastal lagoon entrance dynamics. *Proc. 22nd Intl. Conf. Coastal Eng.*, Delft, The Netherlands, B.L. Edge (Ed.) **3**, Cahp. 218, 2880–2893.

Gourlay, M.R., and Apelt, C.J. (1978) *Coastal Hydraulics and Sediment Transport in a Coastal System.* Lecture Notes, Department of Civil Engineering, University of Queensland, Brisbane, Australia, 250 pages.

Graf, W.H. (1971) *Hydraulics of Sediment Transport.* (McGraw-Hill: New York, USA).

Graf, W.H., and Altinakar, M.S. (1998) *Fluvial Hydraulics. Flow and Transport Processes in Channels of Simple Geometry.* (John Wiley: Chichester, UK), 681 pages.

Gualtieri, C., Gualtieri, P., and Doria, G.P. (2002) Dimensional analysis of reaeration rate in streams. *Jl. Environ. Eng., ASCE*, **128**(1), 12–18.

Gulliver, J.S. (1990) Introduction to air–water mass transfer. *Proc. 2nd Intl. Symp. on Gas Transfer at Water Surfaces, Air–Water Mass Transfer*, ASCE Publication, S.C. Wilhelms and J.S. Gulliver (Eds), Minneapolis MN, USA, pp. 1–7.

Gulliver, J.S., and Halverson, M.J. (1989) Air–water gas transfer in open channels. *Wat. Res. Res.*, **25**(8), 1783–1793.

Gyr, A. (1989) Structure of turbulence and drag reduction. *IUTAM Symp.*, Zürich, Switzerland, (Springer-Verlag Publication: Berlin, Germany).

Hebenstreit, G. (1997) Perspectives on Tsunami hazard reduction. Observations, theory and planning. Kluwer Academic, Dordrecht, The Netherlands, 218 pages (Also *Proc. 17th Intl. Tsunami Symp.*, AGU, Boulder CO, USA, July 1995.)

Helmholtz, H.L.F. (1868) Über discontinuirliche flüssigkeits-bewegungen. *Monatsberichte der Königlich Preussichen Akademie der Wissenschaft zu Berlin*, pp. 215–228 (in German).

Henderson, F.M. (1966) *Open Channel Flow*. (MacMillan Company: New York, USA).

Herringe, R.A. (1973) *A Study of the Structure of Gas–Liquid Mixture Flows*. Ph.D. Thesis, University of New South Wales, Kensington, Australia.

Herschy, R. (2002) The world's maximum observed floods. *Flow Meas. Instrum.*, **13**, 231–235.

Howe, J.W. (1949) Flow Measurement. *Proc. 4th Hydraulic Conf.*, Iowa Institute of Hydraulic Research, John Wiley & Sons Publishers, H. Rouse Ed., June, pp. 177–229.

Hughes, S.A. (1993) Physical models and laboratory techniques in coastal engineering. *Adv. Ser. Ocean Eng.*, **7**, World Scientific Publication, Singapore.

Hunt, B. (1982) Asymptotic solution for dam-break problems. *Jl. Hyd. Div.*, Proc., ASCE, **108**(HY1), 115–126.

Hunt, B. (1984) Perturbation solution for dam break floods. *Jl. Hyd. Eng.*, ASCE, **110**(8), 1058–1071.

Hunt, B. (1988) An asymptotic solution for dam break floods in sloping channels. In *Civil Engineering Practice* (Technomic Publications: Lancaster, PA, USA), P.N. Cheremisinoff, N.P. Cheremisinoff and S.L. Cheng (Eds), Vol. 2, Section 1, Chapter 1, pp. 3–11.

Hunt, B. (1999) Dispersion model for mountain streams. *Jl. Hyd. Eng.*, ASCE, **125**(2), 99–105. Discussion: **126**(3), 224–225.

Hwang, H.Y. (1999) *Taiwan Chi-Chi Earthquake 9.21.99. Bird's Eye View of Cher-Lung-Pu Fault*. Flying Tiger Cultural Publication, Taipei, Taiwan, 150 pages.

Idelchik, I.E. (1986) *Handbook of Hydraulic Resistance*. Hemisphere Publication, 2nd rev. and augm. ed., New York, USA.

Idel'Cik, I.E. (1969) *Mémento des Pertes de Charge*. (*Handbook of Hydraulic Resistance*.) Eyrolles Editor, Collection de la direction des études et recherches d'EDF, Paris, France.

Ippen, A.T. (1966) *Estuary and Coastal Hydrodynamics*. (McGraw-Hill: New York, USA).

Ippen, A.T., and Harleman, R.F. (1956) Verification of theory for oblique standing waves. *Transactions, ASCE*, **121**, 678–694.

Isaacson, E., Stoker, J.J., and Troesch, A. (1954) *Numerical Solution of Flood Prediction and River Regulation Problems. Report II. Numerical Solution of Flood Problems in Simplified Models of the Ohio River and the Junction of the Ohio and Mississippi Rivers*. Report No. IMM-NYU-205, Institute of Mathematical Science, New York University, pp. 1–46.

Isaacson, E., Stoker, J.J., and Troesch, A. (1956) *Numerical Solution of Flood Prediction and River Regulation Problems. Report III. Results of the Numerical Prediction of the 1945 and 1948 Floods on the Ohio River and of the 1947 Flood through the Junction of the Ohio and Mississippi Rivers, and of the Floods of the 1950 and 1948 through Kentucky Reservoir*. Report No. IMM-NYU-235, Institute of Mathematical Science, New York University, pp. 1–70.

Jevons, W.S. (1858) On clouds; their various forms, and producing causes. *Sydney Mag. Sci. Art*, **1**(8), 163–176.

Jones, E. (2003) *Person. Comm.*, 26 March.

Jones, O.C., and Delhaye, J.M. (1976) Transient and statistical measurement techniques for two-phase flows: a critical review. *Intl. Jl. Multiphase Flow*, **3**, 89–116.

Kelvin, Lord (1871) *The Influence of Wind and Waves in Water Supposed Frictionless*. London, Edinburgh and Dublin Phil. Mag. and Jl. Science, Series 4, **42**, 368–374.

Kennedy, J.F. (1963) The mechanics of dunes and antidunes in erodible-bed channels. *Jl. Fluid Mech.*, **16**(4), 521–544 (and 2 plates).

Khan, A.A., Steffler, P.M., and Gerard, R. (2000) Dam-break surges with floating debris. *Jl. Hyd. Eng.*, ASCE, **126**(5), 375–379.

Kjerfve, B., and Ferreira, H.O. (1993) Tidal bores: first ever measurements. *Ciência e Cultura* (*Jl. Brazilian Assoc. Advan. Sci.*), **45**(2), March/April, 135–138.

Kobus, H. (1984) *Proc. of the Intl. Symp. on Scale Effects in Modelling Hydraulic Structures.* IAHR, Esslingen, Germany (Also Institut fur Wasserbau, Universitat Stuttgart, Stuttgart, Germany, 1985).

Kolaini, A.R. (1998) Sound radiation by various types of laboratory breaking waves in fresh and salt water. *Jl. Acoust. Soc. Am.*, **103**(1), 300–308.

Lagrange, J.L. (1781) *Mémoire sur la Théorie du Mouvement des Fluides.* (*Memoir on the Theory of Fluid Motion.*) in Oeuvres de Lagrange, Gauthier-Villars, Paris, France (printed in 1882) (in French).

Lauber, G. (1997) *Experimente zur Talsperrenbruchwelle im Glatten Geneigten Rechteckkanal.* (*Dam Break Wave Experiments in Rectangular Channels.*) Ph.D. Thesis, VAW-ETH, Zürich, Switzerland (in German) (Also Mitt. der Versuchsanstalt fur Wasserbau, Hydrologie und Glaziologie, ETH-Zurich, Switzerland, No. 152).

Leal, J.G.A.B., Ferreira, R.M.L., Franco, A.B., and Cardoso, A.H. (2001) Dam-break waves over movable bed channels. Experimental study. *Proc. 29th IAHR Cong.*, Beijing, China, Theme C, Tsinghua University Press, Beijing, G. Li (Ed.), pp. 232–239.

Leighton, T.G. (1994) *The Acoustic Bubble.* (Academic Press: London, UK).

Levi, E. (1995) *The Science of Water. The Foundation of Modern Hydraulics.* (ASCE Press: New York, USA), 649 pages.

Lewis, R. (1997) *Dispersion in Estuaries and Coastal Waters.* (John Wiley: Chichester, UK), 312 pages.

Liggett, J.A. (1975) Basic equations of unsteady flow. In *Unsteady Flow in Open Channels.* WRP Publication, Fort Collins, USA, K. Mahmood, and V. Yevdjevich (Eds), **1**, 29–62.

Liggett, J.A. (1994) *Fluid Mechanics.* (McGraw-Hill: New York, USA).

Liggett, J.A., and Cunge, J.A. (1975) Numerical methods of solution of the unsteady flow equations. In *Unsteady Flow in Open Channels.* WRP Publication, Fort Collins, USA, K. Mahmood and V. Yevdjevich (Eds), **1**, 89–182.

Liu, D.H.F., Liptak, B.G., and Bouis, P.A. (1997) *Environmental Engineers' Handbook.* (Lewis Publication: Boca Raton, USA), 2nd edn.

Longo, S., Petti, M., and Losada, I.J. (2002) Turbulence in the swash and surf zones: a review. *Coast. Eng.*, **45**(3–4), 129–147.

Lynch, D.K. (1982) Tidal bores. *Sci. Am.*, **247**(4, October), 134–143.

Mahmood, K., and Yevdjevich, V. (1975) *Unsteady Flow in Open Channels.* (WRP Publication: Fort Collins, USA), 3 volumes.

Malandain, J.J. (1988) La seine au temps du mascaret. *Le Chasse-Marée*, **34**, 30–45.

Mariotte, E. (1686) *Traité du Mouvement des Eaux et des Autres Corps Fluides.* (*Treaty on the Motion of Waters and other Fluids.*) (Paris, France) (in French) (Translated by J.T. Desaguliers, Senex and Taylor, London, UK, 1718).

Massau, J. (1889) Appendice au mémoire sur l'intégration graphique. *Annales de l'Association des Ingénieurs Sortis des Ecoles Spéciales de Gand, Belgique*, **12**, 135–444 (in French).

Massau, J. (1900) Mémoire sur l'intégration graphique des equations aux dérivées partielles. *Annales de l'Association des Ingénieurs Sortis des Ecoles Spéciales de Gand, Belgique*, **23**, 95–214 (in French).

Matos, J. (2000) Hydraulic design of stepped spillways over RCC dams. *Intl. Workshop on Hydraulics of Stepped Spillways*, Zürich, Switzerland, H.E. Minor, and W.H. Hager (Eds), Balkema Publication, pp. 187–194.

McCurdy, E. (1956) *The Notebooks of Leonardo da Vinci.* (Jonathan Cape: London, UK), 6th edn, 2 volumes.

McKay, G.R. (1971) *Design of Minimum Energy Culverts.* Research Report, Department of Civil Engineering, University of Queensland, Brisbane, Australia, 29 pages and 7 plates.

McKeogh, E.J. (1978) *A Study of Air Entrainment using Plunging Water Jets.* Ph.D. Thesis, Queen's University of Belfast, UK, 374 pages.

Metcalf, and Eddy (1991) *Wastewater Engineering Treatment, Disposal and Reuse.* (McGraw-Hill: New York), 3rd edn.

Miller, W.A., and Cunge, J.A. (1975) Simplified equations of unsteady flows. In *Unsteady Flow in Open Channels.* (WRP Publication: Fort Collins, USA), K. Mahmood, and V. Yevdjevich (Eds), **1**, 183–257.

Minnaert, M. (1933) On musical air bubbles and the sound of running water. *Phil. Mag.,* **16**, 235–248.

Molchan-Douthit, M. (1998) *Alaska Bore Tales.* National Oceanic and Atmospheric Administration, Anchorage, USA, revised, 2 pages.

Montes, J.S. (1998) *Hydraulics of Open Channel Flow.* (ASCE Press: New-York, USA), 697 pages.

Montes, J.S., and Chanson, H. (1998) Characteristics of undular hydraulic jumps. Results and calculations. *Jl. Hyd. Eng., ASCE,* **124**(2), 192–205.

Murphy, D. (1983) *Pororoca!. Calypso Log, Cousteau Society,* **10**(2, June), 8–11.

Nakai, M., and Arita, M. (2002) An experimental study on prevention of saline wedge intrusion by an air curtain in rivers. *Jl. Hyd. Res., IAHR,* **40**(3), 333–339.

Navier, M. (1823) Mémoire sur les lois du mouvement des fluides. (Memoirs on the laws of fluid motion.) *Mém. Acad. des Sciences* (Paris, France), **6**, 389–416.

Neal, L.S., and Bankoff, S.G. (1963) A high resolution resistivity probe for determination of local void properties in gas–liquid flows. *Am. Inst. Chem. Jl.,* **9**, 49–54.

Nezu, I., and Nakagawa, H. (1993) *Turbulence in Open-Channel Flows.* IAHR Monograph, IAHR Fluid Mechanics Section, (Balkema Publication: Rotterdam, The Netherlands), 281 pages.

Nielsen, P. (1992) Coastal bottom boundary layers and sediment transport. Adv. Ser. Ocean Eng., **4**, World Scientific Publication, Singapore.

Nikora, V.I., Aberle, J., Jowett, I.G., Biggs, B.J.F., and Sykes, J.R.E. (2003) On turbulence effects on fish swimming performance: a case study of the New Zealand Native Fish Galaxias Maculatus (Inanga). *Proc. 30th IAHR Biennial Cong.,* Thessaloniki, Greece, J. Ganoulis, and P. Prinos (Eds), **C2**, 425–432.

Nikuradse, J. (1932) *Gesetzmässigkeit der turbulenten Strömung in glatten Rohren.* (*Laws of Turbulent Pipe Flow in Smooth Pipes.*) (VDI-Forschungsheft: Germany) No. 356 (in German) (Translated in NACA TT F-10, 359).

Nikuradse, J. (1933) *Strömungsgesetze in rauhen Rohren.* (*Laws of Turbulent Pipe Flow in Rough Pipes.*) (VDI-Forschungsheft: Germany) No. 361 (in German) (Translated in NACA Tech. Memo. No. 1292, 1950).

Novak, P., Moffat, A.I.B., Nalluri, C., and Narayanan, R. (2001) *Hydraulic Structures.* (Spon Press: London, UK), 3rd edn, 666 pages.

Nsom, B., Debiane, K., and Piau, J.M. (2000) Bed slope effect on the dam break problem. *Jl. Hyd. Res., IAHR,* **38**(6), 459–464.

Open University Course Team (1995) *Seawater: Its Composition, Properties and Behaviour.* (Butterworth-Heinemann: Oxford, UK), 2nd edn, 168 pages.

Pinto, N.L.de S., and Neidert, S.H. (1982) Model prototype conformity in aerated spillway flow. *Intl. Conf. on the Hydraulic Modelling of Civil Engineering Structures,* BHRA Fluid Engineering, Coventry, UK, pp. 273–284.

Plumptre, G. (1993) *The Water Garden.* (Thames and Hudson: London, UK).

Poiseuille, J.L.M. (1839) Sur le mouvement des liquides dans le tube de très petit diamètre. (On the movement of liquids in the pipe of very small diameter.) *Comptes Rendues de l'Académie des Sciences de Paris,* **9**, 487 (in French).

Pope, S.B. (2000) *Turbulent Flows*. Cambridge University Press, 771 pages.

Prandtl, L. (1904) Über Flussigkeitsbewegung bei sehr kleiner reibung. (On fluid motion with very small friction.) *Verh. III Intl. Math. Kongr.*, Heidelberg, Germany (in German) (Also NACA Tech. Memo. No. 452, 1928).

Prandtl, L. (1925) Über die ausgebildete turbulenz. (On fully developed turbulence.) *Z.A.M.M.*, **5**, 136–139 (in German).

Preissmann, A. (1960) Propagation des intumescences dans les canaux et rivières. *Proc. 1st Cong. of the Association Française de Calcul*, Grenoble, France, pp. 433–442.

Preissmann, A., and Cunge, J.A. (1961) Calculs des intumescences sur machines electroniques. *Proc. 9th IAHR Cong.*, Dubrovnik, Yugoslavia, pp. 656–664 (in French).

Rajaratnam, N. (1976) *Turbulent Jets*. Elsevier Scientific, Development in Water Science, **5**, New York, USA.

Rayleigh, Lord (1883) Investigation on the character of the equilibrium of an incompressible heavy fluid of variable density. *Proc. London Mathematical Society*, **14**, 170–177.

Ré, R. (1946) Etude du lacher instantané d'une retenue d'eau dans un canal par la méthode graphique. (Study of the sudden water release from a reservoir in a channel by a graphical method.) *Jl. Houille Blanche*, **1**(3, May), 181–187 (and 5 plates) (in French).

Rein, M. (1998) Turbulent open-channel flows: drop-generation and self-aeration. *Jl. Hyd. Eng., ASCE*, **124**(1), 98–102. Discussion: **125**(60), 668–670.

Resch, F.J., and Leutheusser, H.J. (1972) Le ressaut hydraulique: mesure de turbulence dans la région diphasique. (The hydraulic jump: turbulence measurements in the two-phase flow region.) *Jl. Houille Blanche*, **4**, 279–293 (in French).

Reynolds, O. (1883) An experimental investigation of the circumstances which determine whether the motion of water shall be direct or sinuous, and the laws of resistance in parallel channels. *Phil. Trans. Roy. Soc. Lond.*, **174**, 935–982.

Richardson, L.F. (1922) *Weather Prediction by Numerical Process*. London, UK.

Richardson, L.F. (1926) Atmospheric diffusion shown on a distance-neighbour graph. *Proc. Roy. Soc. Lond.*, **A110**, 709.

Rijn, L.C.van (1984) Sediment transport, Part I: bed load transport. *Jl. Hyd. Eng., ASCE*. **110**(10), 1431–1456.

Riley, J.P., and Skirrow, G. (1965) *Chemical Oceanography*. (Academic Press: London, UK), 3 volumes.

Ritter, A. (1892) Die fortpflanzung der wasserwellen. *Vereine Deutscher Ingenieure Zeitswchrift*, **36**(2, 33), 13 August, 947–954 (in German).

Rouse, H. (1937) Modern conceptions of the mechanics of turbulence. *Trans., ASCE*, **102**, 463–543.

Rozovskii, I.L. (1957) *Flow of Water in Bends of Open Channels*. Academy of Sciences of Ukrainian SSR, Kiew, USSR, 233 pages (in Russian) (Also Israel Program for Scientific Translations, Jerusalem, Israel, 1961 (in English).)

Rulifson, R.A., and Tull, K.A. (1999) Striped bass spawning in a tidal bore river: the Shubenacadie Estuary, Atlantic Canada. *Trans. Am. Fish. Soc.*, **128**, 613–624.

Runge, C. (1908) Uber eine method die partielle differentialgleichung? $u = $ constant numerisch zu integrieren. *Zeitschrift der Mathematik und Physik*, **56**, 225–232 (in German).

Rutherford, J.C. (1994) *River Mixing*. (John Wiley: Chichester, USA), 347 pages.

Sarrau (1884) *Cours de Mécanique*. (*Lecture Notes in Mechanics*.) Ecole Polytechnique: Paris, France (in French).

Schetz, J.A. (1993) *Boundary Layer Analysis*. (Prentice Hall: Englewood Cliffs, USA).

Schlichting, H. (1979) *Boundary Layer Theory*. (McGraw-Hill: New York, USA), 7th edn.

Schlichting, H., and Gersten, K. (2000) *Boundary Layer Theory*. (Springer Verlag: Berlin, Germany), 8th edn, 707 pages.

Schnitter, N.J. (1994) *A History of Dams: The Useful Pyramids*. (Balkema Publication: Rotterdam, The Netherlands).

Schoklitsch, A. (1917) Über dambruchwellen. *Sitzungberichten der Königliche Akademie der Wissenschaften*, Vienna, **126**(Part Iia), 1489–1514.

Scott, J.C. (1975) The preparation of water for surface clean fluid mechanics. *Jl. Fluid Mech.*, **69**(Part 2), 339–351.

Sene, K.J. (1984) *Aspects of Bubbly Two-Phase Flow*. Ph.D. Thesis, Trinity College, Cambridge, UK, December.

Serizawa, A., Kataoka, I., and Michiyoshi, I. (1975) Turbulence structure of air–water bubbly flows – I. measuring techniques. *Intl. Jl. Multiphase Flow*, **2**(3), 221–233.

Slauenwhite, D.E., and Johnson, B.D. (1999) Bubble shattering: differences in bubble formation in freshwater and seawater. *Jl. Geophy. Res.*, **104**(C2), 3265–3275.

Smith, N. (1971) *A History of Dams*. (The Chaucer Press: Peter Davies, London, UK).

Spiegel, M.R. (1974) *Mathematical Handbook of Formulas and Tables*. (McGraw-Hill Inc.: New York, USA).

Stoker, J.J. (1953) *Numerical Solution of Flood Prediction and River Regulation Problems. Report I. Derivation of Basic Theory and Formulation of Numerical Method of Attack*. Report No. IMM-200, Institute of Mathematical Science, New York University.

Stoker, J.J. (1957) *Water Waves. The Mathematical Theory with Appications*. (Interscience Publishers: New York, USA), 567 pages.

Straub, L.G., and Anderson, A.G. (1958) Experiments on self-aerated flow in open channels. *Jl. Hyd. Div., Proc. ASCE*, **84**(HY7), paper 1890, 1890-1–1890-35.

Stutz, B., and Reboud, J.L. (2000) Measurements within unsteady cavitation. *Exp. Fluids*, **29**, 545–552.

Swanson, W.M. (1961) The magnus effect: a summary of investigations to date. *Jl. Basic Eng., Trans. ASME*, Series D, **83**, 461–470.

Tamburrino, A., and Gulliver, J.S. (1990) Surface renewal due to large flow structures in open channel flow. *Proc. 2nd Intl. Symp. on Gas Transfer at Water Surfaces, Air–Water Mass Transfer, ASCE Publ.*, S.C. Wilhelms, and J.S. Gulliver (Eds), Minneapolis MN, USA, pp. 126–139.

Taylor, G.I. (1953) Dispersion of soluble matter in solvent flowing slowly through a tube. *Proc. Roy. Soc. Lond., Ser. A*, **219**, 186–203.

Taylor, G.I. (1954) The dispersion of matter in turbulent flow through a pipe. *Proc. Roy. Soc. Lond., Ser. A*, **223**, 446–468.

Tessier, B., and Terwindt, J.H.J. (1994) An example of soft-sediment deformations in an intertidal environment – the effect of a tidal bore. *Comptes-Rendus de l'Académie des Sciences*, Série II, **319**, No. 2, Part 2, 217–233 (in French).

Toombes, L. (2002) *Experimental Study of Air–Water Flow Properties on Low-Gradient Stepped Cascades*. Ph.D. Thesis, Department of Civil Engineering, University of Queensland.

Treske, A. (1994) Undular bores (favre-waves) in open channels – experimental studies. *Jl. Hyd. Res., IAHR*, **32**(3), 355–370. Discussion: **33**(3), 274–278.

Tricker, R.A.R. (1965) *Bores, Breakers, Waves and Wakes*. (American Elsevier Publication Company: New York, USA).

Vallentine, H.R. (1969) *Applied Hydrodynamics*. (Butterworths: London, UK), SI edn.

Valentine, E.M., and Wood, I.R. (1979a) Dispersion in rough rectangular channels. *Jl. Hyd. Div., ASCE*, **105**(HY12), 1537–1553.

Valentine, E.M., and Wood, I.R. (1979b) Experiments in longitudinal dispersion with dead zones. *Jl. Hyd. Div., ASCE*, **105**(HY9), 999–1016.

Van de Sande, E., and Smith, J.M. (1973) Surface entrainment of air by high velocity water jets. *Chem. Eng. Sci.*, **28**, 1161–1168.

Vardoulakis, I. (2000). Catastrophic landslides due to frictional heating of the failure plane. *Mech. of Cohesive-Frictional Materials*, Vol. 5, pp. 443–467.

Vischer, D., and Hager, W.H. (1998). *Dam Hydraulics*. (John Wiley: Chichester, UK) 316 pages.

Vischer, D., Volkart, P., and Sigenthaler, A. (1982) Hydraulic modelling of air slots in open chutes spillways. *Intl. Conf. on the Hydraulic Modelling of Civil Engineering Structures*, BHRA Fluid Engineering, Coventry, UK, pp. 239–252.

Visser, P.J., Vrijling, J.K., and Verhagen, H.J. (1990) A field experiment on breach growth in sand-dykes. *Proc. 22nd Intl. Conf. Coastal Eng.*, Delft, The Netherlands, B. Edge (Ed.), **2**, 2097–2100.

Wallis, S.G., Young, P.C., and Beven, K.J. (1989) Experimental investigation of the aggregated dead zone model for longitudinal solute transport in stream channels. *Proc. Instn. Civ. Eng.*, London, **87**(Part 2), 1–22.

Waltham, T., and Sholji, I. (2001) The demise of the aral sea – an environmental disaster. *Geolog. Today*, **17**(6), 218–224.

Wan, Z., and Wang, Z. (1994) *Hyperconcentrated Flow*. Balkema, IAHR Monograph, Rotterdam, The Netherlands, 290 pages.

Webster, D. (2002) China's unknown Gobu: Alashan. *Natl. Geog.*, **201**(1), 48.

Weiss, R.F. (1970) The solubility of nitrogen, oxygen and argon in water and seawater. *Deep-Sea Res.*, **17**, 721–735.

Whitham, G.B. (1955) The effects of hydraulic resistance in the dam-break problem. *Proc. Roy. Soc. of London, Ser. A*, **227**, 399–407.

Witts, C. (1999) *The Mighty Severn Bore*. (Rivern Severn Publications: Gloucester, UK), 84 pages.

Wolanski, E., Moore, K., Spagnol, S., D'Adamo, N., and Pattieratchi, C. (2001) Rapid, human-induced siltation of the macro-tidal Ord River Estuary, Western Australia. *Estuarine, Coastal and Shelf Science*, **53**, 717–732.

Wood, I.R. (1983) Uniform region of self-aerated flow. *Jl. Hyd. Eng., ASCE*, **109**(3), 447–461.

Wood, I.R. (1985) Air water flows. *Proc. 21st IAHR Cong.*, Melbourne, Australia, Keynote address, pp. 18–29.

Wood, I.R. (1991) *Air Entrainment in Free-Surface Flows*. IAHR Hydraulic Structures Design Manual No. 4, Hydraulic Design Considerations, (Balkema Publication: Rotterdam, The Netherlands), 149 pages.

Wood, I.R., Ackers, P., and Loveless, J. (1983). General method for critical point on spillways. *Jl. of Hyd. Eng., ASCE*, **109**(2), 308–312.

Wood, I.R., Bell, R.G., and Wilkinson, D.L. (1993) *Ocean Disposal of Wastewater*. World Scientific, Singapore.

Yasuda, Y., and Ohtsu, I.O. (1999) Flow resistance of skimming flow in stepped channels. *Proc. 28th IAHR Cong.*, Graz, Austria, Session B14, 6 pages.

Yasuda, Y., Ohtsu, I., Koizumi, N., and Junaidi, A. (2002) Installation test of ladder-type fishway at diversion weirs in Japan and Indonesia. *Proc. 13th IAHR-APD Cong.*, Singapore, pp. 947–950.

Yen, B.C. (2002) Open channel flow resistance. *Jl. Hyd. Eng., ASCE*, **128**(1), 20–39.

Yeh, H., Liu, P., and Synolakis, C. (1996) Long-wave runup models. World Scientific, Singapore, 403 pages (Also *Proc. 2nd Intl. Workshop on Long-Wave Runup Models*, Friday Harbour WAS, USA, September 1995.)

Yevdjevich[1], V. (1975) Introduction. in *Unsteady Flow in Open Channels*. WRP Publication, Fort Collins, USA, K. Mahmood and V. Yevdjevich (Eds),**1**, 1–27.

---

[1]The name of V. Yevdjevich is also spelled Jevdjevich or Yevdyevich.

Abbreviations of journals and institutions

AFMC	Australasian Fluid Mechanics Conference
AGU	American Geophysical Union (USA)
AIAA Jl.	Journal of the American Institute of Aeronautics and Astronautics (USA)
ANCOLD	Australian Committee on Large Dams
Ann. Chim. Phys.	Annales de Chimie et Physique, Paris (France)
ANSSR	Academy of Sciences of the USSR, Moscow
APHA	American Public Health Association
ARC	Aeronautical Research Council (UK)
	Australian Research Council
ARC RM	Aeronautical Research Council Reports and Memoranda
ARC CP	Aeronautical Research Council Current Papers
ASAE	American Society of Agricultural Engineers
ASCE	American Society of Civil Engineers
ASME	American Society of Mechanical Engineers
AVA	Aerodynamische Versuchanstalt, Göttingen (Germany)
BHRA	British Hydromechanics Research Association (BHRA Fluid Engineering)
BSI	British Standards Instituion, London
CIRIA	Construction Industry Research and Information Association
EDF	Electricité de France
EPA	Environmental Protection Agency
Ergeb. AVA Göttingen	Ergebnisse Aerodynamische Versuchanstalt, Göttingen (Germany)
Forsch. Ing. Wes.	Forschung auf dem Gebiete des Ingenieur-Wesens (Germany)
Forschunsheft	Research supplement to Forsch. Ing. Wes. (Germany)
Gid. Stroit.	Gidrotekhnicheskoe Stroitel'stvo (Russia) (translated in Hydrotechnical Construction)
IAHR	International Association for Hydraulic Research
IAWQ	International Association for Water Quality
ICOLD	International Committee on Large Dams
IEAust.	Institution of Engineers, Australia
IIHR	Iowa Institute of Hydraulic Research, Iowa City (USA)
Ing. Arch.	Ingenieur-Archiv (Germany)
JAS	Journal of Aeronautical Sciences (USA) (replaced by JASS in 1959)
JASS	Journal of AeroSpace Sciences (USA) (replaced by AIAA Jl. in 1963)
Jl. Fluid Mech.	Journal of Fluid Mechanics (Cambridge, UK)
Jl. Roy. Aero. Soc.	Journal of the Royal Aeronautical Society, London (UK)
JSCE	Japanese Society of Civil Engineers
JSME	Japanese Society of Mechanical Engineers
Luftfahrt-Forsch.	Luftfahrt-Forschung (Germany)
NACA	National Advisory Committee on Aeronautics (USA)
NACA Rep.	NACA Reports (USA)

NACA TM	NACA Technical Memoranda (USA)
NACA TN	NACA Technical Notes (USA)
NASA	National Aeronautics and Space Administration (USA)
NBS	National Bureau of Standards (USA)
ONERA	Office National d'Etudes et de Recherches Aérospatiales (France)
Phil. Mag.	Philosophical Magazine
Phil. Trans. R. Soc. Lond.	Philosophical Transactions of the Royal Society of London (UK)
Proc. Cambridge Phil. Soc.	Proceedings of the Cambridge Philosophical Society (UK)
Proc. Instn. Civ. Engrs.	Proceedings of the Institution of Civil Engineers (UK)
Proc. Roy. Soc.	Proceedings of the Royal Society, London (UK)
Prog. Aero. Sci.	Progress in Aerospace Sciences
Proc. Cambridge Phil. Soc.	Transactions of the Cambridge Philosophical Society (UK)
SAF	St Anthony Falls Hydraulic Laboratory, Minneapolis (USA)
SHF	Société Hydrotechnique de France
SIA	Société des Ingénieurs et Architectes (Switzerland)
Trans. Soc. Nav. Arch. Mar. Eng.	Transactions of the Society of Naval Architects and Marine Engineers
USBR	United States Bureau of Reclamation, Department of the Interior
VDI Forsch.	Verein Deutsche Ingenieure Forschungsheft (Germany)
Wat. Res. Res.	Water Resources Research Journal
WES	US Army Engineer Waterways Experiment Station
Z.A.M.M.	Zeitschrift für Angewandete Mathematik und Mechanik (Germany)
Z.A.M.P.	Zeitschrift für angewandete Mathematik und Physik (Germany)
Z. Ver. Deut. Ingr.	Zeitschrift Verein Deutsche Ingenieure (Germany)

Common bibliographical abbreviations

Conf.	Conference
Cong.	Congress
DEng.	Doctor of Engineering
Intl.	International
Jl.	Journal
Mitt.	Mitteilungen
Ph.D.	Doctor of Philosophy
Proc.	Proceedings
Symp.	Symposium
Trans.	Transactions

Index

Page numbers in **bold** refers to figures and tables.